One Day on the Somme

as reviewed on the Western Front Association's Website
(www.westernfront.co.uk)
by John Whalley

Every now and then a book comes along which makes you think the author has hit the subject spot on, this is such a book. Aimed mainly at the first or second time visitor to the Somme battlefield, this book lists the British battalions and their objectives on 1st July 1916.

The first half of the book lists the main objectives for the first day, the Senior Commanders and the Battalions involved (including their informal names), a section on the IGN maps needed to study the area, and finally a listing of the trenches and battlesites on and behind the front line with the relevant map reference.

The second half of the book, with the help of clearly defined maps, and present-day pictures, covers the attacks made on the day, from Gommecourt in the north to Montauban in the south. And finally the book concludes with a summary of the days events.

For anyone visiting the area for the first time, particularly to examine the battles of the 1st July, this book is a necessity; read it before you go, then take it with you as you survey the battlefields, even for those who know the area there is much to enjoy here.

Highly recommended.

available from:

GMS Enterprises
67 Pyhill, Bretton
Peterborough PE3 8QQ
ENGLAND
TEL 01733 265123
email:
GMSAVIATIONBOOKS@COMPUSERVE.COM

148 DAYS ON THE SOMME

2nd JULY
to
26th NOVEMBER 1916

BARRY CUTTELL

GMS ENTERPRISES

First published 2000
by GMS Enterprises
67 Pyhill, Bretton, Peterborough,
England PE3 8QQ
Tel and Fax (01733) 265123
EMail: GMSAVIATIONBOOKS@ Compuserve.com

ISBN: 1 870384 79 2

Copyright: Barry Cuttell

All rights reserved.
No part of this publication may be reproduced,
stored in a retrieval system, or transmitted,
in any form or by any means, electronic, mechanical,
photocopying, recording or otherwise
without the prior permission of the publishers.

A donation from the sale of this book will be made towards
the maintenance of the privately owned memorial of
Lochnagar Crater at La Boisselle on the Somme

Printed and bound for GMS Enterprises

CONTENTS

List of Abbreviations	5
List of Sketch Maps	6
Introduction	7
How to use this guide	9
Objectives and Code Numbers for the 2nd July to 26th November 1916	11
The men in charge - Corps and Divisional Commanders	14
Transfers between infantry & pioneer units. Battalion amalgamations	16
About the allocation of battalion engagements	17
Disposition of British Infantry Divisions (numerical)	18
Disposition of British Brigades (numerical)	28
Disposition of British Battalions (alphabetical)	37
British Battalions having informal names (alphabetical)	46
British Battalions not engaged from the 2nd July	49
Disposition of Overseas Divisions and Cavalry (numerical)	50
The I.G.N. maps	52
About the trenches	55
List of Trenches and battle sites (alphabetical)	56
About sites to the rear	85
List of sites situated behind the British front line (alphabetical)	85
The infantry attacks	88
Conclusion	230
Acknowledgements	236
Bibliography	238
Appendix 1 - location of the main military cemeteries (alphabetical)	239
Appendix 2 - alphabetical list of Regiments on the Thiepval Memorial	242

To my grandfather George Alfred Reddish 1882 - 1955

who served in the Royal Horse Artillery
at Ypres and the Somme.

LIST OF ABBREVIATIONS

A&SH	: Argyll & Sutherland Highlanders		m	: meter
AB	: Australian Brigade (on maps)		Maj	: Major
B	: British Brigade (on maps)		Maj.Gen	: Major-General
Berks	: Berkshire		Man	: Manchester
Bucks	: Buckinghamshire		Middx	: Middlesex
Cambs	: Cambridge *also* Cambridgeshire		Mon	: Monmouthshire
Capt	: Captain		N	: North
CB	: Canadian Brigade (on maps)		N Staffs	: North Staffordshire
CCS	: Casualty Clearing Station		NF	: Northumberland Fusiliers
cm	: centimetre		NZ	: New Zealand
C.inC	: Commander in Chief		O&BLI	: Oxfordshire & Buckinghamshire Light Infantry
CO	: Commanding officer			
Coy	: Company		Pnrs	: Pioneers
Conn. Rang	: Connaught Rangers		Queen's	: Queen's Royal West Surrey Regiment
Cpl	: Corporal			
CT	: Communication Trench		Rdt	: Redoubt
DCLI	: Duke of Cornwall Light Infantry		Reg	: Regiment
DWR	: Duke of Wellington Regiment		Rif. Brig	: Rifle Brigade
DLI	: Durham Light Infantry		R	: Royal
E	: East		R Berks	: Royal Berkshire
E Lancs	: East Lancashire		R Dub. Fus	: Royal Dublin Fusiliers
E Surrey	: East Surrey		RE	: Royal Engineers
E Yorks	: East Yorkshire		R Fus	: Royal Fusiliers
Fus	: Fusiliers		R Innis. Fus	: Royal Inniskilling Fusiliers
Gen	: General		R Irish	: Royal Irish
Gloucs	: Gloucestershire		R Irish Fus	: Royal Irish Fusiliers
Green Howards	: Green Howards Yorkshire Regiment		R Irish Rif	: Royal Irish Rifles
			RMF	: Royal Munster Fusiliers
Gren. Gds	: Grenadier Guards		R Scots	: Royal Scots
Gds	: Guards		RSF	: Royal Scots Fusiliers
Hants	: Hampshire		R Sussex	: Royal Sussex
Herts	: Hertfordshire		R W Kent	: Queen's Own (Royal West Kent)
HLI	: Highland Light Infantry		R. Warwick	: Royal Warwick
HAC	: Honourable Artillery Company		RWF	: Royal Welsh Fusiliers
IGN	: Institut Géographique National		Scot. Rif	: Scottish Rifles The Cameronians
km	: kilometer		Sgt	: Sergeant
King's	: King's Liverpool		Sher. Fors	: Sherwood Foresters, Nottinghamshire & Derbyshire Regiment
King's Own	: King's Own Royal Lancaster Regiment			
KOSB	: King's Own Scottish Borderers		SLI	: Somerset Light Infantry
KOYLI	: King's Own Yorkshire Light Infantry		S	: South
			S Lancs	: South Lancashire
KRRC	: King's Royal Rifle Corps		S Staffs	: South Staffordshire
KSLI	: King's Shropshire Light Infantry		SWB	: South Wales Borderers
Lancs	: Lancashire		Staffs	: Staffordshire
Lancs Fus	: Lancashire Fusiliers		Tr	: Trench
Leics	: Leicestershire		W	: West
Lt/Lieut	: Lieutenant		W Yorks	: West Yorkshire
Lt.Gen	: Lieutenant-General		Wilts	: Wiltshire
Lincs	: Lincolnshire		Worcs	: Worcestershire
LRB	: London Rifle Brigade		XR	: Cross Roads
Lon. Scot	: London Scottish		Y&L	: York & Lancaster Regiment
LNL	: Loyal North Lancs			

LIST OF SKETCH MAPS SHOWING DIRECTION OF BRIGADE ATTACKS

OBJECTIVE(S)	CODE	PAGE
General map of all objectives	15-36	12/13
Fricourt and north to Birch Tree Wood and north-east to Bottom Wood	15	89
La Boisselle and east to Lincoln Redoubt and south-east to Sausage Redoubt and Horseshoe Trench	16	92
Ovillers	17	95
General map of area from Contalmaison to Trônes Wood	18 & 19	99
Mametz Wood and Contalmaison	18	100
Bernafay Wood, Trônes Wood and Maltzhorn Farm	19	107
The Dawn Attack on 14th July 1916 The Bazentin villages, woods and ridge and east towards Longueval	20	110
General map of the area of Longueval and Delville Wood	21	113
Longueval and Delville Wood	21	114
The infantry attacks at Pozières, the southern sector of objective 22	22	122
The infantry attacks on Mouquet Farm, the northern sector of objective 22	22	128
High Wood, Switch Line Centre, trenches to the south-east and west	23	133
Guillemont, Station, Quarry and trenches north-west towards Waterlot Farm	24	148
General map of area from Waterlot Farm and north-east to the "Alcohol Trenches", Ginchy village and German strongpoints to the east	25 & 26	155
Waterlot Farm, "Alcohol Trenches" and Ginchy	25 & 26 (West)	156
Quadrilateral, Triangle and Straight Trench	26 (East)	158
Flers and surrounding trenches	27	164
Martinpuich and surrounding trenches	28	169
Courcelette, Sugar Factory and trenches north towards Pys and Grandcourt	29	172
Morval, Lesboeufs, "Meteorological Trenches", Gueudecourt and the Gird Lines	30,31 & 32	179
Combles, Bouleaux Wood, Leuze Wood and Falfemont Farm, Wedge Wood	33	191
Thiepval, St. Pierre Divion, trenches north and north-east of Thiepval	34	195
Beaumont Hamel, Beaucourt, Serre and associated trenches	35	208
Le Sars, Destremont Farm and trenches to the east. Eaucourt Abbey and the Butte de Warlencourt	36	222
The development of the British Front Line from the 1st July to the end of November 1916		230

INTRODUCTION

This guide takes up the story of the events on the Somme from the 2nd July to the end of November 1916. It is not a story of resounding success but is, nevertheless, one of unceasing effort and endeavour on the part of the commanders, artillery, infantry, the new war machine called the tank and the many unsung heroes such as pioneers, tunnellers, sappers, labour forces, doctors, stretcher bearers and all those who participate in the execution of war. It is concerned principally with infantry, pioneer, artillery and tank units with occasional reference to other units which, although each played an important role, are generally outside the scope of this guide.

The author has retained a similar format to that used in his guide to the 'Big Push' on the 1st July 1916* that is to say, a number of statistical listings of divisions, brigades and battalions followed by a description of the infantry attacks. The justification for a seemingly large number of statistical lists is quite simply to assist the reader to find information quickly and easily. The order of battle by divisions is sometimes presented as an appendix in books of the Great War and finding information from these sources is relatively easy and useful. The author has found during his reading and research that he often needed to find out, or, quite simply, to be reminded of which division a brigade belonged, or to which brigade and, therefore division, certain battalion were assigned. Having separate numerical listings for divisions and brigades and an alphabetical list of battalions makes this information instantly available, the comb-binding of the book lending itself readily to keeping the book open at the required page.

A list of over one thousand nine hundred trenches and sites incorporates the main objectives assaulted by the infantry during the four and a half months of the campaign. Where possible, the principal trenches have been included on the maps which accompany a description of the infantry attacks. Certain villages and woods became household names in 1916 - Pozières, Ovillers, Guillemont, Ginchy, Thiepval and the infamous Mametz, Trônes, Delville and High Woods. These are sites which can be seen today but what of the trenches situated away from the villages and woods in the vast cratered lunar landscape often knee-deep in mud and where titanic struggles were enacted, sometimes involving close-quarter fighting with bomb and bayonet? Whereas the capture of a village or wood made front-page headlines in the English newspapers a few days after the event, the capture of a trench in the middle of nowhere seldom received such attention. There was no headline such as '..Tank seen in main street Flers going on with large number of troops following.....' And yet, the attacks through the sleet, rain and mud over the open countryside by the 32nd Division on Munich and Frankfort Trenches in November, the assault on Munster Alley and O.G. Lines in July by the Australians, the Canadian attacks on Regina Trench or the innumerable attempts to capture the "Meteorological Trenches" in horrendous conditions by a host of divisions were equally important. Where are these trenches? Some can still be seen and visited today such as those preserved in the Newfoundland Memorial Park to the south of Beaumont Hamel. Others have been immortalised by the Commonwealth War Graves Commission which constructed a number of cemeteries on the sites of old trenches - Frankfort Trench British Cemetery and Regina Trench Cemetery being two examples. A number of trenches were located adjacent to roads, tracks and lanes which, although destroyed during the war were, for the most part, faithfully reconstructed in the nineteen-twenties and can be found today. One such example is Courcelette Trench which headed north out of Courcelette from just west of the Miraumont Road East. The trenches were filled in after the war, the blood-soaked battleground gradually being reclaimed and returned to its original agricultural use. The trench list enables the visitor to locate the sites of such trenches using modern IGN maps. A visit in February or March can be very rewarding, the traces of chalk displacement during the excavation of many trenches can still be seen today after the plough has passed. Each year, farm machinery unearths hundreds of tons of unexploded shells and the usual detritus of war. Such souvenirs should be left untouched if seen by the visitor - they are dangerous. It was reported recently that it would take about eight hundred years before Flanders fields could be declared safe.

It is hoped the statistical information contained in this guide along with the description of the infantry attacks will serve a useful purpose to those who come to trace the path taken by a relative all those years ago. It may be that the presentation of the statistical information may also serve as a useful reference to more experienced visitors.

One other small expression has always intrigued the author. It is the English prepostional use of 'on' when talking about being 'on' the Somme. You can be 'in' the Nord, 'at' Mametz, 'in' the Pas de Calais but one is always 'on' the Somme. The French, like the English, usually use the appropriate preposition for 'in' when speaking of counties and say, 'in' the Marne, 'in' the Aisne as we say 'in' Kent, 'in' Yorkshire and they would say 'in' the Somme, meaning in the county of the Somme. But the English have always talked of being 'on' the Somme. One wonders if that is to keep everyone's feet dry.

It is gratifying to see so many visitors each year on the Somme. Some are first-time visitors, others are those who return more often and then there is a large group which return year after year, drawn by the tumultuous events enacted on this blooded ground. They come to pay homage to a fallen relative or because they are interested in the Great War and wish to learn more or, quite simply, to enjoy the very pleasant undulating wooded countryside and follow the course of the River Somme just a few kilometres to the south. Whatever the reason, they come in their thousands each year in the sure knowledge that they will meet up with other people with the same interest and bonds and friendships are founded today in the same way as they were over eighty years ago. This is indeed a very special relationship.

* see *ONE DAY ON THE SOMME,* B. Cuttell, GMS Enterprises, Peterborough, 1998

The Golden Virgin leaning precariously out over the Basilica in the centre of Albert. The statue was made safe with chains by French Engineers after being struck by a German shell in January 1915. [A.Perret]

HOW TO USE THIS GUIDE

On the 1st July 1916 a division was composed of three brigades, each brigade having four battalions. In addition each division had one battalion of pioneers. In all, a total of thirteen battalions per division. As a battalion consisted of roughly one thousand men, the divisional commander, normally with the rank of major-general, had well over thirteen thousand men under his orders. Battalion commanders usually held the rank of lieutenant-colonel.

If the reader wishes to trace the steps of a relative and knows in which battalion the soldier served then it is quite easy to trace his path by finding the battalion from the alphabetical list. This will reveal the division, brigade and the sector where he was engaged. Each sector, referred to in this guide as an objective area is numerically coded and the reader should then turn to the end section of the guide where he will find a description of the infantry attacks for each objective. A detailed description of each ojective can be found on page 11. The battalion may be known but not the number. In this case, if it is known where he fought, the battalion can be traced by the coded objective number. The statistics have been listed in different forms to help the reader in his search. It may well be of course, that practically little or no information is known regarding the battalion, brigade or division or where he was engaged. If he died during the battle of the Somme, the Commonwealth War Graves Commission at 2 Marlow Road, Maidenhead, Berkshire, SL6 7DX, will be able to provide information on his regiment, where he is buried or, if the person was listed as missing, on which memorial his name is engraved. It is essential to quote the soldier's serial number. If this is not known but the family has medals awarded to the soldier, the serial number can be found on the outer edge of the General Service Medal and the Victory Medal. On the 1914 Star and the 1914/15 Star, the name and serial number are on the obverse side. If no medals are to hand, the search becomes rather more complicated but not impossible. Let us suppose a reader wishes to trace his ancestor who lived in Derby at the outbreak of war. It is very possible he enlisted in one of the battalions of the Notts. & Derbyshire Regiment, known as the Sherwood Foresters and initially, contact could be made at the Regimental H.Q at Nottingham. Without a serial number, it is helpful if he did not have the name of Smith, Taylor, Green, etc. but do not be discouraged - he was known to many people. Can you trace any letters or photographs from the soldier to your family or friends? Although it was strictly forbidden to write place names on letters and most were headed B.E.F. France plus the date, it is surprising what passed the censor's scrutiny. It is also very probable he enlisted with some of his friends. Try to make contact with these families as this is one of the best sources of information. Many newly enlisted soldiers had their photograph taken with their friends before leaving for the Western Front and many of these pictures are now possibly tucked away in family "archives" throughout the country. It is more than likely he joined the same battalion as his friends. Look at copies of the local newspapers in the archives dated July 1916 or after for information on those who died or were wounded or mentioned in despatches. Articles were abundant on the ordinary soldier. He may not have enlisted in the Notts. & Derbyshire Regiment at all but perhaps wished to join a kilted regiment or any other whose name or uniform appealed to him. Harry Fellows of Nottingham for instance joined the Northumberland Fusiliers simply because he fancied a nice long free train ride up north! If the subject of your research survived the war then the chance of finding information more quickly is evidently easier. An advertisement in the local paper may bring some positive result. A further good source of information is the Army Records Office at Hayes in Middlesex. Unfortunately, many records relating to men and women who served in The Great War were destroyed through Luftwaffe bombing during World War II. However, to give the public greater access to what is still available, the existing records have been transferred on to microfilm at the Public Records Office at Kew where they can be seen. The Western Front Association journal, 'Stand To' publishes enquiries from members seeking information. Enquiries concerning membership of the WFA should be addressed to Paul Hanson, 17 Aldrin Way, Cannon Park, Coventry CV4 7DP. A more modern approach is by using Internet or E Mail facilities and the recently issued CDRom Soldiers Died.

As soon as you find some information, look at the contents page and you will find the appropriate page number which will advance your research further. For those readers who are simply on a general visit to the Somme which will probably include a visit to the Newfoundland Park, Thiepval Memorial, Ulster Tower, Lochnagar Crater, High Wood, Delville Wood and other well-known sites, the guide can be used to find more information about a site. For example, on the eastern edge of High Wood, the reader will see on the map the site of the mine crater from where the German defences defied repeated assaults. Although partly hidden by thick growth, the water-filled crater can still be seen from the perimeter track. The action enacted at High Wood is described under Objective 23.

Appendix One lists a selection of military cemeteries using IGN map references accurate to within twenty meters which will enable the reader to locate them easily, but more detailed information has been excluded, there being a number of first-class cemetery guides available to the visitor. Before Endeavours Fade by Rose Coombs MBE has been in print for over twenty years, the 1991 guide The Somme Battlefields by Martin and Mary Middlebrook and Major & Mrs. Holts Battlefield Guide to the Somme are all highly recommended. Excellent walking tours including useful information about the cemeteries can also be found in Paul Reed's Walking the Somme and other books in the Battleground Europe Series.

Appendix Two gives an alphabetical list of Regiments engraved on the many faces and piers of the Thiepval Memorial. It can sometimes take some while as well as causing some discomfort, such as an aching neck, to the visitor while searching the higher sections of the piers and faces in order to find a particular name. Registers are, of course, available but they are much in demand during the main tourist season and, regrettably, some pages are occasionally torn out or complete registers taken away.

Because of the nature and format of this guide, the author has not prepared an Index. The reader should be able to find the information he needs from the contents page.

OBJECTIVE CODE NUMBERS
2nd July to 26th November 1916

The objective code numbers are not official but numbers allocated to facilitate computer data.

CODE   OBJECTIVE

1-14   The attack on the 1st July, the Big Push, from Gommecourt in the north to Montauban at the junction with the French Army (see *One Day on the Somme*)

15   Fricourt village and wood, Lonely Copse, Shelter Wood, Bottom Wood, Fricourt Farm, Lozenge Wood

16   La Boisselle, Sausage Valley, Bloater Trench, Horseshoe Trench, Kaufmann Graben, Sausage Redoubt, Schwaben Höhe, Y Sap crater

17   Ovillers, Mash Valley and trenches north towards Authuille Wood

18   Contalmaison, Quadrangle, Caterpillar Wood, Mametz Wood, Flat Iron Copse

19   Bernafay Wood, Trônes Wood, Maltzhorn Farm

20   The Dawn Attack, 14th July 1916. Bazentin-le-Petit/Bazentin-le-Grand villages and woods, Bazentin Ridge and east towards Longueval

21   Delville Wood, Longueval, Wood Lane South, Carlton Trench, Switch Trench and trenches north towards Flers

22   Pozières village and Ridge, Mouquet Farm, Fabeck Graben, the Elbow, Munster Alley, Skyline Trench, Constance Trench, Zigzag Trench

23   High Wood, Switch Line Centre, Wood Lane North, Intermediate Trench, Hook Trench, Crest Trench

24   Guillemont village, quarry and station, trenches north-west towards Delville Wood, Cochrane Alley, Mount Street, Shropshire Trench, South Street

25   Waterlot Farm, ZZ Trench, "Alcohol" Trenches between Delville Wood and Ginchy

26   Ginchy, Ginchy Telegraph, the Quadrilateral, the Triangle, Straight Trench

27   Flers village, Flers Trench East, Flers Support East, Switch Line East, Gap Trench, Bulls Road

28   Martinpuich, Switch Line West, Starfish Line, Martin Trench, Prue Trench

29   Courcelette, Sugar Factory, MacDonnell Trench, Dyke Trench, Desire Trench, Courcelette Trench, Kenora Trench, Regina Trench East and trenches north towards Grandcourt and Pys

30   Morval and Lesboeufs villages and Le Transloy Ridge

31   The "Meteorological" Trenches between Lesboeufs and Gueudecourt

32   Gueudecourt village, Gird Trench East, Gird Support East, Glebe Street, Seven Dials, Bayonet Trench East

33   Combles, Bouleaux Wood, Leuze Wood, Wedge Wood, Falfemont Farm, Angle Wood, Combles Trench, Loop Trench

34   Thiepval village and ridge, Schwaben Redoubt, St. Pierre Divion, the Mounds, Stuff Trench, Stuff Redoubt, Regina Trench West, Zollern Trench, Zollern Redoubt

35   North of the Ancre - Beaumont Hamel, Frankfort and Munich Trenches, Beaucourt-sur-Ancre, Redan Ridge East and north to Serre

36   Destremont Farm, Le Sars, Gird Trench West, Gird Support West, Flers Trench West, Flers Support West, Eaucourt Abbey, Drop Alley, Goose Alley, Bayonet Trench West, the Nose, Pimple and Tail, Butte de Warlencourt

37   Battalions withdrawn from front line in order to rest, recover and refill ranks after heavy losses or be used in other sectors away from the Somme

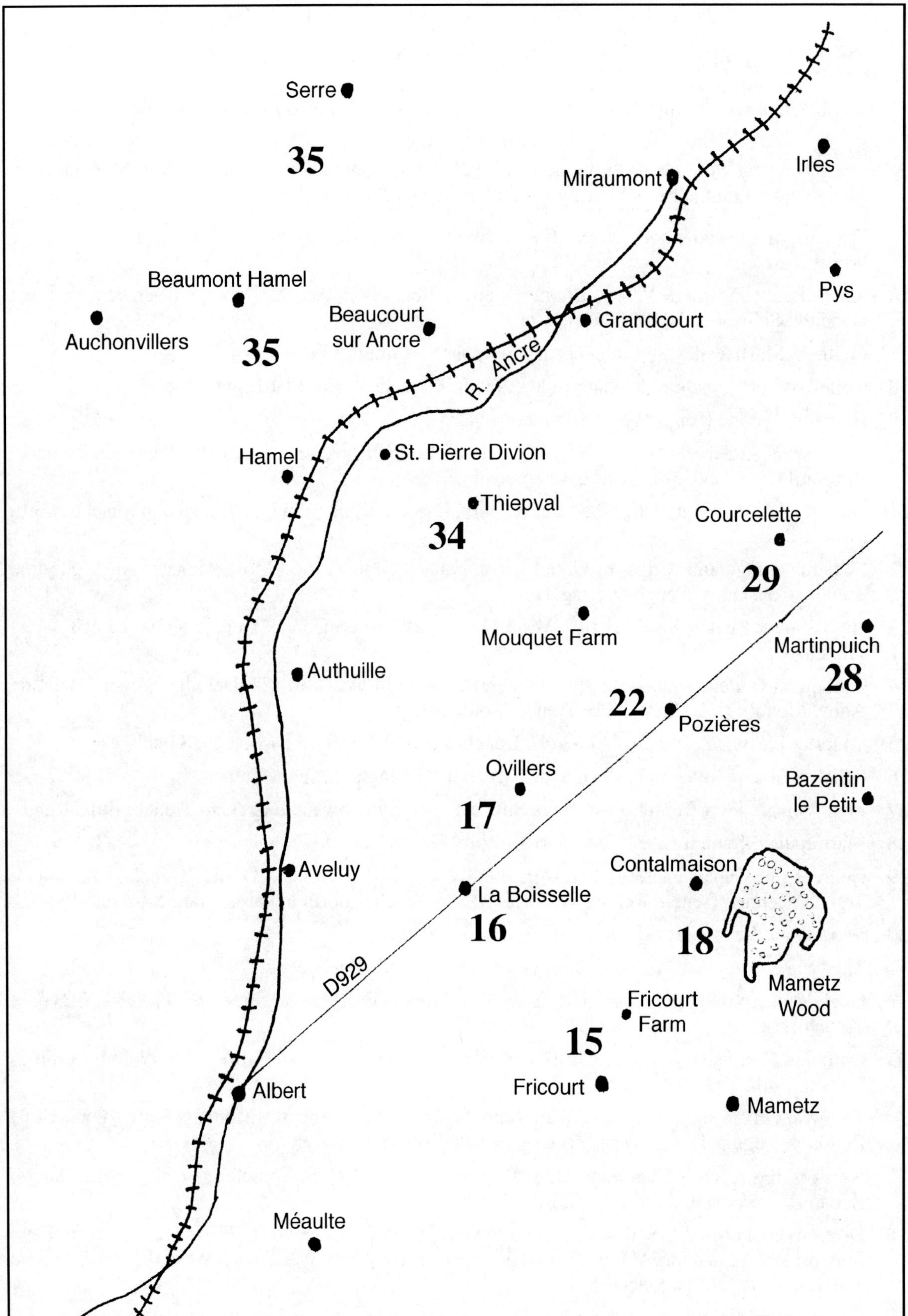
Serre
35
Miraumont
Irles
Beaumont Hamel
Pys
Beaucourt
sur Ancre
Grandcourt
Auchonvillers
35
R. Ancre
Hamel
St. Pierre Divion
Thiepval
Courcelette
34
29
Mouquet Farm
Martinpuich
Authuille
28
22
Pozières
Ovillers
Bazentin
le Petit
17
Contalmaison
Aveluy
La Boisselle
16
18
D929
Mametz
Wood
Fricourt
Farm
15
Albert
Fricourt
Mametz
Méaulte

SKETCH-MAP OF OBJECTIVES 15 TO 36

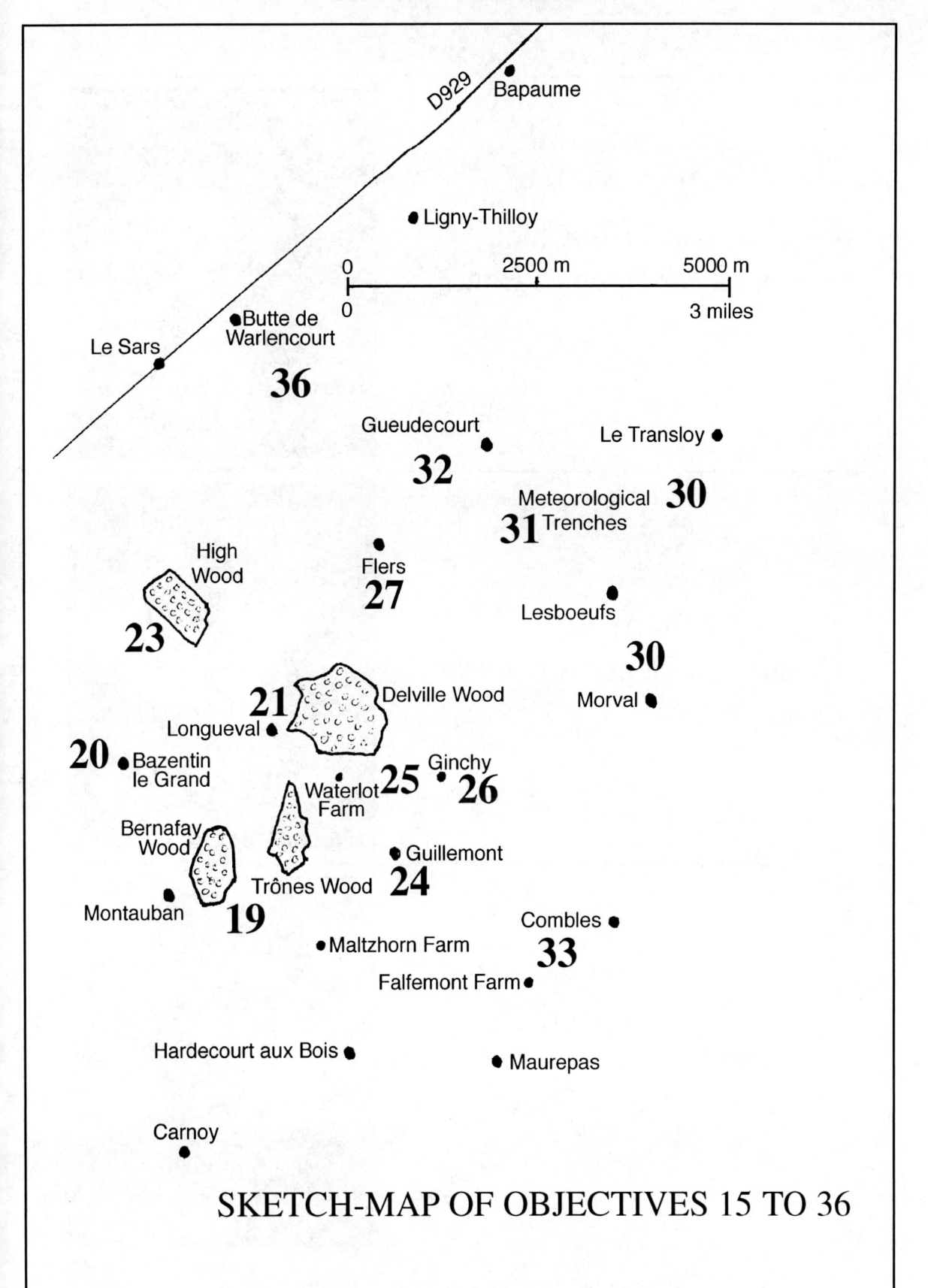

SENIOR COMMANDERS

Commander-in-Chief:
General Sir Douglas Haig

Fifth Army:
General Sir Henry Rawlinson

Reserve Army
(later the Fifth Army):
General Sir H Gough

[All photographs Michelin]

THE CORPS COMMANDERS

II CORPS	Lt.Gen Sir C.W. Jacob
III CORPS	Lt.Gen Sir William Pulteney
V CORPS	Lt.Gen E.A. Fanshawe
VI CORPS	Lt.Gen J.A.L. Haldane
VII CORPS	Lt.Gen Sir Thomas D'O Snow
VIII CORPS	Lt.Gen Sir Aylmer Hunter-Weston
X CORPS	Lt.Gen Sir Thomas Morland
XI CORPS	Lt.Gen Sir Richard Haking
XIII CORPS	Lt.Gen Sir Walter Congreve
	Lt.Gen Earl of Cavan (temporary command)
XIV CORPS	Lt.Gen Earl of Cavan
XV CORPS	Lt.Gen H.S. Horne (to 1st October)
	Lt.Gen J.P. du Cane (from 2nd October)

THE DIVISIONAL COMMANDERS

Guards	Maj.Gen G.P.T. Feilding
1st Division	Maj.Gen E.P. Strickland
2nd Division	Maj.Gen W.G. Walker
3rd Division	Maj.Gen J.A.L. Haldane (later GCO VI Corps)
	Maj.Gen C.J. Deverell (formerly GCO 20th Brigade)
4th Division	Maj.Gen Hon. W. Lambton
5th Division	Maj.Gen R.B. Stephens
6th Division	Maj.Gen C. Ross
7th Division	Maj.Gen H.E. Watts
8th Division	Maj.Gen H. Hudson
9th Division	Maj.Gen W.T. Furse
11th Division	Lt.Gen Sir Charles Woollcombe
12th Division	Maj.Gen A.B. Scott
14th Division	Maj.Gen V.A. Couper
15th Division	Maj.Gen F.W.N. McCracken
16th Division	Maj.Gen W.B. Hickie
17th Division	Maj.Gen T.D. Pilcher
	Maj.Gen P.R. Robertson (from the end of July)
18th Division	Maj.Gen F.I. Maxse
19th Division	Maj.Gen G.T.M. Bridges
20th Division	Maj.Gen W. Douglas Smith
21st Division	Maj.Gen D.G.M. Campbell
23rd Division	Maj.Gen J.M. Babington
24th Division	Maj.Gen J.E. Capper
25th Division	Maj.Gen E.G.T. Bainbridge
29th Division	Maj.Gen H. de B. Lisle
30th Division	Maj.Gen J.S.M. Shea
31st Division	Maj.Gen R. Wanless O'Gowan
32nd Division	Maj.Gen W.H. Rycroft
33rd Division	Maj.Gen H.J.S. Landon
34th Division	Maj.Gen E.C. Ingouville-Williams (killed 22nd July)
	Maj.Gen C.L. Nicholson (from 25th July)
35th Division	Maj.Gen R.J. Pinney
36th Division	Maj.Gen O.S.W. Nugent
37th Division	Maj.Gen S.W. ScraseDickens (to 9th November)
	Maj.Gen H.B. Williams (from 9th November)
38th Division	Maj.Gen I. Philipps (to 11th July)
	Maj.Gen C.G. Blackader (from 12th July)
39th Division	Maj.Gen G.J. Cuthbert
41st Division	Maj.Gen S.T.B. Lawford
47th Division	Maj.Gen Sir C. St. L. Barter (to 15th September)
	Brigadier-General W.H. Greenly (16th September to 1st October)
	Maj.Gen G.F. Gorringe (from 2nd October)
48th Division	Maj.Gen R. Fanshawe
49th Division	Maj.Gen E.M. Perceval
50th Division	Maj.Gen P.S. Wilkinson
51st Division	Maj.Gen G.M. Harper
55th Division	Maj.Gen H.S. Jeudwine
56th Division	Maj.Gen C.P.A. Hull
63rd Division	Maj.Gen Sir Archibald Paris (severely wounded 12th Oct)
	Maj.Gen C.D. Shute (from 12th Oct- formerly GCO 59th Brigade)

AUSTRALIAN

1st Division	Maj.Gen H.B. Walker
2nd Division	Maj.Gen J.G. Legge
4th Division	Maj.Gen Sir Herbert Cox
5th Division	Maj.Gen Hon. J.W. McCay

CANADIAN

1st Division	Maj.Gen A.W. Currie
2nd Division	Maj.Gen R.E.W. Turner
3rd Division	Maj.Gen L.J. Lipsett
4th Division	Maj.Gen D. Watson

NEW ZEALAND

1st Division	Maj.Gen Sir Andrew H. Russell

TRANSFERS BETWEEN UNITS DURING THE SOMME CAMPAIGN

24th Brigade of 8th Division with 23rd Division until 15th July, in exchange for the 70th Brigade, 23rd Division

70th Brigade returned to 23rd Divison 15th July

63rd Brigade of 21st Division exchanged with 110th Brigade of 37th Division on the 7th July

19/Lancashire Fusiliers of 14th Brigade, 32nd Division replaced by a battalion of the Royal Scots Regiment on 29th July. The 19/Lancashire Fusiliers left the front line on the 15th July and a short time later, trained as pioneers. They were allotted in this new role to the 49th (WEST RIDING) Division on 6th August, the original pioneers, 3/Monmouth being disbanded and sent to replace losses in other units

17/Northumberland Fusiliers, pioneers of 32nd Division, replaced by 12/Loyal North Lancs on 19th October

102nd Brigade of 34th Division attached to 37th Division from the 7th July to 21st August and replaced by the 37th Division's 111th Brigade

103rd Brigade of 34th Division attached to 37th Division from the 7th July to 21st August and replaced by the 37th Division's 112th Brigade

18/Northumberland Fusiliers, pioneers of 34th Division attached to 37th Division from the 7th July to 21st August and replaced by the 37th Division's pioneers, the 9/North Staffs

AMALGAMATIONS*

Battalions	Renamed 1916	Date	Shown in this guide as
4th & 5th Black Watch	4th/5th Black Watch	15th March	5/Black Watch
6th & 7th RSF	6th/7th RSF	7th May	7/RSF
8th & 10th Gordon High.	8th/10th Gordon High.	11th May	10/Gordon High.
10th & 11th HLI	10th/11th HLI	14th May	11/HLI
7th & 8th KOSB	7th/8th KOSB	28th May	8/KOSB

* Source: Ray Westlake Archives

ABOUT THE ALLOCATION OF OBJECTIVES TO INFANTRY AND PIONEER BATTALIONS

If this guide is to be of any real use to the reader, it is of great importance that the statistical information given here is accurate and I have done my best to achieve this. I have occasionally come across conflicting information and in these cases, the information has been rechecked and cross-checked and, therefore, any remaining errors or ommisions are solely of my own making.

To meet the needs of those wishing to trace the movements of a single battalion, brigade or division, there were many factors to be considered. The aim of this guide is that each battalion can be traced to specific objectives which necessitated dividing the whole of the Somme battlefield into several sectors, each one relating to a specific part of the campaign. After some considerable thought, the battleground was divided into twenty-two sectors, numbered 15 to 36* allowing the movements of the infantry battalions to be followed more clearly. Details of these sectors or objectives are shown on pages 12 and 13. It is inevitable that there has to be some over-lap of the boundaries of the various objectives and in the section dealing with the description of the infantry attacks, some have been amalgamated to achieve better continuity.

There was then the question of how best to allocate the battalions to the objectives. The leading and support battalions' objectives and roles are clearly defined and documented in many books, making allocation relatively easy. But what of the reserve battalions who brought up supplies, food, water and ammunition, the artillery men, the pioneers who often worked in the most dreadful conditions under fire repairing roads and digging assembly and communication trenches, the stretcher bearers, cooks, tunnellers, engineers, drivers, motor-cyclists and runners? The risk of death was everywhere. There has to be a reasoned criteria to include or exclude battalions engaged in, for example, the capture of High Wood, known in this guide as Objective 23. Did a man have to die in the wood to qualify for the dubious honour of inclusion at Objective 23? I think not. When the waves of infantry climbed towards High Wood protected by the reverse slope, they were not immune to death and many fell long before the German machine guns cut them down as they topped the rise leading to the dark mass of trees. Many more died digging forward trenches to reduce the width of the open land before the wood. And so finally, the adopted criteria was quite simply that any battalion serving the needs of the capture and consolidation of High Wood has been included in that particular objective coding.

In the section starting on page 88 which describes the attacks made by the infantry, it has not been possible to give a full description of those units which, although at the front, did not participate in an actual engagement. The fact that they were on stand-by or on general front line duty has served as justification for inclusion at the objective in question. They were often, like the leading and supporting waves, the subject of a hostile barrage.

* Objectives 1 to14 cover the events of the 1st July 1916 and are the subject of a previous publication, *'One day on the Somme'* (GMS Enterprises)

DISPOSITION BY DIVISION

Div.	Brig.	No. Battalion	Objective
Guards			
	1	2 / Grenadier Guards	26, 30, 31, 32, 35
	1	2 / Coldstream Guards	26, 30, 32, 35
	1	3 / Coldstream Guards	26, 30, 32, 35
	1	1 / Irish Guards	26, 30, 31, 32, 35
	2	3 / Grenadier Guards	26, 30, 35
	2	1 / Coldstream Guards	26, 30, 31
	2	1 / Scots Guards	26, 30, 31, 35
	2	2 / Irish Guards	26, 30, 35
	3	1 / Grenadier Guards	24, 26, 30, 31, 32, 35
	3	4 / Grenadier Guards	26, 30, 31, 32, 33, 35
	3	2 / Scots Guards	26, 30, 31, 32, 35
	3	1 / Welsh Guards	26, 30, 31, 32, 35
		4 / Coldstream Guards (Pnrs)	26, 30, 35
1	1	10 / Gloucestershire	20, 21, 23, 28, 36
1	1	1 / Black Watch	18, 20, 23, 36
1	1	8 / Royal Berkshire	18, 20, 21, 23, 28, 36
1	1	1 / Cameron Highlanders	18, 20, 23, 28, 36
1	2	2 / Royal Sussex	18, 22, 23, 28
1	2	1 / Loyal North Lancashire	20, 22, 23, 36
1	2	1 / Northamptonshire	18, 20, 23, 28, 36
1	2	2 / King's Royal Rifle Corps	18, 20, 23, 28, 36
1	3	1 / South Wales Borderers	18, 20, 21, 22, 23
1	3	1 / Gloucestershire	18, 20, 23
1	3	2 / Welsh	18, 20, 22, 23, 28
1	3	2 / Royal Munster Fusiliers	18, 20, 21, 22, 23, 28
1		6 / Welsh (Pnrs)	18, 20, 22, 23, 28, 34
2	5	17 / Royal Fusiliers	21, 25, 35
2	5	24 / Royal Fusiliers	24, 25, 35
2	5	2 / Oxford & Bucks Light Inf.	24, 25, 35
2	5	2 / Highland Light Infantry	25, 35
2	6	1 / King's Liverpool	21, 24, 35
2	6	2 / South Staffordshire	21, 35
2	6	13 / Essex	21, 24, 25, 35
2	6	17 / Middlesex	21, 24, 25, 35
2	99	22 / Royal Fusiliers	21, 35
2	99	23 / Royal Fusiliers	21, 35
2	99	1 / Royal Berkshire	21, 35
2	99	1 / King's Royal Rifle Corps	21, 35
2		10 / Duke of Cornwall L.I. (Pnrs)	19, 21, 24, 35
3	8	2 / Royal Scots	19, 20, 24, 25, 35
3	8	8 / East Yorkshire	18, 19, 20, 23, 24, 35
3	8	1 / Royal Scots Fusiliers	19, 20, 24, 35
3	8	7 / King's Shropshire L. Inf.	19, 20, 21, 24, 25, 35
3	9	1 / Northumberland Fusiliers	20, 21, 24, 35
3	9	4 / Royal Fusiliers	18, 20, 21, 24, 35
3	9	13 / King's Liverpool	20, 21, 24, 35
3	9	12 / West Yorkshire	18, 20, 21, 24, 35
3	76	8 / King's Own Roy. Lancaster	21, 24, 35
3	76	2 / Suffolk	18, 19, 21, 24, 35
3	76	10 / Royal Welsh Fusiliers	18, 21, 24, 35
3	76	1 / Gordon Highlanders	18, 19, 21, 24, 35
3		20 / King's Royal Rifle C. (Pnrs)	18, 20, 21
4	10	1 / Royal Warwickshire	30, 31, 35
4	10	2 / Seaforth Highlanders	30, 31, 35
4	10	1 / Royal Irish Fusiliers	30, 31, 35
4	10	2 / Royal Dublin Fusiliers	30, 31, 35
4	11	1 / Somerset Light Infantry	30, 31, 35

DISPOSITION BY DIVISION

Div.	Brig.	No.	Battalion	Objective
4	11	1 / East Lancashire		30, 31, 35
4	11	1 / Hampshire		30, 31, 35
4	11	1 / Rifle Brigade		30, 31, 35
4	12	1 / King's Own Roy. Lancaster		26, 30, 31, 35
4	12	2 / Lancashire Fusiliers		30, 31, 35
4	12	2 / Essex		31, 35
4	12	2 / Duke of Wellington		31, 35
4		21 / West Yorkshire (Pnrs)		30, 31, 35
5	13	14 / Royal Warwickshire		20, 21, 23, 24, 30, 33
5	13	15 / Royal Warwickshire		20, 21, 23, 33
5	13	2 / King's Own Scott. Borderers		20, 21, 23, 30, 33
5	13	1 / Royal West Kent-Queen's Own		20, 21, 23, 30, 33
5	15	16 / Royal Warwickshire		21, 30, 33
5	15	1 / Norfolk		21, 23, 30, 33
5	15	1 / Bedfordshire		21, 23, 25, 26, 30, 33
5	15	1 / Cheshire		21, 23, 26, 30, 33
5	95	1 / Devonshire		21, 24, 30, 33
5	95	12 / Gloucestershire		21, 24, 26, 30, 33
5	95	1 / East Surrey		21, 24, 30, 33
5	95	1 / Duke of Cornwall Light Inf.		19, 21, 24, 33
5		6 / Argyll & Sutherland (Pnrs)		23, 24, 30
6	16	1 / Buffs (East Kent)		26, 30, 33, 35
6	16	8 / Bedfordshire		24, 26, 30, 31, 33, 35
6	16	1 / King's Shropshire L. Inf.		24, 26, 30, 31, 32, 35
6	16	2 / York & Lancaster		26, 30, 31, 33, 35
6	18	1 / West Yorkshire		26, 30, 31, 32, 35
6	18	11 / Essex		26, 30, 31, 32, 33, 35
6	18	2 / Durham Light Infantry		26, 30, 31, 32, 35
6	18	14 / Durham Light Infantry		26, 30, 31, 35
6	71	9 / Norfolk		26, 31, 32, 35
6	71	9 / Suffolk		26, 30, 31, 34
6	71	1 / Leicestershire		19, 26, 30, 31, 32, 35
6	71	2 / Sherwood Foresters		26, 30, 31, 35
6		11 / Leicestershire (Pnrs)		24, 35
7	20	8 / Devonshire		18, 20, 23, 25, 26
7	20	9 / Devonshire		18, 20, 23, 25, 26
7	20	2 / Border		18, 20, 24, 26
7	20	2 / Gordon Highlanders		18, 20, 23, 25, 26
7	22	2 / Royal Warwickshire		18, 20, 23, 26
7	22	2 / Royal Irish		15, 18, 20, 23, 25, 26
7	22	1 / Royal Welsh Fusiliers		18, 20, 21, 23, 25, 26
7	22	20 / Manchester		18, 20, 23, 25, 26
7	91	2 / Queen's - Royal West Surrey		18, 21, 23, 25
7	91	1 / South Staffordshire		18, 20, 21, 23, 25
7	91	21 / Manchester		15, 18, 20, 23, 25
7	91	22 / Manchester		18, 21, 23, 25
7		24 / Manchester (Pnrs)		18, 20, 21, 23, 24, 34
8	23	2 / Devonshire		17, 30, 31
8	23	2 / West Yorkshire		17, 30, 31
8	23	2 / Middlesex		17, 30, 31
8	23	2 / Scott. Rifles-Cameronians		17, 31, 36
8	24	1 / Worcestershire		18, 30, 31, 32
8	24	1 / Sherwood Foresters		16, 18, 31
8	24	2 / Northamptonshire		18, 30, 31
8	24	2 / East Lancashire		18, 30, 31, 36
8	25	2 / Lincolnshire		17, 30, 31
8	25	2 / Royal Berkshire		17, 30, 31, 32
8	25	1 / Royal Irish Rifles		17, 30, 31
8	25	2 / Rifle Brigade		17, 30, 31
8		22 / Durham Light Infantry (Pnrs)		17, 30, 31

Div.	Brig.	No.	Battalion	Objective
SCOTTISH				
9	26	8 / Black Watch		19, 20, 21, 36
9	26	7 / Seaforth Highlanders		19, 20, 21, 25, 36
9	26	5 / Cameron Highlanders		20, 21, 25, 36
9	26	10 / Argyll & Sutherland High.		20, 21, 36
9	27	11 / Royal Scots		18, 19, 20, 21, 28, 36
9	27	12 / Royal Scots		18, 19, 20, 21, 36
9	27	6 / King's Own Scott. Borderers		18, 20, 21, 36
9	27	9 / Scott. Rifles-Cameronians		18, 20, 21, 36
9	S.African	1 / Reg. Cape Prov.		21, 25, 36
9	S.African	2 / Reg. Natal & OFS		21, 36
9	S.African	3 / Reg. Trans. & Rhod.		21, 36
9	S.African	4 / Reg. Scottish		19, 21, 25, 36
9		9 / Seaforth Highlanders (Pnrs)		15, 19, 20, 21, 28, 36
11	32	9 / West Yorkshire		22, 34
11	32	6 / Green Howards (Yorks Reg)		34
11	32	8 / Duke of Wellington		34
11	32	6 / York & Lancaster		22, 34, 35
11	33	6 / Lincolnshire		22, 34
11	33	6 / Border		22, 34
11	33	7 / South Staffordshire		34
11	33	9 / Sherwood Foresters		22, 34
11	34	8 / Northumberland Fusiliers		34
11	34	9 / Lancashire Fusiliers		22, 34
11	34	5 / Dorsetshire		22, 34
11	34	11 / Manchester		22, 34
11		6 / East Yorkshire (Pnrs)		22, 34
EASTERN				
12	35	7 / Norfolk		17, 22, 32
12	35	7 / Suffolk		17, 22, 32, 35
12	35	9 / Essex		16, 17, 22, 32
12	35	5 / Royal Berkshire		17, 22, 35, 36
12	36	8 / Royal Fusiliers		17, 22, 31, 32, 36
12	36	9 / Royal Fusiliers		17, 22, 31, 32
12	36	7 / Royal Sussex		17, 22, 32
12	36	11 / Middlesex		17, 22, 34, 36
12	37	6 / Queen's - Royal West Surrey		17, 22, 31, 32
12	37	6 / Buffs (East Kent)		17, 18, 22, 31, 32, 34
12	37	7 / East Surrey		17, 22, 32, 34, 36
12	37	6 / Royal West Kent-Queen's Own		17, 22, 31, 32, 34
12		5 / Northamptonshire (Pnrs)		17, 22, 31, 32
LIGHT				
14	41	7 / King's Royal Rifle Corps		21, 25, 27
14	41	8 / King's Royal Rifle Corps		21, 25, 27
14	41	7 / Rifle Brigade		21, 25, 27
14	41	8 / Rifle Brigade		21, 27
14	42	5 / Oxford & Bucks Light Inf.		21, 25, 27, 32
14	42	5 / King's Shropshire L. Inf.		21, 25, 27, 32
14	42	9 / King's Royal Rifle Corps		21, 25, 27, 32
14	42	9 / Rifle Brigade		21, 25, 27
14	43	6 / Somerset Light Infantry		21, 25, 27, 32
14	43	6 / Duke of Cornwall Light Inf.		21, 25, 27, 32
14	43	6 / King's Own Yorks L.I.		21, 25, 27, 32
14	43	10 / Durham Light Infantry		21, 25, 27, 32
14		11 / King's Liverpool (Pnrs)		21
SCOTTISH				
15	44	9 / Black Watch		22, 23, 28, 36
15	44	8 / Seaforth Highlanders		22, 23, 28, 36
15	44	10 / Gordon Highlanders		20, 28, 36
15	44	7 / Cameron Highlanders		22, 23, 28, 36
15	45	13 / Royal Scots		20, 21, 28, 36

DISPOSITION BY DIVISION

Div.	Brig.	No. Battalion	Objective
15	45	7 / Royal Scots Fusiliers	22, 28, 36
15	45	6 / Cameron Highlanders	21, 22, 28, 36
15	45	11 / Argyll & Sutherland High.	22, 28, 36
15	46	10 / Scott. Rifles-Cameronians	23, 28, 36
15	46	8 / King's Own Scott. Borderers	28, 36
15	46	11 / Highland Light Infantry	22, 23, 28, 36
15	46	12 / Highland Light Infantry	22, 23, 28, 36
15		9 / Gordon Highlanders (Pnrs)	23, 28, 36

IRISH

Div.	Brig.	No. Battalion	Objective
16	47	6 / Royal Irish	24, 26
16	47	6 / Connaught Rangers	24, 26
16	47	7 / Leinster	24, 25, 26
16	47	8 / Royal Munster Fusiliers	19, 24, 26
16	48	7 / Royal Irish Rifles	24, 26
16	48	1 / Royal Munster Fusiliers	19, 26
16	48	8 / Royal Dublin Fusiliers	26
16	48	9 / Royal Dublin Fusiliers	24, 26, 33
16	49	7 / Royal Inniskilling Fus.	24, 26, 33
16	49	8 / Royal Inniskilling Fus.	24, 26, 33
16	49	7 / Royal Irish Fusiliers	24, 26, 33
16	49	8 / Royal Irish Fusiliers	24, 33
16		11 / Hampshire (Pnrs)	19, 24, 26

NORTHERN

Div.	Brig.	No. Battalion	Objective
17	50	10 / West Yorkshire	15, 21, 30, 31, 35
17	50	7 / East Yorkshire	15, 18, 21, 30, 31, 35
17	50	7 / Green Howards (Yorks Reg)	15, 18, 21, 31, 35
17	50	6 / Dorsetshire	15, 18, 21, 31, 35
17	51	7 / Lincolnshire	15, 18, 21, 31, 32, 35
17	51	7 / Border	15, 18, 21, 31, 32.35
17	51	8 / South Staffordshire	15, 18, 21, 35
17	51	10 / Sherwood Foresters	15, 18, 21, 32, 35
17	52	9 / Northumberland Fusiliers	18, 21, 30, 31, 32
17	52	10 / Lancashire Fusiliers	18, 20, 21, 32, 35
17	52	9 / Duke of Wellington	18, 21, 31, 32, 35
17	52	12 / Manchester	18, 21, 30, 35
17		7 / York & Lancaster (Pnrs)	15, 18, 21, 35

EASTERN

Div.	Brig.	No. Battalion	Objective
18	53	8 / Norfolk	18, 21, 29, 34
18	53	8 / Suffolk	18, 19, 21, 29, 34
18	53	10 / Essex	18, 21, 29, 34
18	53	6 / Royal Berkshire	18, 21, 29, 34
18	54	11 / Royal Fusiliers	18, 19, 34, 35
18	54	7 / Bedfordshire	18, 19, 34
18	54	6 / Northamptonshire	18, 19, 22, 29, 34
18	54	12 / Middlesex	18, 19, 34
18	55	7 / Queen's - Royal West Surrey	18, 19, 29, 34
18	55	7 / Buffs (East Kent)	18, 19, 29, 34
18	55	8 / East Surrey	18, 19, 29, 34
18	55	7 / Royal West Kent-Queen's Own	18, 19, 29, 34
18		8 / Royal Sussex (Pnrs)	18, 19, 22, 24, 29, 34

WESTERN

Div.	Brig.	No. Battalion	Objective
19	56	7 / King's Own Roy. Lancaster	16, 17, 18, 20, 23, 34
19	56	7 / East Lancashire	16, 20, 28, 34, 35
19	56	7 / South Lancashire	16, 20, 28, 34, 35
19	56	7 / Loyal North Lancashire	16, 21, 34, 35
19	57	10 / Royal Warwickshire	16, 20, 23, 34
19	57	8 / Gloucestershire	16, 20, 21, 23, 34, 35
19	57	10 / Worcestershire	16, 20, 23, 34
19	57	8 / North Staffordshire	16, 18, 20, 34
19	58	9 / Cheshire	16, 20, 29, 34, 35
19	58	9 / Royal Welsh Fusiliers	16, 18, 20, 34, 35

Div.	Brig.	No.	Battalion	Objective
19	58	9 / Welsh		16, 18, 20, 34, 35
19	58	6 / Wiltshire		16, 18, 20, 34
19		5 / South Wales Borderers (Pnrs)		16, 20, 21, 22, 23, 34
LIGHT				
20	59	10 / King's Royal Rifle Corps		24, 30, 32, 35
20	59	11 / King's Royal Rifle Corps		24, 30, 32, 35
20	59	10 / Rifle Brigade		24, 26, 32, 35
20	59	11 / Rifle Brigade		24, 26, 30, 32, 35
20	60	6 / Oxford & Bucks Light Inf.		24, 25, 30, 31, 35
20	60	6 / King's Shropshire L. Inf.		24, 25, 30, 35
20	60	12 / King's Royal Rifle Corps		21, 24, 25, 30, 31, 32, 35
20	60	12 / Rifle Brigade		24, 26, 30, 31, 32, 35
20	61	7 / Somerset Light Infantry		24, 26, 31, 32, 33, 35
20	61	7 / Duke of Cornwall Light Inf.		24, 26, 30, 31, 32, 35
20	61	7 / King's Own Yorks L.I.		24, 30, 31, 32, 35
20	61	12 / King's Liverpool		24, 26, 30, 31, 32, 35
20		11 / Durham Light Infantry (Pnrs)		19, 24, 26, 30, 31, 35
21	62	12 / Northumberland Fusiliers		15, 18, 27, 32
21	62	13 / Northumberland Fusiliers		15, 18, 27
21	62	1 / Lincolnshire		15, 18, 27, 30, 32
21	62	10 / Green Howards (Yorks Reg)		15, 18, 20, 31, 32
21	63	8 / Lincolnshire		15, 35
21	63	8 / Somerset Light Infantry		15, 35
21	63	4 / Middlesex		15, 35
21	63	10 / York & Lancaster		15, 35
21	64	1 / East Yorkshire		15, 16, 20, 27, 32
21	64	9 / King's Own Yorks L.I.		15, 27, 30, 32
21	64	10 / King's Own Yorks L.I.		15, 20, 27, 30, 32
21	64	15 / Durham Light Infantry		15, 18, 20, 31, 32
21		14 / Northumberland Fus. (Pnrs)		15, 18, 20, 27, 30, 32
23	68	10 / Northumberland Fusiliers		16, 20, 21, 22, 28, 29, 36
23	68	11 / Northumberland Fusiliers		18, 21, 22, 28, 29, 36
23	68	12 / Durham Light Infantry		16, 18, 22, 28, 36
23	68	13 / Durham Light Infantry		18, 22, 23, 28, 36
23	69	11 / West Yorkshire		16, 18, 22, 28, 36
23	69	8 / Green Howards (Yorks Reg)		16, 18, 20, 22, 36
23	69	9 / Green Howards (Yorks Reg)		16, 18, 20, 22, 28, 36
23	69	10 / Duke of Wellington		16, 18, 22, 36
23	70	11 / Sherwood Foresters		17, 28, 34, 36
23	70	8 / King's Own Yorks L.I.		17, 20, 36
23	70	8 / York & Lancaster		17, 28, 36
23	70	9 / York & Lancaster		17, 28, 29, 36
23		9 / South Staffordshire (Pnrs)		18, 22, 28, 36
24	17	8 / Buffs (East Kent)		21, 23, 24, 25
24	17	1 / Royal Fusiliers		21, 24
24	17	12 / Royal Fusiliers		21, 24, 25
24	17	3 / Rifle Brigade		21, 24
24	72	8 / Queen's - Royal West Surrey		21, 24
24	72	9 / East Surrey		21, 24, 25
24	72	8 / Royal West Kent-Queen's Own		19, 21, 24
24	72	1 / North Staffordshire		21, 24, 25
24	73	9 / Royal Sussex		21, 24
24	73	7 / Northamptonshire		21, 24
24	73	13 / Middlesex		21, 24
24	73	2 / Leinster		19, 21, 24
24		12 / Sherwood Foresters (Pnrs)		19, 20, 21, 24
25	7	10 / Cheshire		16, 17, 34, 35
25	7	3 / Worcestershire		16, 17, 34, 35
25	7	8 / Loyal North Lancashire		17, 34, 35
25	7	1 / Wiltshire		17, 22, 34, 35
25	74	11 / Lancashire Fusiliers		17, 22, 29, 34, 35

DISPOSITION BY DIVISION

Div.	Brig.	No. Battalion	Objective
25	74	13 / Cheshire	16, 17, 22, 34, 35
25	74	9 / Loyal North Lancashire	16, 17, 34, 35
25	74	2 / Royal Irish Rifles	16, 17, 34, 35
25	75	11 / Cheshire	17, 34, 35
25	75	8 / Border	17, 34, 35
25	75	2 / South Lancashire	17, 34, 35
25	75	8 / South Lancashire	17, 34, 35
25		6 / South Wales Borderers (Pnrs)	16, 17, 34, 35
29	86	2 / Royal Fusiliers	27, 35, 36
29	86	1 / Lancashire Fusiliers	27, 35
29	86	16 / Middlesex	35, 36
29	86	1 / Royal Dublin Fusiliers	35, 36
29	87	2 / South Wales Borderers	30, 32, 35
29	87	1 / King's Own Scott. Borderers	30, 31, 32, 35
29	87	1 / Royal Inniskilling Fus.	30, 32, 35
29	87	1 / Border	30, 32, 35
29	88	4 / Worcestershire	27, 31, 32, 35
29	88	1 / Essex	31, 32, 35
29	88	1 / Newfoundland Regiment	31, 32, 35
29	88	2 / Hampshire	31, 32, 35
29		2 / Monmouth (Pnrs)	26, 27, 30, 35
30	21	18 / King's Liverpool	18, 19, 36
30	21	2 / Green Howards (Yorks Reg)	18, 19, 21, 24, 36
30	21	2 / Wiltshire	18, 19, 24, 36
30	21	19 / Manchester	18, 19, 24, 36
30	89	17 / King's Liverpool	18, 19, 24, 36
30	89	19 / King's Liverpool	18, 19, 24, 36
30	89	20 / King's Liverpool	18, 19, 24, 32, 36
30	89	2 / Bedfordshire	18, 19, 24, 32, 36
30	90	2 / Royal Scots Fusiliers	18, 19, 24, 36
30	90	16 / Manchester	18, 19, 24, 36
30	90	17 / Manchester	18, 19, 24, 36
30	90	18 / Manchester	18, 19, 24, 36
30		11 / South Lancashire (Pnrs)	18, 19, 30
31	92	10 / East Yorkshire	35
31	92	11 / East Yorkshire	35
31	92	12 / East Yorkshire	35
31	92	13 / East Yorkshire	35
31	93	15 / West Yorkshire	35
31	93	16 / West Yorkshire	35
31	93	18 / West Yorkshire	35
31	93	18 / Durham Light Infantry	35
31	94	11 / East Lancashire	35
31	94	12 / York & Lancaster	35
31	94	13 / York & Lancaster	35
31	94	14 / York & Lancaster	35
31		12 / King's Own Yorks L.I. (Pnrs)	35
32	14	19 / Lancashire Fusiliers	17, 34
32	14	1 / Dorsetshire	17, 20, 34, 35
32	14	2 / Manchester	17, 34, 35
32	14	15 / Highland Light Infantry	17, 34, 35
32	96	16 / Northumberland Fusiliers	17, 34
32	96	15 / Lancashire Fusiliers	17, 34, 35
32	96	16 / Lancashire Fusiliers	17, 34
32	96	2 / Royal Inniskilling Fus.	17, 34, 35
32	97	11 / Border	17, 34, 35
32	97	2 / King's Own Yorks L.I.	17, 34, 35
32	97	16 / Highland Light Infantry	17, 34, 35
32	97	17 / Highland Light Infantry	17, 34, 35
32		17 / Northumberland Fus. (Pnrs)	22, 34, 35
33	19	20 / Royal Fusiliers	20, 21, 23, 30

Div.	Brig.	No. Battalion	Objective
33	19	2 / Royal Welsh Fusiliers	20, 21, 23, 30, 31
33	19	1 / Scott. Rifles-Cameronians	20, 23, 30, 31, 35
33	19	6 / Scott. Rifles-Cameronians	20, 21, 23, 30, 31, 35, 36
33	98	4 / King's Liverpool	20, 21, 23, 30, 31, 35
33	98	4 / Suffolk	20, 21, 23, 31, 34
33	98	1 / Middlesex	20, 21, 23, 28, 31, 35, 36
33	98	2 / Argyll & Sutherland High.	20, 23, 30, 31
33	100	1 / Queen's - Royal West Surrey	20, 21, 23, 28, 31, 35
33	100	2 / Worcestershire	18, 20, 21, 23, 30, 31, 35
33	100	16 / King's Royal Rifle Corps	18, 20, 21, 23, 30, 31, 35
33	100	9 / Highland Light Infantry	20, 21, 23, 28, 30, 31, 35
33		18 / Middlesex (Pnrs)	20, 21, 23
34	101	15 / Royal Scots	16, 20, 21
34	101	16 / Royal Scots	16, 21
34	101	10 / Lincolnshire	16, 20, 21
34	101	11 / Suffolk	16, 20, 23
34	102	20 / Northumberland Fusiliers	16, 37
34	102	21 / Northumberland Fusiliers	16, 37
34	102	22 / Northumberland Fusiliers	16, 37
34	102	23 / Northumberland Fusiliers	16, 37
34	103	24 / Northumberland Fusiliers	16, 28
34	103	25 / Northumberland Fusiliers	16, 28
34	103	26 / Northumberland Fusiliers	16, 28
34	103	27 / Northumberland Fusiliers	16, 28
34		18 / Northumberland Fus. (Pnrs)	16, 28, 35
BANTAM			
35	104	17 / Lancashire Fusiliers	19, 24, 33
35	104	18 / Lancashire Fusiliers	19, 20, 24, 33
35	104	20 / Lancashire Fusiliers	20, 33
35	104	23 / Manchester	19, 20, 24
35	105	15 / Cheshire	19, 24
35	105	16 / Cheshire	19, 24, 25
35	105	14 / Gloucestershire	18, 19, 24
35	105	15 / Sherwood Foresters	19, 24
35	106	17 / Royal Scots	19, 20, 25
35	106	17 / West Yorkshire	18, 19, 21, 24
35	106	19 / Durham Light Infantry	18, 19, 21
35	106	18 / Highland Light Infantry	18, 19, 21
35		19 / Northumberland Fus. (Pnrs)	19, 33, 35
ULSTER			
36	107	8 / Royal Irish Rifles	34, 37
36	107	9 / Royal Irish Rifles	34, 37
36	107	10 / Royal Irish Rifles	34
36	107	15 / Royal Irish Rifles	34
36	108	11 / Royal Irish Rifles	34, 37
36	108	12 / Royal Irish Rifles	34, 37
36	108	13 / Royal Irish Rifles	34, 37
36	108	9 / Royal Irish Fusiliers	34, 37
36	109	9 / Royal Inniskilling Fus.	34
36	109	10 / Royal Inniskilling Fus.	34
36	109	11 / Royal Inniskilling Fus.	34
36	109	14 / Royal Irish Rifles	34, 37
36		16 / Royal Irish Rifles (Pnrs)	34
37	110	6 / Leicestershire	18, 20, 32
37	110	7 / Leicestershire	18, 20, 21, 32
37	110	8 / Leicestershire	18, 20, 32
37	110	9 / Leicestershire	18, 20, 32
37	111	10 / Royal Fusiliers	16, 22, 23, 35
37	111	13 / Royal Fusiliers	17, 22, 35
37	111	13 / King's Royal Rifle Corps	16, 23, 35
37	111	13 / Rifle Brigade	16, 17, 20, 35

Div.	Brig.	No.	Battalion	Objective
37	112	11 / Royal Warwickshire		16, 18, 20, 21, 22, 35
37	112	6 / Bedfordshire		16, 21, 22, 35
37	112	8 / East Lancashire		16, 20, 21, 22, 23, 35
37	112	10 / Loyal North Lancashire		16, 21, 22, 23, 35
37		9 / North Staffordshire (Pnrs)		18, 22, 35

WELSH

Div.	Brig.	No.	Battalion	Objective
38	113	13 / Royal Welsh Fusiliers		18, 35
38	113	14 / Royal Welsh Fusiliers		18, 35
38	113	15 / Royal Welsh Fusiliers		18, 35
38	113	16 / Royal Welsh Fusiliers		18, 35
38	114	10 / Welsh		18, 35
38	114	13 / Welsh		18, 35
38	114	14 / Welsh		18, 35
38	114	15 / Welsh		18, 35
38	115	10 / South Wales Borderers		18, 35
38	115	11 / South Wales Borderers		18, 35
38	115	17 / Royal Welsh Fusiliers		18, 35
38	115	16 / Welsh		18, 35
38		19 / Welsh (Pnrs)		18, 35
39	116	11 / Royal Sussex		34, 35
39	116	12 / Royal Sussex		34, 35
39	116	13 / Royal Sussex		34, 35
39	116	14 / Hampshire		34, 35
39	117	16 / Sherwood Foresters		34, 35
39	117	17 / Sherwood Foresters		34, 35
39	117	17 / King's Royal Rifle Corps		34, 35
39	117	16 / Rifle Brigade		34, 35
39	118	6 / Cheshire		34, 35
39	118	1 / Cambridgeshire		34, 35
39	118	1 / Hertfordshire		34, 35
39	118	5 / Black Watch		34
39		13 / Gloucestershire (Pnrs)		34, 35
41	122	12 / East Surrey		21, 27, 31, 32, 36
41	122	15 / Hampshire		21, 27, 32, 36
41	122	11 / Royal West Kent-Queen's Own		21, 27, 36
41	122	18 / King's Royal Rifle Corps		21, 27, 32, 36
41	123	11 / Queen's - Royal West Surrey		21, 27, 32, 36
41	123	10 / Royal West Kent-Queen's Own		21, 27, 32
41	123	23 / Middlesex		21, 27, 32, 36
41	123	20 / Durham Light Infantry		18, 19, 21, 27, 32, 36
41	124	10 / Queen's - Royal West Surrey		21, 27, 32, 36
41	124	26 / Royal Fusiliers		21, 27, 32, 36
41	124	32 / Royal Fusiliers		24, 27, 32, 36
41	124	21 / King's Royal Rifle Corps		21, 27, 32, 36
41		19 / Middlesex (Pnrs)		21, 27, 30, 32

NORTH MIDLAND

Div.	Brig.	No.	Battalion	Objective
46	137	5 / South Staffordshire		35, 37
46	137	6 / South Staffordshire		35, 37
46	137	5 / North Staffordshire		35, 37
46	137	6 / North Staffordshire		35, 37
46	138	4 / Lincolnshire		35
46	138	5 / Lincolnshire		35
46	138	4 / Leicestershire		35
46	138	5 / Leicestershire		35
46	139	5 / Sherwood Foresters		35, 37
46	139	6 / Sherwood Foresters		35, 37
46	139	7 / Sherwood Foresters		35, 37
46	139	8 / Sherwood Foresters		35, 37
46		1 / Monmouth (Pnrs)		35, 37

Div.	Brig.	No.	Battalion	Objective
2nd LONDON				
47	140	6 / London		23, 36
47	140	7 / London		23, 28, 36
47	140	8 / London		23, 36
47	140	15 / London		23, 36
47	141	17 / London		23, 28, 36
47	141	18 / London		23, 36
47	141	19 / London		23, 36
47	141	20 / London		23, 36
47	142	21 / London		23, 28, 36
47	142	22 / London		23, 28, 36
47	142	23 / London		23, 28, 36
47	142	24 / London		23, 28, 36
47		4 / Royal Welsh Fusiliers (Pnrs)		23, 36
SOUTH MIDLAND				
48	143	5 / Royal Warwickshire		17, 22, 34, 35, 36
48	143	6 / Royal Warwickshire		17, 22, 34, 35, 36
48	143	7 / Royal Warwickshire		17, 22, 34, 35, 36
48	143	8 / Royal Warwickshire		17, 22, 35, 36
48	144	4 / Gloucestershire		17, 22, 34, 35, 36
48	144	6 / Gloucestershire		17, 22, 34, 35, 36
48	144	7 / Worcestershire		17, 20, 22, 35, 36
48	144	8 / Worcestershire		16, 35, 36
48	145	5 / Gloucestershire		17, 22, 34, 35, 36
48	145	4 / Oxford & Bucks Light Inf.		22, 28, 34, 35, 36
48	145	1 / Buckinghamshire		22, 34, 35, 36
48	145	4 / Royal Berkshire		22, 35, 36
48		5 / Royal Sussex (Pnrs)		17, 22, 34
WEST RIDING				
49	146	5 / West Yorkshire		34, 35
49	146	6 / West Yorkshire		34, 35
49	146	7 / West Yorkshire		34, 35
49	146	8 / West Yorkshire		34, 35
49	147	4 / Duke of Wellington		34, 35
49	147	5 / Duke of Wellington		34, 35
49	147	6 / Duke of Wellington		34, 35
49	147	7 / Duke of Wellington		34, 35
49	148	4 / King's Own Yorks L.I.		34, 35
49	148	5 / King's Own Yorks L.I.		34, 35
49	148	4 / York & Lancaster		34, 35
49	148	5 / York & Lancaster		34, 35
49		3 / Monmouth (Pnrs)		16, 34, 37
NORTHUMBRIAN				
50	149	4 / Northumberland Fusiliers		23, 28, 36
50	149	5 / Northumberland Fusiliers		23, 28, 36
50	149	6 / Northumberland Fusiliers		23, 28, 36
50	149	7 / Northumberland Fusiliers		23, 28, 36
50	150	4 / East Yorkshire		23, 28, 36
50	150	4 / Green Howards (Yorks Reg)		28, 36
50	150	5 / Green Howards (Yorks Reg)		28, 36
50	150	5 / Durham Light Infantry		28, 36
50	151	5 / Border		23, 28, 36
50	151	6 / Durham Light Infantry		28, 36
50	151	8 / Durham Light Infantry		28, 36
50	151	9 / Durham Light Infantry		28, 36
50		7 / Durham Light Infantry (Pnrs)		28, 36
HIGHLAND				
51	152	5 / Seaforth Highlanders		20, 23, 35
51	152	6 / Seaforth Highlanders		20, 23, 35
51	152	6 / Gordon Highlanders		23, 35
51	152	8 / Argyll & Sutherland High.		23, 35

DISPOSITION BY DIVISION

Div.	Brig.	No. Battalion	Objective
51	153	6 / Black Watch	20, 23, 35
51	153	7 / Black Watch	23, 35
51	153	5 / Gordon Highlanders	20, 23, 35
51	153	7 / Gordon Highlanders	20, 23, 35
51	154	9 / Royal Scots	20, 23, 35
51	154	4 / Seaforth Highlanders	23, 35
51	154	4 / Gordon Highlanders	20, 23, 35
51	154	7 / Argyll & Sutherland High.	23, 35
51		8 / Royal Scots (Pnrs)	23, 35

WEST LANCASHIRE

Div.	Brig.	No. Battalion	Objective
55	164	4 / King's Own Roy. Lancaster	21, 24, 25, 32
55	164	8 / King's Liverpool	21, 24, 25, 27, 32, 36
55	164	5 / Lancashire Fusiliers	21, 24, 25, 26, 27
55	164	4 / Loyal North Lancashire	24, 25, 27, 32
55	165	5 / King's Liverpool	19, 21, 24, 27, 32
55	165	6 / King's Liverpool	19, 21, 24, 27, 32
55	165	7 / King's Liverpool	19, 21, 24, 27, 32
55	165	9 / King's Liverpool	21, 24, 27, 32
55	166	5 / King's Own Roy. Lancaster	21, 24, 27
55	166	10 / King's Liverpool	21, 24
55	166	5 / South Lancashire	19, 21, 25, 27
55	166	5 / Loyal North Lancashire	19, 21, 24
55		4 / South Lancashire (Pnrs)	24, 26, 27, 30, 32

1st LONDON

Div.	Brig.	No. Battalion	Objective
56	167	1 / London	24, 30, 31, 33, 35
56	167	3 / London	30, 31, 33, 35, 36
56	167	7 / Middlesex	26, 30, 31, 33, 35
56	167	8 / Middlesex	26, 30, 31, 33, 35
56	168	4 / London	24, 30, 31, 33, 35
56	168	12 / London	24, 26, 30, 31, 33, 35
56	168	13 / London	33, 35
56	168	14 / London	26, 30, 31, 33, 35
56	169	2 / London	30, 31, 33, 35, 36
56	169	5 / London	30, 31, 33, 35
56	169	9 / London	30, 31, 33, 35
56	169	16 / London	30, 33, 35
56		5 / Cheshire (Pnrs)	30, 33, 35

ROYAL NAVY (Serving as Infantry Battalions)

Div.	Brig.	No. Battalion	Objective
63	188	0 / Anson Bn.	35
63	188	0 / Howe Bn.	35
63	188	1 / Royal Marine	35
63	188	2 / Royal Marine	35
63	189	0 / Hood Bn.	35
63	189	0 / Nelson Bn.	35
63	189	0 / Hawke Bn.	35
63	189	0 / Drake Bn.	35
63	190	1 / Honourable Artillery Coy.	35
63	190	7 / Royal Fusiliers	35
63	190	4 / Bedfordshire	35
63	190	10 / Royal Dublin Fusiliers	35
63		14 / Worcestershire (Pnrs)	35

DISPOSITION BY BRIGADE

Div.	Brig.	No.	Battalion	Objective
GDS	1	Gds.Reg.	2 / Grenadier Guards	26, 30, 31, 32, 35
GDS	1	Gds.Reg.	2 / Coldstream Guards	26, 30, 32, 35
GDS	1	Gds.Reg.	3 / Coldstream Guards	26, 30, 32, 35
GDS	1	Gds.Reg.	1 / Irish Guards	26, 30, 31, 32, 35
1	1	Service	10 / Gloucestershire	20, 21, 23, 28, 36
1	1	Regular	1 / Black Watch	18, 20, 23, 36
1	1	Service	8 / Royal Berkshire	18, 20, 21, 23, 28, 36
1	1	Regular	1 / Cameron Highlanders	18, 20, 23, 28, 36
GDS	2	Gds.Reg.	3 / Grenadier Guards	26, 30, 35
GDS	2	Gds.Reg.	1 / Coldstream Guards	26, 30, 31
GDS	2	Gds.Reg.	1 / Scots Guards	26, 30, 31, 35
GDS	2	Gds.Reg.	2 / Irish Guards	26, 30, 35
1	2	Regular	2 / Royal Sussex	18, 22, 23, 28
1	2	Regular	1 / Loyal North Lancashire	20, 22, 23, 36
1	2	Regular	1 / Northamptonshire	18, 20, 23, 28, 36
1	2	Regular	2 / King's Royal Rifle Corps	18, 20, 23, 28, 36
GDS	3	Gds.Reg.	1 / Grenadier Guards	24, 26, 30, 31, 32, 35
GDS	3	Gds.Reg.	4 / Grenadier Guards	26, 30, 31, 32, 33, 35
GDS	3	Gds.Reg.	2 / Scots Guards	26, 30, 31, 32, 35
GDS	3	Gds.Reg.	1 / Welsh Guards	26, 30, 31, 32, 35
1	3	Regular	1 / South Wales Borderers	18, 20, 21, 22, 23
1	3	Regular	1 / Gloucestershire	18, 20, 23
1	3	Regular	2 / Welsh	18, 20, 22, 23, 28
1	3	Regular	2 / Royal Munster Fusiliers	18, 20, 21, 22, 23, 28
2	5	Service	17 / Royal Fusiliers	21, 25, 35
2	5	Service	24 / Royal Fusiliers	24, 25, 35
2	5	Regular	2 / Oxford & Bucks Light Inf.	24, 25, 35
2	5	Regular	2 / Highland Light Infantry	25, 35
2	6	Regular	1 / King's Liverpool	21, 24, 35
2	6	Regular	2 / South Staffordshire	21, 35
2	6	Service	13 / Essex	21, 24, 25, 35
2	6	Service	17 / Middlesex	21, 24, 25, 35
25	7	Service	10 / Cheshire	16, 17, 34, 35
25	7	Regular	3 / Worcestershire	16, 17, 34, 35
25	7	Service	8 / Loyal North Lancashire	17, 34, 35
25	7	Regular	1 / Wiltshire	17, 22, 34, 35
3	8	Regular	2 / Royal Scots	19, 20, 24, 25, 35
3	8	Service	8 / East Yorkshire	18, 19, 20, 23, 24, 35
3	8	Regular	1 / Royal Scots Fusiliers	19, 20, 24, 35
3	8	Service	7 / King's Shropshire L. Inf.	19, 20, 21, 24, 25, 35
3	9	Regular	1 / Northumberland Fusiliers	20, 21, 24, 35
3	9	Regular	4 / Royal Fusiliers	18, 20, 21, 24, 35
3	9	Service	13 / King's Liverpool	20, 21, 24, 35
3	9	Service	12 / West Yorkshire	18, 20, 21, 24, 35
4	10	Regular	1 / Royal Warwickshire	30, 31, 35
4	10	Regular	2 / Seaforth Highlanders	30, 31, 35
4	10	Regular	1 / Royal Irish Fusiliers	30, 31, 35
4	10	Regular	2 / Royal Dublin Fusiliers	30, 31, 35
4	11	Regular	1 / Somerset Light Infantry	30, 31, 35
4	11	Regular	1 / East Lancashire	30, 31, 35
4	11	Regular	1 / Hampshire	30, 31, 35
4	11	Regular	1 / Rifle Brigade	30, 31, 35
4	12	Regular	1 / King's Own Roy. Lancaster	26, 30, 31, 35
4	12	Regular	2 / Lancashire Fusiliers	30, 31, 35
4	12	Regular	2 / Essex	31, 35
4	12	Regular	2 / Duke of Wellington	31, 35
5	13	Service	14 / Royal Warwickshire	20, 21, 23, 24, 30, 33
5	13	Service	15 / Royal Warwickshire	20, 21, 23, 33
5	13	Regular	2 / King's Own Scott. Borderers	20, 21, 23, 30, 33
5	13	Regular	1 / Royal West Kent-Queen's Own	20, 21, 23, 30, 33
32	14	Service	19 / Lancashire Fusiliers	17, 34

DISPOSITION BY BRIGADE

Div.	Brig.	No.	Battalion	Objective
32	14	Regular	1 / Dorsetshire	17, 20, 34, 35
32	14	Regular	2 / Manchester	17, 34, 35
32	14	Service	15 / Highland Light Infantry	17, 34, 35
5	15	Service	16 / Royal Warwickshire	21, 30, 33
5	15	Regular	1 / Norfolk	21, 23, 30, 33
5	15	Regular	1 / Bedfordshire	21, 23, 25, 26, 30, 33
5	15	Regular	1 / Cheshire	21, 23, 26, 30, 33
6	16	Regular	1 / Buffs (East Kent)	26, 30, 33, 35
6	16	Service	8 / Bedfordshire	24, 26, 30, 31, 33, 35
6	16	Regular	1 / King's Shropshire L. Inf.	24, 26, 30, 31, 32, 35
6	16	Regular	2 / York & Lancaster	26, 30, 31, 33, 35
24	17	Service	8 / Buffs (East Kent)	21, 23, 24, 25
24	17	Regular	1 / Royal Fusiliers	21, 24
24	17	Service	12 / Royal Fusiliers	21, 24, 25
24	17	Regular	3 / Rifle Brigade	21, 24
6	18	Regular	1 / West Yorkshire	26, 30, 31, 32, 35
6	18	Service	11 / Essex	26, 30, 31, 32, 33, 35
6	18	Regular	2 / Durham Light Infantry	26, 30, 31, 32, 35
6	18	Service	14 / Durham Light Infantry	26, 30, 31, 35
33	19	Service	20 / Royal Fusiliers	20, 21, 23, 30
33	19	Regular	2 / Royal Welsh Fusiliers	20, 21, 23, 30, 31
33	19	Regular	1 / Scott. Rifles-Cameronians	20, 23, 30, 31, 35
33	19	T.F.	6 / Scott. Rifles-Cameronians	20, 21, 23, 30, 31, 35, 36
7	20	Service	8 / Devonshire	18, 20, 23, 25, 26
7	20	Service	9 / Devonshire	18, 20, 23, 25, 26
7	20	Regular	2 / Border	18, 20, 24, 26
7	20	Regular	2 / Gordon Highlanders	18, 20, 23, 25, 26
30	21	Service	18 / King's Liverpool	18, 19, 36
30	21	Regular	2 / Green Howards (Yorks Reg)	18, 19, 21, 24, 36
30	21	Regular	2 / Wiltshire	18, 19, 24, 36
30	21	Service	19 / Manchester	18, 19, 24, 36
7	22	Regular	2 / Royal Warwickshire	18, 20, 23, 26
7	22	Regular	2 / Royal Irish	15, 18, 20, 23, 25, 26
7	22	Regular	1 / Royal Welsh Fusiliers	18, 20, 21, 23, 25, 26
7	22	Service	20 / Manchester	18, 20, 23, 25, 26
8	23	Regular	2 / Devonshire	17, 30, 31
8	23	Regular	2 / West Yorkshire	17, 30, 31
8	23	Regular	2 / Middlesex	17, 30, 31
8	23	Regular	2 / Scott. Rifles-Cameronians	17, 31, 36
8	24	Regular	1 / Worcestershire	18, 30, 31, 32
8	24	Regular	1 / Sherwood Foresters	16, 18, 31
8	24	Regular	2 / Northamptonshire	18, 30, 31
8	24	Regular	2 / East Lancashire	18, 30, 31, 36
8	25	Regular	2 / Lincolnshire	17, 30, 31
8	25	Regular	2 / Royal Berkshire	17, 30, 31, 32
8	25	Regular	1 / Royal Irish Rifles	17, 30, 31
8	25	Regular	2 / Rifle Brigade	17, 30, 31
9	26	Service	8 / Black Watch	19, 20, 21, 36
9	26	Service	7 / Seaforth Highlanders	19, 20, 21, 25, 36
9	26	Service	5 / Cameron Highlanders	20, 21, 25, 36
9	26	Service	10 / Argyll & Sutherland High.	20, 21, 36
9	27	Service	11 / Royal Scots	18, 19, 20, 21, 28, 36
9	27	Service	12 / Royal Scots	18, 19, 20, 21, 36
9	27	Service	6 / King's Own Scott. Borderers	18, 20, 21, 36
9	27	Service	9 / Scott. Rifles-Cameronians	18, 20, 21, 36
11	32	Service	9 / West Yorkshire	22, 34
11	32	Service	6 / Green Howards (Yorks Reg)	34
11	32	Service	8 / Duke of Wellington	34
11	32	Service	6 / York & Lancaster	22, 34, 35
11	33	Service	6 / Lincolnshire	22, 34
11	33	Service	6 / Border	22, 34
11	33	Service	7 / South Staffordshire	34
11	33	Service	9 / Sherwood Foresters	22, 34

Div.	Brig.	No.	Battalion	Objective
11	34	Service	8 / Northumberland Fusiliers	34
11	34	Service	9 / Lancashire Fusiliers	22, 34
11	34	Service	5 / Dorsetshire	22, 34
11	34	Service	11 / Manchester	22, 34
12	35	Service	7 / Norfolk	17, 22, 32
12	35	Service	7 / Suffolk	17, 22, 32, 35
12	35	Service	9 / Essex	16, 17, 22, 32
12	35	Service	5 / Royal Berkshire	17, 22, 35, 36
12	36	Service	8 / Royal Fusiliers	17, 22, 31, 32, 36
12	36	Service	9 / Royal Fusiliers	17, 22, 31, 32
12	36	Service	7 / Royal Sussex	17, 22, 32
12	36	Service	11 / Middlesex	17, 22, 34, 36
12	37	Service	6 / Queen's - Royal West Surrey	17, 22, 31, 32
12	37	Service	6 / Buffs (East Kent)	17, 18, 22, 31, 32, 34
12	37	Service	7 / East Surrey	17, 22, 32, 34, 36
12	37	Service	6 / Royal West Kent-Queen's Own	17, 22, 31, 32, 34
14	41	Service	7 / King's Royal Rifle Corps	21, 25, 27
14	41	Service	8 / King's Royal Rifle Corps	21, 25, 27
14	41	Service	7 / Rifle Brigade	21, 25, 27
14	41	Service	8 / Rifle Brigade	21, 27
14	42	Service	5 / Oxford & Bucks Light Inf.	21, 25, 27, 32
14	42	Service	5 / King's Shropshire L. Inf.	21, 25, 27, 32
14	42	Service	9 / King's Royal Rifle Corps	21, 25, 27, 32
14	42	Service	9 / Rifle Brigade	21, 25, 27
14	43	Service	6 / Somerset Light Infantry	21, 25, 27, 32
14	43	Service	6 / Duke of Cornwall Light Inf.	21, 25, 27, 32
14	43	Service	6 / King's Own Yorks L.I.	21, 25, 27, 32
14	43	Service	10 / Durham Light Infantry	21, 25, 27, 32
15	44	Service	9 / Black Watch	22, 23, 28, 36
15	44	Service	8 / Seaforth Highlanders	22, 23, 28, 36
15	44	Service	10 / Gordon Highlanders	20, 28, 36
15	44	Service	7 / Cameron Highlanders	22, 23, 28, 36
15	45	Service	13 / Royal Scots	20, 21, 28, 36
15	45	Service	7 / Royal Scots Fusiliers	22, 28, 36
15	45	Service	6 / Cameron Highlanders	21, 22, 28, 36
15	45	Service	11 / Argyll & Sutherland High.	22, 28, 36
15	46	Service	10 / Scott. Rifles-Cameronians	23, 28, 36
15	46	Service	8 / King's Own Scott. Borderers	28, 36
15	46	Service	11 / Highland Light Infantry	22, 23, 28, 36
15	46	Service	12 / Highland Light Infantry	22, 23, 28, 36
16	47	Service	6 / Royal Irish	24, 26
16	47	Service	6 / Connaught Rangers	24, 26
16	47	Service	7 / Leinster	24, 25, 26
16	47	Service	8 / Royal Munster Fusiliers	19, 24, 26
16	48	Service	7 / Royal Irish Rifles	24, 26
16	48	Regular	1 / Royal Munster Fusiliers	19, 26
16	48	Service	8 / Royal Dublin Fusiliers	26
16	48	Service	9 / Royal Dublin Fusiliers	24, 26, 33
16	49	Service	7 / Royal Inniskilling Fus.	24, 26, 33
16	49	Service	8 / Royal Inniskilling Fus.	24, 26, 33
16	49	Service	7 / Royal Irish Fusiliers	24, 26, 33
16	49	Service	8 / Royal Irish Fusiliers	24, 33
17	50	Service	10 / West Yorkshire	15, 21, 30, 31, 35
17	50	Service	7 / East Yorkshire	15, 18, 21, 30, 31, 35
17	50	Service	7 / Green Howards (Yorks Reg)	15, 18, 21, 31, 35
17	50	Service	6 / Dorsetshire	15, 18, 21, 31, 35
17	51	Service	7 / Lincolnshire	15, 18, 21, 31, 32, 35
17	51	Service	7 / Border	15, 18, 21, 31, 32.35
17	51	Service	8 / South Staffordshire	15, 18, 21, 35
17	51	Service	10 / Sherwood Foresters	15, 18, 21, 32, 35
17	52	Service	9 / Northumberland Fusiliers	18, 21, 30, 31, 32
17	52	Service	10 / Lancashire Fusiliers	18, 20, 21, 32, 35
17	52	Service	9 / Duke of Wellington	18, 21, 31, 32, 35
17	52	Service	12 / Manchester	18, 21, 30, 35

DISPOSITION BY BRIGADE

Div.	Brig.	No.	Battalion	Objective
18	53	Service	8 / Norfolk	18, 21, 29, 34
18	53	Service	8 / Suffolk	18, 19, 21, 29, 34
18	53	Service	10 / Essex	18, 21, 29, 34
18	53	Service	6 / Royal Berkshire	18, 21, 29, 34
18	54	Service	11 / Royal Fusiliers	18, 19, 34, 35
18	54	Service	7 / Bedfordshire	18, 19, 34
18	54	Service	6 / Northamptonshire	18, 19, 22, 29, 34
18	54	Service	12 / Middlesex	18, 19, 34
18	55	Service	7 / Queen's - Royal West Surrey	18, 19, 29, 34
18	55	Service	7 / Buffs (East Kent)	18, 19, 29, 34
18	55	Service	8 / East Surrey	18, 19, 29, 34
18	55	Service	7 / Royal West Kent-Queen's Own	18, 19, 29, 34
19	56	Service	7 / King's Own Roy. Lancaster	16, 17, 18, 20, 23, 34
19	56	Service	7 / East Lancashire	16, 20, 28, 34, 35
19	56	Service	7 / South Lancashire	16, 20, 28, 34, 35
19	56	Service	7 / Loyal North Lancashire	16, 21, 34, 35
19	57	Service	10 / Royal Warwickshire	16, 20, 23, 34
19	57	Service	8 / Gloucestershire	16, 20, 21, 23, 34, 35
19	57	Service	10 / Worcestershire	16, 20, 23, 34
19	57	Service	8 / North Staffordshire	16, 18, 20, 34
19	58	Service	9 / Cheshire	16, 20, 29, 34, 35
19	58	Service	9 / Royal Welsh Fusiliers	16, 18, 20, 34, 35
19	58	Service	9 / Welsh	16, 18, 20, 34, 35
19	58	Service	6 / Wiltshire	16, 18, 20, 34
20	59	Service	10 / King's Royal Rifle Corps	24, 30, 32, 35
20	59	Service	11 / King's Royal Rifle Corps	24, 30, 32, 35
20	59	Service	10 / Rifle Brigade	24, 26, 32, 35
20	59	Service	11 / Rifle Brigade	24, 26, 30, 32, 35
20	60	Service	6 / Oxford & Bucks Light Inf.	24, 25, 30, 31, 35
20	60	Service	6 / King's Shropshire L. Inf.	24, 25, 30, 35
20	60	Service	12 / King's Royal Rifle Corps	21, 24, 25, 30, 31, 32, 35
20	60	Service	12 / Rifle Brigade	24, 26, 30, 31, 32, 35
20	61	Service	7 / Somerset Light Infantry	24, 26, 31, 32, 33, 35
20	61	Service	7 / Duke of Cornwall Light Inf.	24, 26, 30, 31, 32, 35
20	61	Service	7 / King's Own Yorks L.I.	24, 30, 31, 32, 35
20	61	Service	12 / King's Liverpool	24, 26, 30, 31, 32, 35
21	62	Service	12 / Northumberland Fusiliers	15, 18, 27, 32
21	62	Service	13 / Northumberland Fusiliers	15, 18, 27
21	62	Regular	1 / Lincolnshire	15, 18, 27, 30, 32
21	62	Service	10 / Green Howards (Yorks Reg)	15, 18, 20, 31, 32
21	63	Service	8 / Lincolnshire	15, 35
21	63	Service	8 / Somerset Light Infantry	15, 35
21	63	Regular	4 / Middlesex	15, 35
21	63	Service	10 / York & Lancaster	15, 35
21	64	Regular	1 / East Yorkshire	15, 16, 20, 27, 32
21	64	Service	9 / King's Own Yorks L.I.	15, 27, 30, 32
21	64	Service	10 / King's Own Yorks L.I.	15, 20, 27, 30, 32
21	64	Service	15 / Durham Light Infantry	15, 18, 20, 31, 32
23	68	Service	10 / Northumberland Fusiliers	16, 20, 21, 22, 28, 29, 36
23	68	Service	11 / Northumberland Fusiliers	18, 21, 22, 28, 29, 36
23	68	Service	12 / Durham Light Infantry	16, 18, 22, 28, 36
23	68	Service	13 / Durham Light Infantry	18, 22, 23, 28, 36
23	69	Service	11 / West Yorkshire	16, 18, 22, 28, 36
23	69	Service	8 / Green Howards (Yorks Reg)	16, 18, 20, 22, 36
23	69	Service	9 / Green Howards (Yorks Reg)	16, 18, 20, 22, 28, 36
23	69	Service	10 / Duke of Wellington	16, 18, 22, 36
23	70	Service	11 / Sherwood Foresters	17, 28, 34, 36
23	70	Service	8 / King's Own Yorks L.I.	17, 20, 36
23	70	Service	8 / York & Lancaster	17, 28, 36
23	70	Service	9 / York & Lancaster	17, 28, 29, 36
6	71	Service	9 / Norfolk	26, 31, 32, 35
6	71	Service	9 / Suffolk	26, 30, 31, 34
6	71	Regular	1 / Leicestershire	19, 26, 30, 31, 32, 35
6	71	Regular	2 / Sherwood Foresters	26, 30, 31, 35

Div.	Brig.	No.	Battalion	Objective
24	72	Service	8 / Queen's - Royal West Surrey	21, 24
24	72	Service	9 / East Surrey	21, 24, 25
24	72	Service	8 / Royal West Kent-Queen's Own	19, 21, 24
24	72	Regular	1 / North Staffordshire	21, 24, 25
24	73	Service	9 / Royal Sussex	21, 24
24	73	Service	7 / Northamptonshire	21, 24
24	73	Service	13 / Middlesex	21, 24
24	73	Regular	2 / Leinster	19, 21, 24
25	74	Service	11 / Lancashire Fusiliers	17, 22, 29, 34, 35
25	74	Service	13 / Cheshire	16, 17, 22, 34, 35
25	74	Service	9 / Loyal North Lancashire	16, 17, 34, 35
25	74	Regular	2 / Royal Irish Rifles	16, 17, 34, 35
25	75	Service	11 / Cheshire	17, 34, 35
25	75	Service	8 / Border	17, 34, 35
25	75	Regular	2 / South Lancashire	17, 34, 35
25	75	Service	8 / South Lancashire	17, 34, 35
3	76	Service	8 / King's Own Roy. Lancaster	21, 24, 35
3	76	Regular	2 / Suffolk	18, 19, 21, 24, 35
3	76	Service	10 / Royal Welsh Fusiliers	18, 21, 24, 35
3	76	Regular	1 / Gordon Highlanders	18, 19, 21, 24, 35
29	86	Regular	2 / Royal Fusiliers	27, 35, 36
29	86	Regular	1 / Lancashire Fusiliers	27, 35
29	86	Service	16 / Middlesex	35, 36
29	86	Regular	1 / Royal Dublin Fusiliers	35, 36
29	87	Regular	2 / South Wales Borderers	30, 32, 35
29	87	Regular	1 / King's Own Scott. Borderers	30, 31, 32, 35
29	87	Regular	1 / Royal Inniskilling Fus.	30, 32, 35
29	87	Regular	1 / Border	30, 32, 35
29	88	Regular	4 / Worcestershire	27, 31, 32, 35
29	88	Regular	1 / Essex	31, 32, 35
29	88	Colonial	1 / Newfoundland Regiment	31, 32, 35
29	88	Regular	2 / Hampshire	31, 32, 35
30	89	Service	17 / King's Liverpool	18, 19, 24, 36
30	89	Service	19 / King's Liverpool	18, 19, 24, 36
30	89	Service	20 / King's Liverpool	18, 19, 24, 32, 36
30	89	Regular	2 / Bedfordshire	18, 19, 24, 32, 36
30	90	Regular	2 / Royal Scots Fusiliers	18, 19, 24, 36
30	90	Service	16 / Manchester	18, 19, 24, 36
30	90	Service	17 / Manchester	18, 19, 24, 36
30	90	Service	18 / Manchester	18, 19, 24, 36
7	91	Regular	2 / Queen's - Royal West Surrey	18, 21, 23, 25
7	91	Regular	1 / South Staffordshire	18, 20, 21, 23, 25
7	91	Service	21 / Manchester	15, 18, 20, 23, 25
7	91	Service	22 / Manchester	18, 21, 23, 25
31	92	Service	10 / East Yorkshire	35
31	92	Service	11 / East Yorkshire	35
31	92	Service	12 / East Yorkshire	35
31	92	Service	13 / East Yorkshire	35
31	93	Service	15 / West Yorkshire	35
31	93	Service	16 / West Yorkshire	35
31	93	Service	18 / West Yorkshire	35
31	93	Service	18 / Durham Light Infantry	35
31	94	Service	11 / East Lancashire	35
31	94	Service	12 / York & Lancaster	35
31	94	Service	13 / York & Lancaster	35
31	94	Service	14 / York & Lancaster	35
5	95	Regular	1 / Devonshire	21, 24, 30, 33
5	95	Service	12 / Gloucestershire	21, 24, 26, 30, 33
5	95	Regular	1 / East Surrey	21, 24, 30, 33
5	95	Regular	1 / Duke of Cornwall Light Inf.	19, 21, 24, 33
32	96	Service	16 / Northumberland Fusiliers	17, 34
32	96	Service	15 / Lancashire Fusiliers	17, 34, 35
32	96	Service	16 / Lancashire Fusiliers	17, 34
32	96	Regular	2 / Royal Inniskilling Fus.	17, 34, 35

DISPOSITION BY BRIGADE

Div.	Brig.	No.	Battalion	Objective
32	97	Service	11 / Border	17, 34, 35
32	97	Regular	2 / King's Own Yorks L.I.	17, 34, 35
32	97	Service	16 / Highland Light Infantry	17, 34, 35
32	97	Service	17 / Highland Light Infantry	17, 34, 35
33	98	Ex. Res	4 / King's Liverpool	20, 21, 23, 30, 31, 35
33	98	T.F.	4 / Suffolk	20, 21, 23, 31, 34
33	98	Regular	1 / Middlesex	20, 21, 23, 28, 31, 35, 36
33	98	Regular	2 / Argyll & Sutherland High.	20, 23, 30, 31
2	99	Service	22 / Royal Fusiliers	21, 35
2	99	Service	23 / Royal Fusiliers	21, 35
2	99	Regular	1 / Royal Berkshire	21, 35
2	99	Regular	1 / King's Royal Rifle Corps	21, 35
33	100	Regular	1 / Queen's - Royal West Surrey	20, 21, 23, 28, 31, 35
33	100	Regular	2 / Worcestershire	18, 20, 21, 23, 30, 31, 35
33	100	Service	16 / King's Royal Rifle Corps	18, 20, 21, 23, 30, 31, 35
33	100	T.F.	9 / Highland Light Infantry	20, 21, 23, 28, 30, 31, 35
34	101	Service	15 / Royal Scots	16, 20, 21
34	101	Service	16 / Royal Scots	16, 21
34	101	Service	10 / Lincolnshire	16, 20, 21
34	101	Service	11 / Suffolk	16, 20, 23
34	102	Service	20 / Northumberland Fusiliers	16, 37
34	102	Service	21 / Northumberland Fusiliers	16, 37
34	102	Service	22 / Northumberland Fusiliers	16, 37
34	102	Service	23 / Northumberland Fusiliers	16, 37
34	103	Service	24 / Northumberland Fusiliers	16, 28
34	103	Service	25 / Northumberland Fusiliers	16, 28
34	103	Service	26 / Northumberland Fusiliers	16, 28
34	103	Service	27 / Northumberland Fusiliers	16, 28
35	104	Service	17 / Lancashire Fusiliers	19, 24, 33
35	104	Service	18 / Lancashire Fusiliers	19, 20, 24, 33
35	104	Service	20 / Lancashire Fusiliers	20, 33
35	104	Service	23 / Manchester	19, 20, 24
35	105	Service	15 / Cheshire	19, 24
35	105	Service	16 / Cheshire	19, 24, 25
35	105	Service	14 / Gloucestershire	18, 19, 24
35	105	Service	15 / Sherwood Foresters	19, 24
35	106	Service	17 / Royal Scots	19, 20, 25
35	106	Service	17 / West Yorkshire	18, 19, 21, 24
35	106	Service	19 / Durham Light Infantry	18, 19, 21
35	106	Service	18 / Highland Light Infantry	18, 19, 21
36	107	Service	8 / Royal Irish Rifles	34, 37
36	107	Service	9 / Royal Irish Rifles	34, 37
36	107	Service	10 / Royal Irish Rifles	34
36	107	Service	15 / Royal Irish Rifles	34
36	108	Service	11 / Royal Irish Rifles	34, 37
36	108	Service	12 / Royal Irish Rifles	34, 37
36	108	Service	13 / Royal Irish Rifles	34, 37
36	108	Service	9 / Royal Irish Fusiliers	34, 37
36	109	Service	9 / Royal Inniskilling Fus.	34
36	109	Service	10 / Royal Inniskilling Fus.	34
36	109	Service	11 / Royal Inniskilling Fus.	34
36	109	Service	14 / Royal Irish Rifles	34, 37
37	110	Service	6 / Leicestershire	18, 20, 32
37	110	Service	7 / Leicestershire	18, 20, 21, 32
37	110	Service	8 / Leicestershire	18, 20, 32
37	110	Service	9 / Leicestershire	18, 20, 32
37	111	Service	10 / Royal Fusiliers	16, 22, 23, 35
37	111	Service	13 / Royal Fusiliers	17, 22, 35
37	111	Service	13 / King's Royal Rifle Corps	16, 23, 35
37	111	Service	13 / Rifle Brigade	16, 17, 20, 35
37	112	Service	11 / Royal Warwickshire	16, 18, 20, 21, 22, 35
37	112	Service	6 / Bedfordshire	16, 21, 22, 35
37	112	Service	8 / East Lancashire	16, 20, 21, 22, 23, 35
37	112	Service	10 / Loyal North Lancashire	16, 21, 22, 23, 35

Div.	Brig.	No.	Battalion	Objective
38	113	Service	13 / Royal Welsh Fusiliers	18, 35
38	113	Service	14 / Royal Welsh Fusiliers	18, 35
38	113	Service	15 / Royal Welsh Fusiliers	18, 35
38	113	Service	16 / Royal Welsh Fusiliers	18, 35
38	114	Service	10 / Welsh	18, 35
38	114	Service	13 / Welsh	18, 35
38	114	Service	14 / Welsh	18, 35
38	114	Service	15 / Welsh	18, 35
38	115	Service	10 / South Wales Borderers	18, 35
38	115	Service	11 / South Wales Borderers	18, 35
38	115	Service	17 / Royal Welsh Fusiliers	18, 35
38	115	Service	16 / Welsh	18, 35
39	116	Service	11 / Royal Sussex	34, 35
39	116	Service	12 / Royal Sussex	34, 35
39	116	Service	13 / Royal Sussex	34, 35
39	116	Service	14 / Hampshire	34, 35
39	117	Service	16 / Sherwood Foresters	34, 35
39	117	Service	17 / Sherwood Foresters	34, 35
39	117	Service	17 / King's Royal Rifle Corps	34, 35
39	117	Service	16 / Rifle Brigade	34, 35
39	118	T.F.	6 / Cheshire	34, 35
39	118	T.F.	1 / Cambridgeshire	34, 35
39	118	T.F.	1 / Hertfordshire	34, 35
39	118	T.F.	5 / Black Watch	34
41	122	Service	12 / East Surrey	21, 27, 31, 32, 36
41	122	Service	15 / Hampshire	21, 27, 32, 36
41	122	Service	11 / Royal West Kent-Queen's Own	21, 27, 36
41	122	Service	18 / King's Royal Rifle Corps	21, 27, 32, 36
41	123	Service	11 / Queen's - Royal West Surrey	21, 27, 32, 36
41	123	Service	10 / Royal West Kent-Queen's Own	21, 27, 32
41	123	Service	23 / Middlesex	21, 27, 32, 36
41	123	Service	20 / Durham Light Infantry	18, 19, 21, 27, 32, 36
41	124	Service	10 / Queen's - Royal West Surrey	21, 27, 32, 36
41	124	Service	26 / Royal Fusiliers	21, 27, 32, 36
41	124	Service	32 / Royal Fusiliers	24, 27, 32, 36
41	124	Service	21 / King's Royal Rifle Corps	21, 27, 32, 36
46	137	T.F.	5 / South Staffordshire	35, 37
46	137	T.F.	6 / South Staffordshire	35, 37
46	137	T.F.	5 / North Staffordshire	35, 37
46	137	T.F.	6 / North Staffordshire	35, 37
46	138	T.F.	4 / Lincolnshire	35
46	138	T.F.	5 / Lincolnshire	35
46	138	T.F.	4 / Leicestershire	35
46	138	T.F.	5 / Leicestershire	35
46	139	T.F.	5 / Sherwood Foresters	35, 37
46	139	T.F.	6 / Sherwood Foresters	35, 37
46	139	T.F.	7 / Sherwood Foresters	35, 37
46	139	T.F.	8 / Sherwood Foresters	35, 37
47	140	T.F.	6 / London	23, 36
47	140	T.F.	7 / London	23, 28, 36
47	140	T.F.	8 / London	23, 36
47	140	T.F.	15 / London	23, 36
47	141	T.F.	17 / London	23, 28, 36
47	141	T.F.	18 / London	23, 36
47	141	T.F.	19 / London	23, 36
47	141	T.F.	20 / London	23, 36
47	142	T.F.	21 / London	23, 28, 36
47	142	T.F.	22 / London	23, 28, 36
47	142	T.F.	23 / London	23, 28, 36
47	142	T.F.	24 / London	23, 28, 36
48	143	T.F.	5 / Royal Warwickshire	17, 22, 34, 35, 36
48	143	T.F.	6 / Royal Warwickshire	17, 22, 34, 35, 36
48	143	T.F.	7 / Royal Warwickshire	17, 22, 34, 35, 36
48	143	T.F.	8 / Royal Warwickshire	17, 22, 35, 36

DISPOSITION BY BRIGADE

Div.	Brig.	No.	Battalion	Objective
48	144	T.F.	4 / Gloucestershire	17, 22, 34, 35, 36
48	144	T.F.	6 / Gloucestershire	17, 22, 34, 35, 36
48	144	T.F.	7 / Worcestershire	17, 20, 22, 35, 36
48	144	T.F.	8 / Worcestershire	16, 35, 36
48	145	T.F.	5 / Gloucestershire	17, 22, 34, 35, 36
48	145	T.F.	4 / Oxford & Bucks Light Inf.	22, 28, 34, 35, 36
48	145	T.F.	1 / Buckinghamshire	22, 34, 35, 36
48	145	T.F.	4 / Royal Berkshire	22, 35, 36
49	146	T.F.	5 / West Yorkshire	34, 35
49	146	T.F.	6 / West Yorkshire	34, 35
49	146	T.F.	7 / West Yorkshire	34, 35
49	146	T.F.	8 / West Yorkshire	34, 35
49	147	T.F.	4 / Duke of Wellington	34, 35
49	147	T.F.	5 / Duke of Wellington	34, 35
49	147	T.F.	6 / Duke of Wellington	34, 35
49	147	T.F.	7 / Duke of Wellington	34, 35
49	148	T.F.	4 / King's Own Yorks L.I.	34, 35
49	148	T.F.	5 / King's Own Yorks L.I.	34, 35
49	148	T.F.	4 / York & Lancaster	34, 35
49	148	T.F.	5 / York & Lancaster	34, 35
50	149	T.F.	4 / Northumberland Fusiliers	23, 28, 36
50	149	T.F.	5 / Northumberland Fusiliers	23, 28, 36
50	149	T.F.	6 / Northumberland Fusiliers	23, 28, 36
50	149	T.F.	7 / Northumberland Fusiliers	23, 28, 36
50	150	T.F.	4 / East Yorkshire	23, 28, 36
50	150	T.F.	4 / Green Howards (Yorks Reg)	28, 36
50	150	T.F.	5 / Green Howards (Yorks Reg)	28, 36
50	150	T.F.	5 / Durham Light Infantry	28, 36
50	151	T.F.	5 / Border	23, 28, 36
50	151	T.F.	6 / Durham Light Infantry	28, 36
50	151	T.F.	8 / Durham Light Infantry	28, 36
50	151	T.F.	9 / Durham Light Infantry	28, 36
51	152	T.F.	5 / Seaforth Highlanders	20, 23, 35
51	152	T.F.	6 / Seaforth Highlanders	20, 23, 35
51	152	T.F.	6 / Gordon Highlanders	23, 35
51	152	T.F.	8 / Argyll & Sutherland High.	23, 35
51	153	T.F.	6 / Black Watch	20, 23, 35
51	153	T.F.	7 / Black Watch	23, 35
51	153	T.F.	5 / Gordon Highlanders	20, 23, 35
51	153	T.F.	7 / Gordon Highlanders	20, 23, 35
51	154	T.F.	9 / Royal Scots	20, 23, 35
51	154	T.F.	4 / Seaforth Highlanders	23, 35
51	154	T.F.	4 / Gordon Highlanders	20, 23, 35
51	154	T.F.	7 / Argyll & Sutherland High.	23, 35
55	164	T.F.	4 / King's Own Roy. Lancaster	21, 24, 25, 32
55	164	T.F.	8 / King's Liverpool	21, 24, 25, 27, 32, 36
55	164	T.F.	5 / Lancashire Fusiliers	21, 24, 25, 26, 27
55	164	T.F.	4 / Loyal North Lancashire	24, 25, 27, 32
55	165	T.F.	5 / King's Liverpool	19, 21, 24, 27, 32
55	165	T.F.	6 / King's Liverpool	19, 21, 24, 27, 32
55	165	T.F.	7 / King's Liverpool	19, 21, 24, 27, 32
55	165	T.F.	9 / King's Liverpool	21, 24, 27, 32
55	166	T.F.	5 / King's Own Roy. Lancaster	21, 24, 27
55	166	T.F.	10 / King's Liverpool	21, 24
55	166	T.F.	5 / South Lancashire	19, 21, 25, 27
55	166	T.F.	5 / Loyal North Lancashire	19, 21, 24
56	167	T.F.	1 / London	24, 30, 31, 33, 35
56	167	T.F.	3 / London	30, 31, 33, 35, 36
56	167	T.F.	7 / Middlesex	26, 30, 31, 33, 35
56	167	T.F.	8 / Middlesex	26, 30, 31, 33, 35
56	168	T.F.	4 / London	24, 30, 31, 33, 35
56	168	T.F.	12 / London	24, 26, 30, 31, 33, 35
56	168	T.F.	13 / London	33, 35
56	168	T.F.	14 / London	26, 30, 31, 33, 35

Div.	Brig.	No.	Battalion	Objective
56	169	T.F.	2 / London	30, 31, 33, 35, 36
56	169	T.F.	5 / London	30, 31, 33, 35
56	169	T.F.	9 / London	30, 31, 33, 35
56	169	T.F.	16 / London	30, 33, 35
63	188	Naval	0 / Anson Bn.	35
63	188	Naval	0 / Howe Bn.	35
63	188	Naval	1 / Royal Marine	35
63	188	Naval	2 / Royal Marine	35
63	189	Naval	0 / Hood Bn.	35
63	189	Naval	0 / Nelson Bn.	35
63	189	Naval	0 / Hawke Bn.	35
63	189	Naval	0 / Drake Bn.	35
63	190	T.F.	1 / Honourable Artillery Coy.	35
63	190	Ex. Res.	7 / Royal Fusiliers	35
63	190	Ex. Res.	4 / Bedfordshire	35
63	190	Service	10 / Royal Dublin Fusiliers	35
		Gds.Reg.	4 / Coldstream Guards (Pnrs)	26, 30, 35
1		T.F.	6 / Welsh (Pnrs.)	18, 20, 22, 23, 28, 34
2		Service	10 / Duke of Cornwall L.I. (Pnrs)	19, 21, 24, 35
3		Service	20 / King's Royal Rifle C. (Pnrs)	18, 20, 21
4		Service	21 / West Yorkshire (Pnrs)	30, 31, 35
5		T.F.	6 / Argyll & Sutherland (Pnrs)	23, 24, 30
6		Service	11 / Leicestershire (Pnrs)	24, 35
7		Service	24 / Manchester (Pnrs)	18, 20, 21, 23, 24, 34
8		Service	22 / Durham Light Infantry (Pnrs)	17, 30, 31
9		S.African	1 / Reg. Cape Prov.	21, 25, 36
9		S.African	2 / Reg. Natal & OFS	21, 36
9		S.African	3 / Reg. Trans. & Rhod.	21, 36
9		S.African	4 / Reg. Scottish	19, 21, 25, 36
9		Service	9 / Seaforth Highlanders (Pnrs)	15, 19, 20, 21, 28, 36
11		Service	6 / East Yorkshire (Pnrs)	22, 34
12		Service	5 / Northamptonshire (Pnrs)	17, 22, 27, 31, 32
14		Service	11 / King's Liverpool (Pnrs)	21
15		Service	9 / Gordon Highlanders(Pnrs)	23, 28, 36
16		Service	11 / Hampshire (Pnrs)	19, 24, 26
17		Service	7 / York & Lancaster (Pnrs)	15, 18, 21, 35
18		Service	8 / Royal Sussex (Pnrs.)	18, 19, 22, 24, 29, 34
19		Service	5 / South Wales Borderers (Pnrs)	16, 20, 21, 22, 23, 34
20		Service	11 / Durham Light Infantry (Pnrs)	19, 24, 26, 30, 31, 35
21		Service	14 / Northumberland Fus. (Pnrs)	15, 18, 20, 27, 30, 32
23		Service	9 / South Staffordshire (Pnrs)	18, 22, 28, 36
24		Service	12 / Sherwood Foresters (Pnrs)	19, 20, 21, 24
25		Service	6 / South Wales Borderers (Pnrs)	16, 17, 34, 35
29		T.F.	2 / Monmouth (Pnrs)	26, 27, 30, 35
30		Service	11 / South Lancashire (Pnrs)	18, 19, 30
31		Service	12 / King's Own Yorks L.I. (Pnrs)	35
32		Service	17 / Northumberland Fus. (Pnrs)	22, 34, 35
33		Service	18 / Middlesex (Pnrs)	20, 21, 23
34		Service	18 / Northumberland Fus. (Pnrs)	16, 28, 35
35		Service	19 / Northumberland Fus. (Pnrs)	19, 33, 35
36		Service	16 / Royal Irish Rifles (Pnrs)	34
37		Service	9 / North Staffordshire (Pnrs)	18, 22, 35
38		Service	19 / Welsh (Pnrs)	18, 35
39		Service	13 / Gloucestershire (Pnrs)	34, 35
41		Service	19 / Middlesex (Pnrs)	21, 27, 30, 32
46		T.F.	1 / Monmouth (Pnrs)	35, 37
47		T.F.	4 / Royal Welsh Fusiliers (Pnrs)	23, 36
48		T.F.	5 / Royal Sussex (Pnrs)	17, 22, 34
49		T.F.	3 / Monmouth (Pnrs)	16, 34, 37
50		T.F.	7 / Durham Light Infantry (Pnrs)	28, 36
51		T.F.	8 / Royal Scots (Pnrs)	23, 35
55		T.F.	4 / South Lancashire (Pnrs)	24, 26, 27, 30, 32
56		T.F.	5 / Cheshire (Pnrs)	30, 33, 35
63		Service	14 / Worcestershire (Pnrs)	35

ORDER OF BATTLE - INFANTRY AND PIONEER BATTALIONS

Div.	Brig.	Type	Battalion	Objective
63	188	Naval	0 / Anson Bn.	35
33	98	Regular	2 / Argyll & Sutherland High.	20, 23, 30, 31
5		T.F.	6 / Argyll & Sutherland High. (Pnrs)	23, 24, 30
51	154	T.F.	7 / Argyll & Sutherland High.	23, 35
51	152	T.F.	8 / Argyll & Sutherland High.	23, 35
9	26	Service	10 / Argyll & Sutherland High.	20, 21, 36
15	45	Service	11 / Argyll & Sutherland High.	22, 28, 36
5	15	Regular	1 / Bedfordshire	21, 23, 25, 26, 30, 33
30	89	Regular	2 / Bedfordshire	18, 19, 24, 32, 36
63	190	Ex. Res.	4 / Bedfordshire	35
37	112	Service	6 / Bedfordshire	16, 21, 22, 35
18	54	Service	7 / Bedfordshire	18, 19, 34
6	16	Service	8 / Bedfordshire	24, 26, 30, 31, 33, 35
1	1	Regular	1 / Black Watch	18, 20, 23, 36
39	118	T.F.	5 / Black Watch	34
51	153	T.F.	6 / Black Watch	20, 23, 35
51	153	T.F.	7 / Black Watch	23, 35
9	26	Service	8 / Black Watch	19, 20, 21, 36
15	44	Service	9 / Black Watch	22, 23, 28, 36
29	87	Regular	1 / Border	30, 32, 35
7	20	Regular	2 / Border	18, 20, 24, 26
50	151	T.F.	5 / Border	23, 28, 36
11	33	Service	6 / Border	22, 34
17	51	Service	7 / Border	15, 18, 21, 31, 32.35
25	75	Service	8 / Border	17, 34, 35
32	97	Service	11 / Border	17, 34, 35
48	145	T.F.	1 / Buckinghamshire	22, 34, 35, 36
6	16	Regular	1 / Buffs (East Kent)	26, 30, 33, 35
12	37	Service	6 / Buffs (East Kent)	17, 18, 22, 31, 32, 34
18	55	Service	7 / Buffs (East Kent)	18, 19, 29, 34
24	17	Service	8 / Buffs (East Kent)	21, 23, 24, 25
39	118	T.F.	1 / Cambridgeshire	34, 35
1	1	Regular	1 / Cameron Highlanders	18, 20, 23, 28, 36
9	26	Service	5 / Cameron Highlanders	20, 21, 25, 36
15	45	Service	6 / Cameron Highlanders	21, 22, 28, 36
15	44	Service	7 / Cameron Highlanders	22, 23, 28, 36
5	15	Regular	1 / Cheshire	21, 23, 26, 30, 33
56		T.F.	5 / Cheshire (Pnrs)	30, 33, 35
39	118	T.F.	6 / Cheshire	34, 35
19	58	Service	9 / Cheshire	16, 20, 29, 34, 35
25	7	Service	10 / Cheshire	16, 17, 34, 35
25	75	Service	11 / Cheshire	17, 34, 35
25	74	Service	13 / Cheshire	16, 17, 22, 34, 35
35	105	Service	15 / Cheshire	19, 24
35	105	Service	16 / Cheshire	19, 24, 25
	2	Gds.Reg.	1 / Coldstream Guards	26, 30, 31
	1	Gds.Reg.	2 / Coldstream Guards	26, 30, 32, 35
	1	Gds.Reg.	3 / Coldstream Guards	26, 30, 32, 35
		Gds.Reg.	4 / Coldstream Guards (Pnrs)	26, 30, 35
16	47	Service	6 / Connaught Rangers	24, 26
5	95	Regular	1 / Devonshire	21, 24, 30, 33
8	23	Regular	2 / Devonshire	17, 30, 31
7	20	Service	8 / Devonshire	18, 20, 23, 25, 26
7	20	Service	9 / Devonshire	18, 20, 23, 25, 26
32	14	Regular	1 / Dorsetshire	17, 20, 34, 35
11	34	Service	5 / Dorsetshire	22, 34
17	50	Service	6 / Dorsetshire	15, 18, 21, 31, 35
63	189	Naval	0 / Drake Bn.	35
5	95	Regular	1 / Duke of Cornwall Light Inf.	19, 21, 24, 33
14	43	Service	6 / Duke of Cornwall Light Inf.	21, 25, 27, 32
20	61	Service	7 / Duke of Cornwall Light Inf.	24, 26, 30, 31, 32, 35

Div.	Brig.	Type	Battalion	Objective
2		Service	10 / Duke of Cornwall L.I. (Pnrs)	19, 21, 24, 35
4	12	Regular	2 / Duke of Wellington	31, 35
49	147	T.F.	4 / Duke of Wellington	34, 35
49	147	T.F.	5 / Duke of Wellington	34, 35
49	147	T.F.	6 / Duke of Wellington	34, 35
49	147	T.F.	7 / Duke of Wellington	34, 35
11	32	Service	8 / Duke of Wellington	34
17	52	Service	9 / Duke of Wellington	18, 21, 31, 32, 35
23	69	Service	10 / Duke of Wellington	16, 18, 22, 36
6	18	Regular	2 / Durham Light Infantry	26, 30, 31, 32, 35
50	150	T.F.	5 / Durham Light Infantry	28, 36
50	151	T.F.	6 / Durham Light Infantry	28, 36
50		T.F.	7 / Durham Light Infantry (Pnrs)	28, 36
50	151	T.F.	8 / Durham Light Infantry	28, 36
50	151	T.F.	9 / Durham Light Infantry	28, 36
14	43	Service	10 / Durham Light Infantry	21, 25, 27, 32
20		Service	11 / Durham Light Infantry (Pnrs)	19, 24, 26, 30, 31, 35
23	68	Service	12 / Durham Light Infantry	16, 18, 22, 28, 36
23	68	Service	13 / Durham Light Infantry	18, 22, 23, 28, 36
6	18	Service	14 / Durham Light Infantry	26, 30, 31, 35
21	64	Service	15 / Durham Light Infantry	15, 18, 20, 31, 32
31	93	Service	18 / Durham Light Infantry	35
35	106	Service	19 / Durham Light Infantry	18, 19, 21
41	123	Service	20 / Durham Light Infantry	18, 19, 21, 27, 32, 36
8		Service	22 / Durham Light Infantry (Pnrs)	17, 30, 31
4	11	Regular	1 / East Lancashire	30, 31, 35
8	24	Regular	2 / East Lancashire	18, 30, 31, 36
19	56	Service	7 / East Lancashire	16, 20, 28, 34, 35
37	112	Service	8 / East Lancashire	16, 20, 21, 22, 23, 35
31	94	Service	11 / East Lancashire	35
5	95	Regular	1 / East Surrey	21, 24, 30, 33
12	37	Service	7 / East Surrey	17, 22, 32, 34, 36
18	55	Service	8 / East Surrey	18, 19, 29, 34
24	72	Service	9 / East Surrey	21, 24, 25
41	122	Service	12 / East Surrey	21, 27, 31, 32, 36
21	64	Regular	1 / East Yorkshire	15, 16, 20, 27, 32
50	150	T.F.	4 / East Yorkshire	23, 28, 36
11		Service	6 / East Yorkshire (Pnrs)	22, 34
17	50	Service	7 / East Yorkshire	15, 18, 21, 30, 31, 35
3	8	Service	8 / East Yorkshire	18, 19, 20, 23, 24, 35
31	92	Service	10 / East Yorkshire	35
31	92	Service	11 / East Yorkshire	35
31	92	Service	12 / East Yorkshire	35
31	92	Service	13 / East Yorkshire	35
29	88	Regular	1 / Essex	31, 32, 35
4	12	Regular	2 / Essex	31, 35
12	35	Service	9 / Essex	16, 17, 22, 32
18	53	Service	10 / Essex	18, 21, 29, 34
6	18	Service	11 / Essex	26, 30, 31, 32, 33, 35
2	6	Service	13 / Essex	21, 24, 25, 35
1	3	Regular	1 / Gloucestershire	18, 20, 23
48	144	T.F.	4 / Gloucestershire	17, 22, 34, 35, 36
48	145	T.F.	5 / Gloucestershire	17, 22, 34, 35, 36
48	144	T.F.	6 / Gloucestershire	17, 22, 34, 35, 36
19	57	Service	8 / Gloucestershire	16, 20, 21, 23, 34, 35
1	1	Service	10 / Gloucestershire	20, 21, 23, 28, 36
5	95	Service	12 / Gloucestershire	21, 24, 26, 30, 33
39		Service	13 / Gloucestershire (Pnrs)	34, 35
35	105	Service	14 / Gloucestershire	18, 19, 24
3	76	Regular	1 / Gordon Highlanders	18, 19, 21, 24, 35
7	20	Regular	2 / Gordon Highlanders	18, 20, 23, 25, 26
51	154	T.F.	4 / Gordon Highlanders	20, 23, 35
51	153	T.F.	5 / Gordon Highlanders	20, 23, 35
51	152	T.F.	6 / Gordon Highlanders	23, 35

ORDER OF BATTLE - INFANTRY AND PIONEER BATTALIONS

Div.	Brig.	Type	Battalion	Objective
51	153	T.F.	7 / Gordon Highlanders	20, 23, 35
15		Service	9 / Gordon Highlanders(Pnrs)	23, 28, 36
15	44	Service	10 / Gordon Highlanders	20, 28, 36
30	21	Regular	2 / Green Howards (Yorks Reg)	18, 19, 21, 24, 36
50	150	T.F.	4 / Green Howards (Yorks Reg)	28, 36
50	150	T.F.	5 / Green Howards (Yorks Reg)	28, 36
11	32	Service	6 / Green Howards (Yorks Reg)	34
17	50	Service	7 / Green Howards (Yorks Reg)	15, 18, 21, 31, 35
23	69	Service	8 / Green Howards (Yorks Reg)	16, 18, 20, 22, 36
23	69	Service	9 / Green Howards (Yorks Reg)	16, 18, 20, 22, 28, 36
21	62	Service	10 / Green Howards (Yorks Reg)	15, 18, 20, 31, 32
	3	Gds.Reg.	1 / Grenadier Guards	24, 26, 30, 31, 32, 35
	1	Gds.Reg.	2 / Grenadier Guards	26, 30, 31, 32, 35
	2	Gds.Reg.	3 / Grenadier Guards	26, 30, 35
	3	Gds.Reg.	4 / Grenadier Guards	26, 30, 31, 32, 33, 35
4	11	Regular	1 / Hampshire	30, 31, 35
29	88	Regular	2 / Hampshire	31, 32, 35
16		Service	11 / Hampshire (Pnrs)	19, 24, 26
39	116	Service	14 / Hampshire	34, 35
41	122	Service	15 / Hampshire	21, 27, 32, 36
63	189	Naval	0 / Hawke Bn.	35
39	118	T.F.	1 / Hertfordshire	34, 35
2	5	Regular	2 / Highland Light Infantry	25, 35
33	100	T.F.	9 / Highland Light Infantry	20, 21, 23, 28, 30, 31, 35
15	46	Service	11 / Highland Light Infantry	22, 23, 28, 36
15	46	Service	12 / Highland Light Infantry	22, 23, 28, 36
32	14	Service	15 / Highland Light Infantry	17, 34, 35
32	97	Service	16 / Highland Light Infantry	17, 34, 35
32	97	Service	17 / Highland Light Infantry	17, 34, 35
35	106	Service	18 / Highland Light Infantry	18, 19, 21
63	190	T.F.	1 / Honourable Artillery Coy.	35
63	189	Naval	0 / Hood Bn.	35
63	188	Naval	0 / Howe Bn.	35
	1	Gds.Reg.	1 / Irish Guards	26, 30, 31, 32, 35
	2	Gds.Reg.	2 / Irish Guards	26, 30, 35
2	6	Regular	1 / King's Liverpool	21, 24, 35
33	98	Ex. Res.	4 / King's Liverpool	20, 21, 23, 30, 31, 35
55	165	T.F.	5 / King's Liverpool	19, 21, 24, 27, 32
55	165	T.F.	6 / King's Liverpool	19, 21, 24, 27, 32
55	165	T.F.	7 / King's Liverpool	19, 21, 24, 27, 32
55	164	T.F.	8 / King's Liverpool	21, 24, 25, 27, 32, 36
55	165	T.F.	9 / King's Liverpool	21, 24, 27, 32
55	166	T.F.	10 / King's Liverpool	21, 24
14		Service	11 / King's Liverpool (Pnrs)	21
20	61	Service	12 / King's Liverpool	24, 26, 30, 31, 32, 35
3	9	Service	13 / King's Liverpool	20, 21, 24, 35
30	89	Service	17 / King's Liverpool	18, 19, 24, 36
30	21	Service	18 / King's Liverpool	18, 19, 36
30	89	Service	19 / King's Liverpool	18, 19, 24, 36
30	89	Service	20 / King's Liverpool	18, 19, 24, 32, 36
4	12	Regular	1 / King's Own Roy. Lancaster	26, 30, 31, 35
55	164	T.F.	4 / King's Own Roy. Lancaster	21, 24, 25, 32
55	166	T.F.	5 / King's Own Roy. Lancaster	21, 24, 27
19	56	Service	7 / King's Own Roy. Lancaster	16, 17, 18, 20, 23, 34
3	76	Service	8 / King's Own Roy. Lancaster	21, 24, 35
29	87	Regular	1 / King's Own Scott. Borderers	30, 31, 32, 35
5	13	Regular	2 / King's Own Scott. Borderers	20, 21, 23, 30, 33
9	27	Service	6 / King's Own Scott. Borderers	18, 20, 21, 36
15	46	Service	8 / King's Own Scott. Borderers	28, 36
32	97	Regular	2 / King's Own Yorks L.I.	17, 34, 35
49	148	T.F.	4 / King's Own Yorks L.I.	34, 35
49	148	T.F.	5 / King's Own Yorks L.I.	34, 35
14	43	Service	6 / King's Own Yorks L.I.	21, 25, 27, 32
20	61	Service	7 / King's Own Yorks L.I.	24, 30, 31, 32, 35

Div.	Brig.	Type	Battalion	Objective
23	70	Service	8 / King's Own Yorks L.I.	17, 20, 36
21	64	Service	9 / King's Own Yorks L.I.	15, 27, 30, 32
21	64	Service	10 / King's Own Yorks L.I.	15, 20, 27, 30, 32
31		Service	12 / King's Own Yorks L.I. (Pnrs)	35
2	99	Regular	1 / King's Royal Rifle Corps	21, 35
1	2	Regular	2 / King's Royal Rifle Corps	18, 20, 23, 28, 36
14	41	Service	7 / King's Royal Rifle Corps	21, 25, 27
14	41	Service	8 / King's Royal Rifle Corps	21, 25, 27
14	42	Service	9 / King's Royal Rifle Corps	21, 25, 27, 32
20	59	Service	10 / King's Royal Rifle Corps	24, 30, 32, 35
20	59	Service	11 / King's Royal Rifle Corps	24, 30, 32, 35
20	60	Service	12 / King's Royal Rifle Corps	21, 24, 25, 30, 31, 32, 35
37	111	Service	13 / King's Royal Rifle Corps	16, 23, 35
33	100	Service	16 / King's Royal Rifle Corps	18, 20, 21, 23, 30, 31, 35
39	117	Service	17 / King's Royal Rifle Corps	34, 35
41	122	Service	18 / King's Royal Rifle Corps	21, 27, 32, 36
3		Service	20 / King's Royal Rifle Corps (Pnrs)	18, 20, 21
41	124	Service	21 / King's Royal Rifle Corps	21, 27, 32, 36
6	16	Regular	1 / King's Shropshire L. Inf.	24, 26, 30, 31, 32, 35
14	42	Service	5 / King's Shropshire L. Inf.	21, 25, 27, 32
20	60	Service	6 / King's Shropshire L. Inf.	24, 25, 30, 35
3	8	Service	7 / King's Shropshire L. Inf.	19, 20, 21, 24, 25, 35
29	86	Regular	1 / Lancashire Fusiliers	27, 35
4	12	Regular	2 / Lancashire Fusiliers	30, 31, 35
55	164	T.F.	5 / Lancashire Fusiliers	21, 24, 25, 26, 27
11	34	Service	9 / Lancashire Fusiliers	22, 34
17	52	Service	10 / Lancashire Fusiliers	18, 20, 21, 32, 35
25	74	Service	11 / Lancashire Fusiliers	17, 22, 29, 34, 35
32	96	Service	15 / Lancashire Fusiliers	17, 34, 35
32	96	Service	16 / Lancashire Fusiliers	17, 34
35	104	Service	17 / Lancashire Fusiliers	19, 24, 33
35	104	Service	18 / Lancashire Fusiliers	19, 20, 24, 33
32	14	Service	19 / Lancashire Fusiliers	17, 34
35	104	Service	20 / Lancashire Fusiliers	20, 33
6	71	Regular	1 / Leicestershire	19, 26, 30, 31, 32, 35
46	138	T.F.	4 / Leicestershire	35
46	138	T.F.	5 / Leicestershire	35
37	110	Service	6 / Leicestershire	18, 20, 32
37	110	Service	7 / Leicestershire	18, 20, 21, 32
37	110	Service	8 / Leicestershire	18, 20, 32
37	110	Service	9 / Leicestershire	18, 20, 32
6		Service	11 / Leicestershire (Pnrs)	24, 35
24	73	Regular	2 / Leinster	19, 21, 24
16	47	Service	7 / Leinster	24, 25, 26
21	62	Regular	1 / Lincolnshire	15, 18, 27, 30, 32
8	25	Regular	2 / Lincolnshire	17, 30, 31
46	138	T.F.	4 / Lincolnshire	35
46	138	T.F.	5 / Lincolnshire	35
11	33	Service	6 / Lincolnshire	22, 34
17	51	Service	7 / Lincolnshire	15, 18, 21, 31, 32, 35
21	63	Service	8 / Lincolnshire	15, 35
34	101	Service	10 / Lincolnshire	16, 20, 21
56	167	T.F.	1 / London	24, 30, 31, 33, 35
56	169	T.F.	2 / London	30, 31, 33, 35, 36
56	167	T.F.	3 / London	30, 31, 33, 35, 36
56	168	T.F.	4 / London	24, 30, 31, 33, 35
56	169	T.F.	5 / London	30, 31, 33, 35
47	140	T.F.	6 / London	23, 36
47	140	T.F.	7 / London	23, 28, 36
47	140	T.F.	8 / London	23, 36
56	169	T.F.	9 / London	30, 31, 33, 35
56	168	T.F.	12 / London	24, 26, 30, 31, 33, 35
56	168	T.F.	13 / London	33, 35
56	168	T.F.	14 / London	26, 30, 31, 33, 35

ORDER OF BATTLE - INFANTRY AND PIONEER BATTALIONS

Div.	Brig.	Type	Battalion	Objective
47	140	T.F.	15 / London	23, 36
56	169	T.F.	16 / London	30, 33, 35
47	141	T.F.	17 / London	23, 28, 36
47	141	T.F.	18 / London	23, 36
47	141	T.F.	19 / London	23, 36
47	141	T.F.	20 / London	23, 36
47	142	T.F.	21 / London	23, 28, 36
47	142	T.F.	22 / London	23, 28, 36
47	142	T.F.	23 / London	23, 28, 36
47	142	T.F.	24 / London	23, 28, 36
1	2	Regular	1 / Loyal North Lancashire	20, 22, 23, 36
55	164	T.F.	4 / Loyal North Lancashire	24, 25, 27, 32
55	166	T.F.	5 / Loyal North Lancashire	19, 21, 24
19	56	Service	7 / Loyal North Lancashire	16, 21, 34, 35
25	7	Service	8 / Loyal North Lancashire	17, 34, 35
25	74	Service	9 / Loyal North Lancashire	16, 17, 34, 35
37	112	Service	10 / Loyal North Lancashire	16, 21, 22, 23, 35
32	14	Regular	2 / Manchester	17, 34, 35
11	34	Service	11 / Manchester	22, 34
17	52	Service	12 / Manchester	18, 21, 30, 35
30	90	Service	16 / Manchester	18, 19, 24, 36
30	90	Service	17 / Manchester	18, 19, 24, 36
30	90	Service	18 / Manchester	18, 19, 24, 36
30	21	Service	19 / Manchester	18, 19, 24, 36
7	22	Service	20 / Manchester	18, 20, 23, 25, 26
7	91	Service	21 / Manchester	15, 18, 20, 23, 25
7	91	Service	22 / Manchester	18, 21, 23, 25
35	104	Service	23 / Manchester	19, 20, 24
7		Service	24 / Manchester (Pnrs)	18, 20, 21, 23, 24, 34
33	98	Regular	1 / Middlesex	20, 21, 23, 31, 35
8	23	Regular	2 / Middlesex	17, 30, 31
21	63	Regular	4 / Middlesex	15, 35
56	167	T.F.	7 / Middlesex	26, 30, 31, 33, 35
56	167	T.F.	8 / Middlesex	26, 30, 31, 33, 35
12	36	Service	11 / Middlesex	17, 22, 34, 36
18	54	Service	12 / Middlesex	18, 19, 34
24	73	Service	13 / Middlesex	21, 24
29	86	Service	16 / Middlesex	35, 36
2	6	Service	17 / Middlesex	21, 24, 25, 35
33		Service	18 / Middlesex (Pnrs)	20, 21, 23
41		Service	19 / Middlesex (Pnrs)	21, 27, 30, 32
41	123	Service	23 / Middlesex	21, 27, 32, 36
46		T.F.	1 / Monmouth (Pnrs)	35, 37
29		T.F.	2 / Monmouth (Pnrs)	26, 27, 30, 35
49		T.F.	3 / Monmouth (Pnrs)	16, 34, 37
63	189	Naval	0 / Nelson Bn.	35
29	88	Colonial	1 / Newfoundland Regiment	31, 32, 35
5	15	Regular	1 / Norfolk	21, 23, 30, 33
12	35	Service	7 / Norfolk	17, 22, 32
18	53	Service	8 / Norfolk	18, 21, 29, 34
6	71	Service	9 / Norfolk	26, 31, 32, 35
24	72	Regular	1 / North Staffordshire	21, 24, 25
46	137	T.F.	5 / North Staffordshire	35, 37
46	137	T.F.	6 / North Staffordshire	35, 37
19	57	Service	8 / North Staffordshire	16, 18, 20, 34
37		Service	9 / North Staffordshire (Pnrs)	18, 22, 35
1	2	Regular	1 / Northamptonshire	18, 20, 23, 28, 36
8	24	Regular	2 / Northamptonshire	18, 30, 31
12		Service	5 / Northamptonshire (Pnrs)	17, 22, 31, 32
18	54	Service	6 / Northamptonshire	18, 19, 22, 29, 34
24	73	Service	7 / Northamptonshire	21, 24
3	9	Regular	1 / Northumberland Fusiliers	20, 21, 24, 35
50	149	T.F.	4 / Northumberland Fusiliers	23, 28, 36
50	149	T.F.	5 / Northumberland Fusiliers	23, 28, 36

Div.	Brig.	Type	Battalion	Objective
50	149	T.F.	6 / Northumberland Fusiliers	23, 28, 36
50	149	T.F.	7 / Northumberland Fusiliers	23, 28, 36
11	34	Service	8 / Northumberland Fusiliers	34
17	52	Service	9 / Northumberland Fusiliers	18, 21, 30, 31, 32
23	68	Service	10 / Northumberland Fusiliers	16, 20, 21, 22, 28, 29, 36
23	68	Service	11 / Northumberland Fusiliers	18, 21, 22, 28, 29, 36
21	62	Service	12 / Northumberland Fusiliers	15, 18, 27, 32
21	62	Service	13 / Northumberland Fusiliers	15, 18, 27
21		Service	14 / Northumberland Fus. (Pnrs)	15, 18, 20, 27, 30, 32
32	96	Service	16 / Northumberland Fusiliers	17, 34
32		Service	17 / Northumberland Fus. (Pnrs)	22, 34, 35
34		Service	18 / Northumberland Fus. (Pnrs)	16, 28, 35
35		Service	19 / Northumberland Fus. (Pnrs)	19, 33, 35
34	102	Service	20 / Northumberland Fusiliers	16, 37
34	102	Service	21 / Northumberland Fusiliers	16, 37
34	102	Service	22 / Northumberland Fusiliers	16, 37
34	102	Service	23 / Northumberland Fusiliers	16, 37
34	103	Service	24 / Northumberland Fusiliers	16, 28
34	103	Service	25 / Northumberland Fusiliers	16, 28
34	103	Service	26 / Northumberland Fusiliers	16, 28
34	103	Service	27 / Northumberland Fusiliers	16, 28
2	5	Regular	2 / Oxford & Bucks Light Inf.	24, 25, 35
48	145	T.F.	4 / Oxford & Bucks Light Inf.	22, 28, 34, 35, 36
14	42	Service	5 / Oxford & Bucks Light Inf.	21, 25, 27, 32
20	60	Service	6 / Oxford & Bucks Light Inf.	24, 25, 30, 31, 35
33	100	Regular	1 / Queen's - Royal West Surrey	20, 21, 23, 28, 31, 35
7	91	Regular	2 / Queen's - Royal West Surrey	18, 21, 23, 25
12	37	Service	6 / Queen's - Royal West Surrey	17, 22, 31, 32
18	55	Service	7 / Queen's - Royal West Surrey	18, 19, 29, 34
24	72	Service	8 / Queen's - Royal West Surrey	21, 24
41	124	Service	10 / Queen's - Royal West Surrey	21, 27, 32, 36
41	123	Service	11 / Queen's - Royal West Surrey	21, 27, 32, 36
9		S.African	1 / Reg. Cape Prov.	21, 25, 36
9		S.African	2 / Reg. Natal & OFS	21, 36
9		S.African	4 / Reg. Scottish	19, 21, 25, 36
9		S.African	3 / Reg. Trans. & Rhod.	21, 36
4	11	Regular	1 / Rifle Brigade	30, 31, 35
8	25	Regular	2 / Rifle Brigade	17, 30, 31
24	17	Regular	3 / Rifle Brigade	21, 24
14	41	Service	7 / Rifle Brigade	21, 25, 27
14	41	Service	8 / Rifle Brigade	21, 27
14	42	Service	9 / Rifle Brigade	21, 25, 27
20	59	Service	10 / Rifle Brigade	24, 26, 32, 35
20	59	Service	11 / Rifle Brigade	24, 26, 30, 32, 35
20	60	Service	12 / Rifle Brigade	24, 26, 30, 31, 32, 35
37	111	Service	13 / Rifle Brigade	16, 17, 20, 35
39	117	Service	16 / Rifle Brigade	34, 35
2	99	Regular	1 / Royal Berkshire	21, 35
8	25	Regular	2 / Royal Berkshire	17, 30, 31, 32
48	145	T.F.	4 / Royal Berkshire	22, 35, 36
12	35	Service	5 / Royal Berkshire	17, 22, 35, 36
18	53	Service	6 / Royal Berkshire	18, 21, 29, 34
1	1	Service	8 / Royal Berkshire	18, 20, 21, 23, 28, 36
29	86	Regular	1 / Royal Dublin Fusiliers	35, 36
4	10	Regular	2 / Royal Dublin Fusiliers	30, 31, 35
16	48	Service	8 / Royal Dublin Fusiliers	26
16	48	Service	9 / Royal Dublin Fusiliers	24, 26, 33
63	190	Service	10 / Royal Dublin Fusiliers	35
24	17	Regular	1 / Royal Fusiliers	21, 24
29	86	Regular	2 / Royal Fusiliers	27, 35, 36
3	9	Regular	4 / Royal Fusiliers	18, 20, 21, 24, 35
63	190	Ex. Res.	7 / Royal Fusiliers	35
12	36	Service	8 / Royal Fusiliers	17, 22, 31, 32, 36
12	36	Service	9 / Royal Fusiliers	17, 22, 31, 32

ORDER OF BATTLE - INFANTRY AND PIONEER BATTALIONS 43

Div.	Brig.	Type	Battalion	Objective
37	111	Service	10 / Royal Fusiliers	16, 22, 23, 35
18	54	Service	11 / Royal Fusiliers	18, 19, 34, 35
24	17	Service	12 / Royal Fusiliers	21, 24, 25
37	111	Service	13 / Royal Fusiliers	17, 22, 35
2	5	Service	17 / Royal Fusiliers	21, 25, 35
33	19	Service	20 / Royal Fusiliers	20, 21, 23, 30
2	99	Service	22 / Royal Fusiliers	21, 35
2	99	Service	23 / Royal Fusiliers	21, 35
2	5	Service	24 / Royal Fusiliers	24, 25, 35
41	124	Service	26 / Royal Fusiliers	21, 27, 32, 36
41	124	Service	32 / Royal Fusiliers	24, 27, 32, 36
29	87	Regular	1 / Royal Inniskilling Fus.	30, 32, 35
32	96	Regular	2 / Royal Inniskilling Fus.	17, 34, 35
16	49	Service	7 / Royal Inniskilling Fus.	24, 26, 33
16	49	Service	8 / Royal Inniskilling Fus.	24, 26, 33
36	109	Service	9 / Royal Inniskilling Fus.	34
36	109	Service	10 / Royal Inniskilling Fus.	34
36	109	Service	11 / Royal Inniskilling Fus.	34
7	22	Regular	2 / Royal Irish	15, 18, 20, 23, 25, 26
16	47	Service	6 / Royal Irish	24, 26
4	10	Regular	1 / Royal Irish Fusiliers	30, 31, 35
16	49	Service	7 / Royal Irish Fusiliers	24, 26, 33
16	49	Service	8 / Royal Irish Fusiliers	24, 33
36	108	Service	9 / Royal Irish Fusiliers	34, 37
8	25	Regular	1 / Royal Irish Rifles	17, 30, 31
25	74	Regular	2 / Royal Irish Rifles	16, 17, 34, 35
16	48	Service	7 / Royal Irish Rifles	24, 26
36	107	Service	8 / Royal Irish Rifles	34, 37
36	107	Service	9 / Royal Irish Rifles	34, 37
36	107	Service	10 / Royal Irish Rifles	34
36	108	Service	11 / Royal Irish Rifles	34, 37
36	108	Service	12 / Royal Irish Rifles	34, 37
36	108	Service	13 / Royal Irish Rifles	34, 37
36	109	Service	14 / Royal Irish Rifles	34, 37
36	107	Service	15 / Royal Irish Rifles	34
36		Service	16 / Royal Irish Rifles (Pnrs)	34
63	188	Naval	1 / Royal Marine	35
63	188	Naval	2 / Royal Marine	35
16	48	Regular	1 / Royal Munster Fusiliers	19, 26
1	3	Regular	2 / Royal Munster Fusiliers	18, 20, 21, 22, 23, 28
16	47	Service	8 / Royal Munster Fusiliers	19, 24, 26
3	8	Regular	2 / Royal Scots	19, 20, 24, 25, 35
51		T.F.	8 / Royal Scots (Pnrs)	23, 35
51	154	T.F.	9 / Royal Scots	20, 23, 35
9	27	Service	11 / Royal Scots	18, 19, 20, 21, 28, 36
9	27	Service	12 / Royal Scots	18, 19, 20, 21, 36
15	45	Service	13 / Royal Scots	20, 21, 28, 36
34	101	Service	15 / Royal Scots	16, 20, 21
34	101	Service	16 / Royal Scots	16, 21
35	106	Service	17 / Royal Scots	19, 20, 25
3	8	Regular	1 / Royal Scots Fusiliers	19, 20, 24, 35
30	90	Regular	2 / Royal Scots Fusiliers	18, 19, 24, 36
15	45	Service	7 / Royal Scots Fusiliers	22, 28, 36
1	2	Regular	2 / Royal Sussex	18, 22, 23, 28
48		T.F.	5 / Royal Sussex (Pnrs)	17, 22, 34
12	36	Service	7 / Royal Sussex	17, 22, 32
18		Service	8 / Royal Sussex (Pnrs)	18, 19, 22, 24, 29, 34
24	73	Service	9 / Royal Sussex	21, 24
39	116	Service	11 / Royal Sussex	34, 35
39	116	Service	12 / Royal Sussex	34, 35
39	116	Service	13 / Royal Sussex	34, 35
4	10	Regular	1 / Royal Warwickshire	30, 31, 35
7	22	Regular	2 / Royal Warwickshire	18, 20, 23, 26
48	143	T.F.	5 / Royal Warwickshire	17, 22, 34, 35, 36

Div.	Brig.	Type	Battalion	Objective
48	143	T.F.	6 / Royal Warwickshire	17, 22, 34, 35, 36
48	143	T.F.	7 / Royal Warwickshire	17, 22, 34, 35, 36
48	143	T.F.	8 / Royal Warwickshire	17, 22, 35, 36
19	57	Service	10 / Royal Warwickshire	16, 20, 23, 34
37	112	Service	11 / Royal Warwickshire	16, 18, 20, 21, 22, 35
5	13	Service	14 / Royal Warwickshire	20, 21, 23, 24, 30, 33
5	13	Service	15 / Royal Warwickshire	20, 21, 23, 33
5	15	Service	16 / Royal Warwickshire	21, 30, 33
7	22	Regular	1 / Royal Welsh Fusiliers	18, 20, 21, 23, 25, 26
33	19	Regular	2 / Royal Welsh Fusiliers	20, 21, 23, 30, 31
47		T.F.	4 / Royal Welsh Fusiliers (Pnrs)	23, 36
19	58	Service	9 / Royal Welsh Fusiliers	16, 18, 20, 34, 35
3	76	Service	10 / Royal Welsh Fusiliers	18, 21, 24, 35
38	113	Service	13 / Royal Welsh Fusiliers	18, 35
38	113	Service	14 / Royal Welsh Fusiliers	18, 35
38	113	Service	15 / Royal Welsh Fusiliers	18, 35
38	113	Service	16 / Royal Welsh Fusiliers	18, 35
38	115	Service	17 / Royal Welsh Fusiliers	18, 35
5	13	Regular	1 / Royal West Kent-Queen's Own	20, 21, 23, 30, 33
12	37	Service	6 / Royal West Kent-Queen's Own	17, 22, 31, 32, 34
18	55	Service	7 / Royal West Kent-Queen's Own	18, 19, 29, 34
24	72	Service	8 / Royal West Kent-Queen's Own	19, 21, 24
41	123	Service	10 / Royal West Kent-Queen's Own	21, 27, 32
41	122	Service	11 / Royal West Kent-Queen's Own	21, 27, 36
	2	Gds.Reg.	1 / Scots Guards	26, 30, 31, 35
	3	Gds.Reg.	2 / Scots Guards	26, 30, 31, 32, 35
33	19	Regular	1 / Scott. Rifles-Cameronians	20, 23, 30, 31, 35
8	23	Regular	2 / Scott. Rifles-Cameronians	17, 31, 36
33	19	T.F.	6 / Scott. Rifles-Cameronians	20, 21, 23, 30, 31, 35, 36
9	27	Service	9 / Scott. Rifles-Cameronians	18, 20, 21, 36
15	46	Service	10 / Scott. Rifles-Cameronians	23, 28, 36
4	10	Regular	2 / Seaforth Highlanders	30, 31, 35
51	154	T.F.	4 / Seaforth Highlanders	23, 35
51	152	T.F.	5 / Seaforth Highlanders	20, 23, 35
51	152	T.F.	6 / Seaforth Highlanders	20, 23, 35
9	26	Service	7 / Seaforth Highlanders	19, 20, 21, 25, 36
15	44	Service	8 / Seaforth Highlanders	22, 23, 28, 36
9		Service	9 / Seaforth Highlanders (Pnrs)	15, 19, 20, 21, 28, 36
8	24	Regular	1 / Sherwood Foresters	16, 18, 31
6	71	Regular	2 / Sherwood Foresters	26, 30, 31, 35
46	139	T.F.	5 / Sherwood Foresters	35, 37
46	139	T.F.	6 / Sherwood Foresters	35, 37
46	139	T.F.	7 / Sherwood Foresters	35, 37
46	139	T.F.	8 / Sherwood Foresters	35, 37
11	33	Service	9 / Sherwood Foresters	22, 34
17	51	Service	10 / Sherwood Foresters	15, 18, 21, 32, 35
23	70	Service	11 / Sherwood Foresters	17, 28, 34, 36
24		Service	12 / Sherwood Foresters (Pnrs)	19, 20, 21, 24
35	105	Service	15 / Sherwood Foresters	19, 24
39	117	Service	16 / Sherwood Foresters	34, 35
39	117	Service	17 / Sherwood Foresters	34, 35
4	11	Regular	1 / Somerset Light Infantry	30, 31, 35
14	43	Service	6 / Somerset Light Infantry	21, 25, 27, 32
20	61	Service	7 / Somerset Light Infantry	24, 26, 31, 32, 33, 35
21	63	Service	8 / Somerset Light Infantry	15, 35
25	75	Regular	2 / South Lancashire	17, 34, 35
55		T.F.	4 / South Lancashire (Pnrs)	24, 26, 27, 30, 32
55	166	T.F.	5 / South Lancashire	19, 21, 25, 27
19	56	Service	7 / South Lancashire	16, 20, 28, 34, 35
25	75	Service	8 / South Lancashire	17, 34, 35
30		Service	11 / South Lancashire (Pnrs)	18, 19, 30
7	91	Regular	1 / South Staffordshire	18, 20, 21, 23, 25
2	6	Regular	2 / South Staffordshire	21, 35
46	137	T.F.	5 / South Staffordshire	35, 37

ORDER OF BATTLE - INFANTRY AND PIONEER BATTALIONS

Div.	Brig.	Type	Battalion	Objective
46	137	T.F.	6 / South Staffordshire	35, 37
11	33	Service	7 / South Staffordshire	34
17	51	Service	8 / South Staffordshire	15, 18, 21, 35
23		Service	9 / South Staffordshire (Pnrs)	18, 22, 28, 36
1	3	Regular	1 / South Wales Borderers	18, 20, 21, 22, 23
29	87	Regular	2 / South Wales Borderers	30, 32, 35
19		Service	5 / South Wales Borderers (Pnrs)	16, 20, 21, 22, 23, 34
25		Service	6 / South Wales Borderers (Pnrs)	16, 17, 34, 35
38	115	Service	10 / South Wales Borderers	18, 35
38	115	Service	11 / South Wales Borderers	18, 35
3	76	Regular	2 / Suffolk	18, 19, 21, 24, 35
33	98	T.F.	4 / Suffolk	20, 21, 23, 31, 34
12	35	Service	7 / Suffolk	17, 22, 32, 35
18	53	Service	8 / Suffolk	18, 19, 21, 29, 34
6	71	Service	9 / Suffolk	26, 30, 31, 34
34	101	Service	11 / Suffolk	16, 20, 23
1	3	Regular	2 / Welsh	18, 20, 22, 23, 28
1		T.F.	6 / Welsh (Pnrs)	18, 20, 22, 23, 28, 34
19	58	Service	9 / Welsh	16, 18, 20, 34, 35
38	114	Service	10 / Welsh	18, 35
38	114	Service	13 / Welsh	18, 35
38	114	Service	14 / Welsh	18, 35
38	114	Service	15 / Welsh	18, 35
38	115	Service	16 / Welsh	18, 35
38		Service	19 / Welsh (Pnrs)	18, 35
	3	Gds.Reg.	1 / Welsh Guards	26, 30, 31, 32, 35
6	18	Regular	1 / West Yorkshire	26, 30, 31, 32, 35
8	23	Regular	2 / West Yorkshire	17, 30, 31
49	146	T.F.	5 / West Yorkshire	34, 35
49	146	T.F.	6 / West Yorkshire	34, 35
49	146	T.F.	7 / West Yorkshire	34, 35
49	146	T.F.	8 / West Yorkshire	34, 35
11	32	Service	9 / West Yorkshire	22, 34
17	50	Service	10 / West Yorkshire	15, 21, 30, 31, 35
23	69	Service	11 / West Yorkshire	16, 18, 22, 28, 36
3	9	Service	12 / West Yorkshire	18, 20, 21, 24, 35
31	93	Service	15 / West Yorkshire	35
31	93	Service	16 / West Yorkshire	35
35	106	Service	17 / West Yorkshire	18, 19, 21, 24
31	93	Service	18 / West Yorkshire	35
4		Service	21 / West Yorkshire (Pnrs)	30, 31, 35
25	7	Regular	1 / Wiltshire	17, 22, 34, 35
30	21	Regular	2 / Wiltshire	18, 19, 24, 36
19	58	Service	6 / Wiltshire	16, 18, 20, 34
8	24	Regular	1 / Worcestershire	18, 30, 31, 32
33	100	Regular	2 / Worcestershire	18, 20, 21, 23, 30, 31, 35
25	7	Regular	3 / Worcestershire	16, 17, 34, 35
29	88	Regular	4 / Worcestershire	27, 31, 32, 35
48	144	T.F.	7 / Worcestershire	17, 20, 22, 35, 36
48	144	T.F.	8 / Worcestershire	16, 35, 36
19	57	Service	10 / Worcestershire	16, 20, 23, 34
63		Service	14 / Worcestershire (Pnrs)	35
6	16	Regular	2 / York & Lancaster	26, 30, 31, 33, 35
49	148	T.F.	4 / York & Lancaster	34, 35
49	148	T.F.	5 / York & Lancaster	34, 35
11	32	Service	6 / York & Lancaster	22, 34, 35
17		Service	7 / York & Lancaster (Pnrs)	15, 18, 21, 35
23	70	Service	8 / York & Lancaster	17, 28, 36
23	70	Service	9 / York & Lancaster	17, 28, 29, 36
21	63	Service	10 / York & Lancaster	15, 35
31	94	Service	12 / York & Lancaster	35
31	94	Service	13 / York & Lancaster	35
31	94	Service	14 / York & Lancaster	35

BATTALIONS HAVING INFORMAL/OTHER NAMES

Div.	Brig.	No.	Familiar/Other name		Official Name
31	94		Accrington Pals	11 /	East Lancashire
51	152		Argyllshire	8 /	Argyll & Sutherland High.
36	108		Armagh, Monaghan & Cavan	9 /	Royal Irish Fusiliers
41	122		Arts & Crafts	18 /	King's Royal Rifle Corps
51	152		Banff & Donside	6 /	Gordon Highlanders
41	124		Bankers	26 /	Royal Fusiliers
31	94	1st	Barnsley Pals	13 /	York & Lancaster
31	94	2nd	Barnsley Pals	14 /	York & Lancaster
41	124		Battersea	10 /	Queen's - Royal West Surrey
36	109		Belfast Young Citizens	14 /	Royal Irish Rifles
41	122		Bermondsey	12 /	East Surrey
35	105	1st	Birkenhead	15 /	Cheshire
35	105	2nd	Birkenhead	16 /	Cheshire
5	13	1st	Birmingham Pals	14 /	Royal Warwickshire
5	13	2nd	Birmingham Pals	15 /	Royal Warwickshire
5	15	3rd	Birmingham Pals	16 /	Royal Warwickshire
47	141		Blackheath & Woolwich	20 /	London
31	93	1st	Bradford Pals	16 /	West Yorkshire
31	93	2nd	Bradford Pals	18 /	West Yorkshire
5	95		Bristol	12 /	Gloucestershire
3			Brit. Emp. League (Pnrs.)	20 /	King's Royal Rifle C. (Pnrs)
39	117		British Empire League	17 /	King's Royal Rifle Corps
51	153		Buchan & Formartin	5 /	Gordon Highlanders
34	101		Cambridgeshire	11 /	Suffolk
38	115		Cardiff City	16 /	Welsh
38	114		Carmarthenshire	15 /	Welsh
36	108		Central Antrim	12 /	Royal Irish Rifles
39	117		Chatsworth Rifles	16 /	Sherwood Foresters
33	100		Church Lads Brigade	16 /	King's Royal Rifle Corps
48			Cinque Ports	5 /	Royal Sussex (Pnrs)
48	144		City of Bristol	4 /	Gloucestershire
47	140		City of London	7 /	London
47	140		City of London Rifles	6 /	London
47	140		Civil Service Rifles	15 /	London
2			Cornwall Pnrs.	10 /	Duke of Cornwall L.I. (Pnrs)
31	93	1st	County	18 /	Durham Light Infantry
35	106	2nd	County	19 /	Durham Light Infantry
8		3rd	County	22 /	Durham Light Infantry (Pnrs)
36	109		County Derry	10 /	Royal Inniskilling Fus.
36	108	1st	County Down	13 /	Royal Irish Rifles
36		2nd	County Down	16 /	Royal Irish Rifles (Pnrs)
36	109		County Tyrone	9 /	Royal Inniskilling Fus.
50	151		Cumberland	5 /	Border
51	153		Deeside Highland	7 /	Gordon Highlanders
47			Denbighshire	4 /	Royal Welsh Fusiliers (Pnrs)
36	109		Donegal & Fermanagh	11 /	Royal Inniskilling Fus.
36	107		East Belfast	8 /	Royal Irish Rifles
41	124		East Ham	32 /	Royal Fusiliers
6	16		East Kent	1 /	Buffs (East Kent)
12	37		East Kent	6 /	Buffs (East Kent)
18	55		East Kent	7 /	Buffs (East Kent)
24	17		East Kent	8 /	Buffs (East Kent)
34	101	1st	Edinburgh City	15 /	Royal Scots
34	101	2nd	Edinburgh City	16 /	Royal Scots
2	5		Empire	17 /	Royal Fusiliers
51	153		Fife	7 /	Black Watch
47	142		First Surrey Rifles	21 /	London
2	6	1st	Football	17 /	Middlesex
41	123	2nd	Football	23 /	Middlesex
39			Forest of Dean Pnrs.	13 /	Gloucestershire (Pnrs)
1			Glamorgan Pnrs.	6 /	Welsh (Pnrs.)

Div.	Brig.	No.	Familiar/Other name		Official Name
38			Glamorgan Pnrs.	19 /	Welsh (Pnrs.)
35	106	4th	Glasgow	18 /	Highland Light Infantry
32	97		Glasgow Boys Brigade	16 /	Highland Light Infantry
32	97		Glasgow Commercials	17 /	Highland Light Infantry
33	100		Glasgow Highland	9 /	Highland Light Infantry
32	14		Glasgow Tramways	15 /	Highland Light Infantry
34	101		Grimsby Chums	10 /	Lincolnshire
38	115	1st	Gwent	10 /	South Wales Borderers
38	115	2nd	Gwent	11 /	South Wales Borderers
31			Halifax Pals Pnrs.	12 /	King's Own Yorks L.I. (Pnrs)
49	148		Hallamshire	4 /	York & Lancaster
31	92		Hull Commercials	10 /	East Yorkshire
31	92		Hull Sportsmen	12 /	East Yorkshire
31	92		Hull Tradesmen	11 /	East Yorkshire
55	164		Irish	8 /	King's Liverpool
2	99		Kensington	22 /	Royal Fusiliers
56	168		Kensington	13 /	London
41	123		Kent County	10 /	Royal West Kent-Queen's Own
41	123		Lambeth	11 /	Queen's - Royal West Surrey
31	93	1st	Leeds Pals	15 /	West Yorkshire
35	106	2nd	Leeds Pals	17 /	West Yorkshire
49	146		Leeds Rifles	7 /	West Yorkshire
49	146		Leeds Rifles	8 /	West Yorkshire
41	122		Lewisham	11 /	Royal West Kent-Queen's Own
55	164		Liverpool Irish	8 /	King's Liverpool
30	89	1st	Liverpool Pals	17 /	King's Liverpool
30	21	2nd	Liverpool Pals	18 /	King's Liverpool
30	89	3rd	Liverpool Pals	19 /	King's Liverpool
30	89	4th	Liverpool Pals	20 /	King's Liverpool
47	141		London Irish Rifles	18 /	London
56	169	1st	London Rifle Brigade	5 /	London
56	168	1st	London Scottish	14 /	London
38	113	1st	London Welsh	15 /	Royal Welsh Fusiliers
32	97		Lonsdale	11 /	Border
30	90	1st	Manchester Pals	16 /	Manchester
30	90	2nd	Manchester Pals	17 /	Manchester
30	90	3rd	Manchester Pals	18 /	Manchester
30	21	4th	Manchester Pals	19 /	Manchester
7	22	5th	Manchester Pals	20 /	Manchester
7	91	6th	Manchester Pals	21 /	Manchester
7	91	7th	Manchester Pals	22 /	Manchester
35	104	8th	Manchester Pals	23 /	Manchester
6			Midland Pioneers	11 /	Leicestershire (Pnrs)
51	152		Morayshire	6 /	Seaforth Highlanders
32	96		Newcastle Commercials	16 /	Northumberland Fusiliers
32			Newcastle Railway Pals	17 /	Northumberland Fus. (Pnrs)
36	107		North Belfast	15 /	Royal Irish Rifles
38	113	1st	North Wales	13 /	Royal Welsh Fusiliers
38	115	2nd	North Wales	17 /	Royal Welsh Fusiliers
35	105		Nottingham	15 /	Sherwood Foresters
7			Oldham Pals Pioneers	24 /	Manchester (Pnrs)
51	153		Perthshire	6 /	Black Watch
47	141		Poplar & Stepney Rifles	17 /	London
39	116	1st	Portsmouth	14 /	Hampshire
41	122	2nd	Portsmouth	15 /	Hampshire
47	140		Post Office Rifles	8 /	London
29	86		Public Schools	16 /	Middlesex
33	19	3rd	Public Schools	20 /	Royal Fusiliers
33		1st	Public Works Pioneers	18 /	Middlesex (Pnrs)
41		2nd	Public Works Pioneers	19 /	Middlesex (Pnrs)
56	169		Queen Victoria's Rifles	9 /	London
56	169		Queen's Westminster Rifles	16 /	London
56	168		Rangers	12 /	London

Div.	Brig.	No.	Familiar/Other name		Official Name
5			Renfrewshire	6 /	Argyll & Sutherland (Pnrs)
38	114	1st	Rhondda	10 /	Welsh
38	114	2nd	Rhondda	13 /	Welsh
55	165		Rifle	6 /	King's Liverpool
46	139		Robin Hood Rifles	7 /	Sherwood Foresters
35	106		Rosebury	17 /	Royal Scots
51	154		Ross Highland	4 /	Seaforth Highlanders
56	167		Royal Fusiliers	1 /	London
56	167		Royal Fusiliers	3 /	London
56	168		Royal Fusiliers	4 /	London
56	169		Royal Fusiliers	2 /	London
32	96	1st	Salford Pals	15 /	Lancashire Fusiliers
32	96	2nd	Salford Pals	16 /	Lancashire Fusiliers
32	14	3rd	Salford Pals	19 /	Lancashire Fusiliers
35	104	4th	Salford Pals	20 /	Lancashire Fusiliers
55	166		Scottish	10 /	King's Liverpool
63			Severn Valley Pioneers	14 /	Worcestershire (Pnrs)
31	94		Sheffield City	12 /	York & Lancaster
36	108		South Antrim	11 /	Royal Irish Rifles
36	107		South Belfast	10 /	Royal Irish Rifles
39	116	1st	South Down	11 /	Royal Sussex
39	116	2nd	South Down	12 /	Royal Sussex
39	116	3rd	South Down	13 /	Royal Sussex
35	104	1st	South East Lancs	17 /	Lancashire Fusiliers
35	104	2nd	South East Lancs	18 /	Lancashire Fusiliers
2	99	1st	Sportsman's	23 /	Royal Fusiliers
2	5	2nd	Sportsman's	24 /	Royal Fusiliers
30			St. Helens Pioneers	11 /	South Lancashire (Pnrs)
39	117		St. Pancras	16 /	Rifle Brigade
47	141		St. Pancras	19 /	London
51	152		Sutherland & Caithness	5 /	Seaforth Highlanders
38	114		Swansea	14 /	Welsh
31	92		T'Others	13 /	East Yorkshire
47	142		The Queen's	22 /	London
47	142		The Queen's	24 /	London
34	103	1st	Tyneside Irish	24 /	Northumberland Fusiliers
34	103	2nd	Tyneside Irish	25 /	Northumberland Fusiliers
34	103	3rd	Tyneside Irish	26 /	Northumberland Fusiliers
34	103	4th	Tyneside Irish	27 /	Northumberland Fusiliers
34		1st	Tyneside Pioneers	18 /	Northumberland Fus. (Pnrs)
35		2nd	Tyneside Pioneers	19 /	Northumberland Fus. (Pnrs)
34	102	1st	Tyneside Scottish	20 /	Northumberland Fusiliers
34	102	2nd	Tyneside Scottish	21 /	Northumberland Fusiliers
34	102	3rd	Tyneside Scottish	22 /	Northumberland Fusiliers
34	102	4th	Tyneside Scottish	23 /	Northumberland Fusiliers
41	123		Wearside	20 /	Durham Light Infantry
39	117		Welbeck Rangers	17 /	Sherwood Foresters
36	107		West Belfast	9 /	Royal Irish Rifles
2	6		West Ham	13 /	Essex
35	105		West of England	14 /	Gloucestershire
4			Wool Textile Pioneers	21 /	West Yorkshire (Pnrs)
41	124		Yeoman Rifles	21 /	King's Royal Rifle Corps

INFANTRY BATTALIONS NOT ENGAGED FROM THE 2ND JULY

A small number of battalions which had been engaged in the 'Big Push' on the 1st July have been allocated to Objective 37, which, as the explanation of the objective codes on page 11 implies, were withdrawn from the front early on the 2nd July to rest, recover, have their losses replaced or to be moved away to other sectors. The following is a list of these battalions.

DIV.	BRIG.	BATTALION	WHERE ENGAGED
34	102	20/Northumberland Fusiliers	La Boisselle
34	102	21/Northumberland Fusiliers	La Boisselle
34	102	22/Northumberland Fusiliers	La Boisselle
34	102	23/Northumberland Fusiliers	La Boisselle
36	107	8/Royal Irish Rifles	North of Thiepval
36	107	9/Royal Irish Rifles	North of Thiepval
36	108	9/Royal Irish Fusiliers	North of Thiepval
36	108	11/Royal Irish Rifles	North of Thiepval
36	108	12/Royal Irish Rifles	North of Thiepval
36	108	13/Royal Irish Rifles	North of Thiepval
36	109	14/Royal Irish Rifles	North of Thiepval
46	137	5/North Staffordshire	Gommecourt
46	137	6/North Staffordshire	Gommecourt
46	137	5/South Staffordshire	Gommecourt
46	137	6/South Staffordshire	Gommecourt
46	139	5/Sherwood Foresters	Gommecourt
46	139	6/Sherwood Foresters	Gommecourt
46	139	7/Sherwood Foresters	Gommecourt
46	139	8/Sherwood Foresters	Gommecourt

Following the failed attack on the 1st July against the German defences from La Boisselle north as far as Gommecourt, it was necessary to make immediate plans to continue the offensive and to relieve the exhausted and depleted battalions*. The so-called 'Gommecourt Diversion' with almost seven thousand casualties had only partially achieved its aim, namely to attract enemy fire away from the main British attack further south. From the 2nd July there was little action at the Gommecout Salient and relief of the units was effected without delay. Under the protection of a red-cross flag on the front of the 46th Division, both German and British wounded were collected. The leading brigades (137th and 139th) were withdrawn on the 2nd July, the 138th Brigade following a few days later. The division was then transferred out of the area and did not return to the Somme. The 56th Division at Gommecourt was also withdrawn to recover from its losses but remained on the Somme, participating in a number of later actions. The 36th Division, after its valiant attack on the Schwaben Redoubt, was withdrawn and did not participate in further actions on the Somme.

Whereas the above battalions were quickly withdrawn, others were less fortunate, some battalions having to wait a number of days before being relieved, some being engaged in mid-July at only half their normal establishment. It took several weeks before the losses of the 1st July could be replaced. The battalions were eventually brought up to strength with good men, but many were inexperienced, some arriving at the front with hardly any training at all. What they lacked in experience, they made up with a surfeit of will and determination.

* The casualties, extracted from the Official History Volume I, in the divisions on the 1st July were:-

34th Division	6,380	La Boisselle		7th Division	3,410	Mametz
29th Division	5,240	Beaumont Hamel		18th Division	3,115	East of Mametz & west of Montauban
8th Division	5,121	Ovillers				
36th Division	5,104	North of Thiepval		30th Division	3,011	Montauban
4th Division	4,692	North of Beaumont Hamel & south of Serre		46th Division	2,455	Gommecourt
				17th Division	1,115	Fricourt (3 Bns. of 50th Brigade only)
56th Division	4,314	Gommecourt				
21st Division	4,256	North of Fricourt		48th Division	1,060	6th & 8th Royal Warwicks attached to 4th Division
32nd Division	3,949	Thiepval				
31st Division	3,600	Serre		49th Division	590	Mostly in reserve at Thiepval

DOMINION/OVERSEAS INFANTRY, PIONEERS & CAVALRY

Division	Brig	Battalion	Objectives
1 (Australian)	1	1 / New South Wales	22, 32, 34, 36
1 (Australian)	1	2 / New South Wales	22, 32, 34, 36
1 (Australian)	1	3 / New South Wales	22, 32, 34, 36
1 (Australian)	1	4 / New South Wales	22, 32, 34, 36
1 (Australian)	2	5 / Victoria	22, 32, 34
1 (Australian)	2	6 / Victoria	22, 32, 34
1 (Australian)	2	7 / Victoria	22, 32, 34
1 (Australian)	2	8 / Victoria	22, 32, 34
1 (Australian)	3	9 / Queensland	22, 34
1 (Australian)	3	10 / S. Australian	22, 34
1 (Australian)	3	11 / W. Australian	22, 34
1 (Australian)	3	12 / S.& W.Australia	22, 34
1 (Australian)		1 / Australian Pnrs.	22, 34
2 (Australian)	5	17 / New South Wales	22, 34, 36
2 (Australian)	5	18 / New South Wales	22, 34, 36
2 (Australian)	5	19 / New South Wales	22, 34, 36
2 (Australian)	5	20 / New South Wales	22, 34, 36
2 (Australian)	6	21 / Victoria	22, 34
2 (Australian)	6	22 / Victoria	22, 34
2 (Australian)	6	23 / Victoria	22, 34
2 (Australian)	6	24 / Victoria	22, 34
2 (Australian)	7	25 / Queensland	22, 31, 36
2 (Australian)	7	26 / Queensland & Tasmania	22, 31, 36
2 (Australian)	7	27 / S. Australia	22, 31, 36
2 (Australian)	7	28 / W. Australia	22, 31, 36
2 (Australian)		2 / Australian Pnr. Bn.	22, 31, 36
4 (Australian)	4	13 / New South Wales	22, 34
4 (Australian)	4	14 / Victoria	22, 34
4 (Australian)	4	15 / Queensland (Tas.)	22, 34
4 (Australian)	4	16 / S. & W. Australia	22, 34
4 (Australian)	12	45 / New South Wales	22
4 (Australian)	12	46 / Victoria	22
4 (Australian)	12	47 / Queensland & Tas.	22
4 (Australian)	12	48 / S. & W. Australia	22
4 (Australian)	13	49 / Queensland	22, 34
4 (Australian)	13	50 / S. Australia	22, 34
4 (Australian)	13	51 / W. Australia	22, 34
4 (Australian)	13	52 / S & W Australia & Tasmania	22, 34
4 (Australian)		4 / Australian Pnrs.	22
5 (Australian)	8	29 / Victoria	36
5 (Australian)	8	30 / New South Wales	36
5 (Australian)	8	31 / Queensland	36
5 (Australian)	8	32 / S. & W. Australia	36
5 (Australian)	14	53 / New South Wales	36
5 (Australian)	14	54 / New South Wales	36
5 (Australian)	14	55 / New South Wales	36
5 (Australian)	14	56 / New South Wales	36
5 (Australian)	15	57 / Victoria	36
5 (Australian)	15	58 / Victoria	36
5 (Australian)	15	59 / Victoria	36
5 (Australian)	15	60 / Victoria	36
5 (Australian)		5 / Australian Pnrs.	36
1 (Canadian)	1	1 / Ontario	22, 29
1 (Canadian)	1	2 / East Ontario	22, 29
1 (Canadian)	1	3 / Toronto	22, 29
1 (Canadian)	1	4 / Toronto	22, 29
1 (Canadian)	2	5 / Western Cav.	29, 34
1 (Canadian)	2	7 / 1st British Columbia	29, 34
1 (Canadian)	2	8 / 90th Rifles	29, 34
1 (Canadian)	2	10 / Battalion	29, 34

Division	Brig	Battalion	Objectives
1 (Canadian)	3	13 / Royal Highlanders	22, 29, 34
1 (Canadian)	3	14 / R. Montreal Reg.	22, 29, 34
1 (Canadian)	3	15 / 48th Highlanders	22, 29, 34
1 (Canadian)	3	16 / Canadian Scottish	22, 29, 34
1 (Canadian)		1 / Canadian Pnr. Bn.	29, 34
2 (Canadian)	4	18 / W. Ontario	28, 29
2 (Canadian)	4	19 / Central Ontario	28, 29, 36
2 (Canadian)	4	20 / Central Ontario	28, 29
2 (Canadian)	4	21 / E. Ontario	28, 29
2 (Canadian)	5	22 / Canadien Francais	29
2 (Canadian)	5	24 / Victoria Rifles	29
2 (Canadian)	5	25 / Nova Scotia Rifles	29, 36
2 (Canadian)	5	26 / New Brunswick	29, 36
2 (Canadian)	6	27 / City of Winnipeg	29, 36
2 (Canadian)	6	28 / North-West	29, 36
2 (Canadian)	6	29 / Vancouver	29
2 (Canadian)	6	31 / Alberta	29
2 (Canadian)		2 / Canadian Pnrs.	29
3 (Canadian)	7	0 / PPCLI	29, 34
3 (Canadian)	7	0 / R. Canadian Reg.	29, 34
3 (Canadian)	7	42 / R. Highlanders	29, 34
3 (Canadian)	7	49 / Edmonton	29, 34
3 (Canadian)	8	1 / Cdn.M.R.	22, 29, 34
3 (Canadian)	8	2 / Cdn.M.R.	22, 29, 34
3 (Canadian)	8	4 / Cdn.M.R.	22, 29, 34
3 (Canadian)	8	5 / Cdn.M.R.	22, 29, 34
3 (Canadian)	9	43 / Cameron Highlanders	29, 34
3 (Canadian)	9	52 / New Ontario	29, 34
3 (Canadian)	9	58 / Battalion	29, 34
3 (Canadian)	9	60 / Victoria Rifles	29, 34
3 (Canadian)		3 / Canadian Pnrs.	29, 34
4 (Canadian)	10	44 / Battalion	29, 34
4 (Canadian)	10	46 / S. Saskatchewan	29, 34
4 (Canadian)	10	47 / British Columbia	29, 34
4 (Canadian)	10	50 / Calgary	29, 34
4 (Canadian)	11	54 / Kootenay	29, 34
4 (Canadian)	11	75 / Mississauga	29, 34
4 (Canadian)	11	87 / Cdn. Grenadier Guards	29, 34
4 (Canadian)	11	102 / Battalion	29, 34
4 (Canadian)	12	38 / Ottawa	29, 34
4 (Canadian)	12	72 / Seaforth Highlanders	29, 34
4 (Canadian)	12	73 / Royal Highlanders	29, 34
4 (Canadian)	12	78 / Winnipeg Grenadiers	29, 34
4 (Canadian)		67 / Canadian Pnrs.	29, 34
(New Zealand)	1	1 / Auckland	27, 32, 36
(New Zealand)	1	1 / Canterbury	27, 32, 36
(New Zealand)	1	1 / Otago	27, 32, 36
(New Zealand)	1	1 / Wellington	27, 32, 36
(New Zealand)	2	2 / Auckland	23, 27, 36
(New Zealand)	2	2 / Canterbury	21, 27, 36
(New Zealand)	2	2 / Otago	23, 27, 36
(New Zealand)	2	2 / Wellington	21, 27, 36
(New Zealand)	3	1 / N.Z. Rifle Brigade	27, 32, 36
(New Zealand)	3	2 / N.Z. Rifle Brigade	27, 32, 36
(New Zealand)	3	3 / N.Z. Rifle Brigade	27, 32, 36
(New Zealand)	3	4 / N.Z. Rifle Brigade	27, 32, 36
(New Zealand)		0 / N.Z. Pioneer Bn.	21, 27, 36
1 CAV. (CAN.)		19 / Lancers Sialkot Cav. Brig.	31, 32
CAVALRY		0 / S. Irish Horse	31, 32
CAV. (IND.)		7 / Dragoon Guards	23
CAV. (IND.)		20 / Deccan Horse	23

THE I.G.N. MAPS AND REFERENCE NUMBERS

Four I.G.N. (Institut Géographique National) maps on a scale of 1:25 000, i.e. one centimetre to one kilometer or approximately two and a half inches to one mile, cover the whole of the battle area from the 1st July to 26th November when the campaign became bogged down in severe winter conditions.

    2407 Ouest (West) which is shown simply as 7 W on the lists

    2407 Est (East)   "   7 E   "

    2408 Ouest (West)   "   8 W   "

    2408 Est (East)   "   8 E   "

They are readily available from Monsieur Brunet's Maison de la Presse opposite the Basilica in Albert. At the time of writing the cost was 46 French francs each. If you are well out of the Somme area, the maps may have to be ordered. In England, they can be obtained from a number of large booksellers. These large-scale maps with so much detail are really most useful when touring the area and when compared with those printed during 1916, it is very pleasing to note that, in spite of the complete devastation of much of the terrain, the roads follow their original course and few seem to have been lost during the post-war reconstruction.

    The map references quoted on lists have been done in the traditional way, i.e. Eastings followed by Northings. Most references to trenches refer to one kilometer square, but smaller sites, such as farms and craters, etc have been given references to one hundred meters and, in some cases, i.e. all the cemteries in Appendix One, to an accuracy of twenty meters. The reference numbers on the maps are shown in two colours, black and blue. ALL the references quoted on the lists are taken from the BLUE figures (known as the Lambert Zone Two). Personally, I have ruled the maps into one kilometer squares (always with the blue figures) and I find this extremely useful - the references can be found in an instant on the appropriate map.

    A number of references have been given a + sign after the Eastings or/and Northings reference. For example, on page 59:, the reference on Map 7 W for Brawn Trench is 624+/2561 which means that the trench extends EAST into the next kilometer line, 625. Similarly, on page 82, the reference 630+/2559+ on Map 8 E for Upper Road means that Upper Road extends EAST into line 631 and NORTH into the next kilometer line which will be 2560. All distances have been measured in a straight line, usually from centre to centre of two points.

    When the maps of the 1st World War were being compiled for the army, some changes were made to the names of certain sites and the Bois d'Authuille, just west of Thiepval is always cited as Thiepval Wood on British military maps of the period. Similarly, the Bois de la Haie is known as Authuille Wood. All references in this guide to these woods use the English names found on British trench maps. Some interesting transpositional errors were made by cartographers and the village of Foncquevillers at the northern end of the line is often listed on British maps as Fonquevillers and the Bois des Troncs, is always shown on British maps as Trônes Wood. Pronunciation of French proper nouns has always presented particular difficulties to foreign visitors to the Somme or indeed, anywhere in France. Villages such as Ginchy, Thiepval and Aveluy when pronounced phonetically by an Englishman would not be understood by his French counterpart. The troops avoided such problems by finding their own names. Delville Wood was quickly baptised Devil's Wood which seems more appropiate than the original aristocratic title. It would not have been wise to officially change the names of villages and so the men had to cope with some strange names and very quickly, Auchonvillers became known as "Ocean Villas" to the men and other places took rather odd names when Anglicized.

    Whereas readers will be familiar with the four cardinal points of the compass, it may be some while since they were called upon to use a sixteen-point compass. Perhaps the compass and the few examples of IGN references on the following page may be useful.

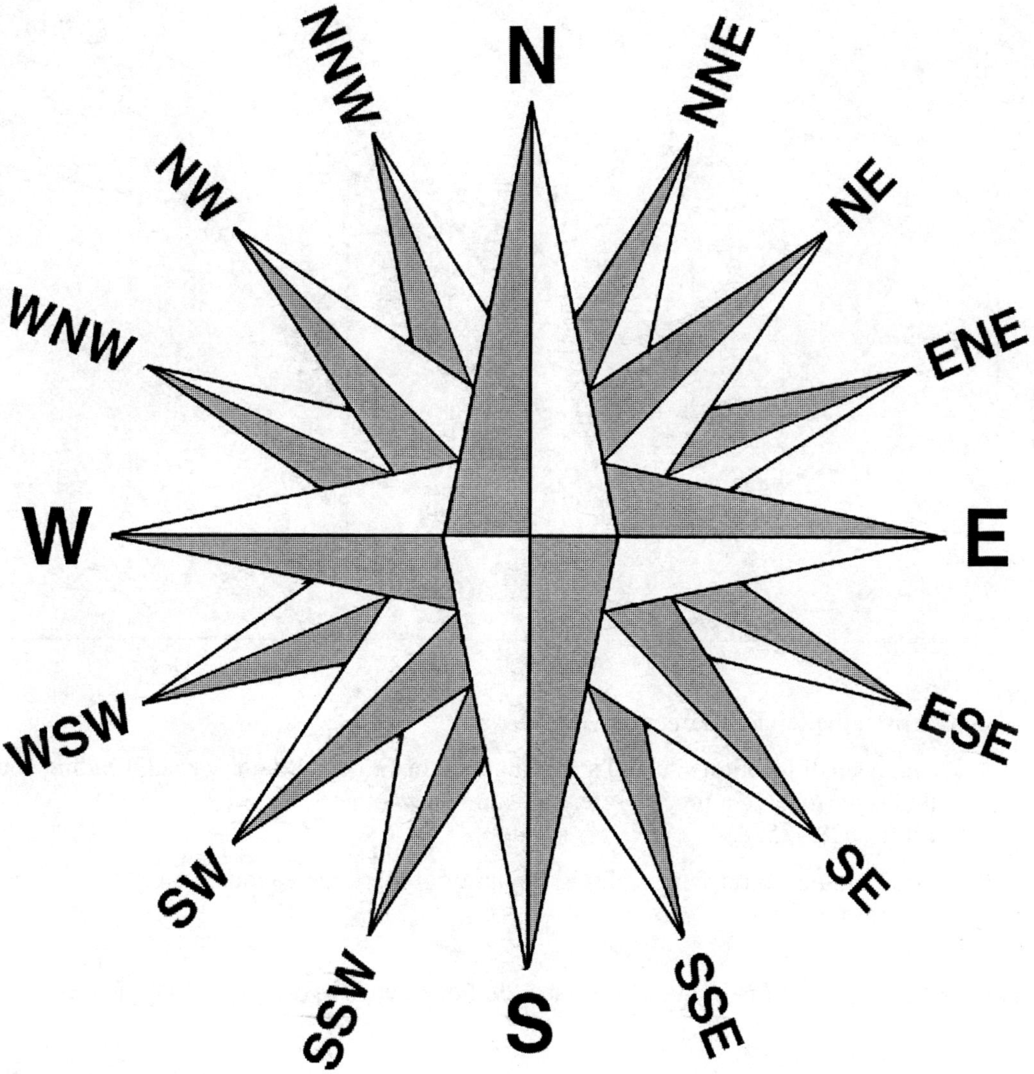
N
NNE
NE
ENE
E
ESE
SE
SSE
S
SSW
SW
WSW
W
WNW
NW
NNW

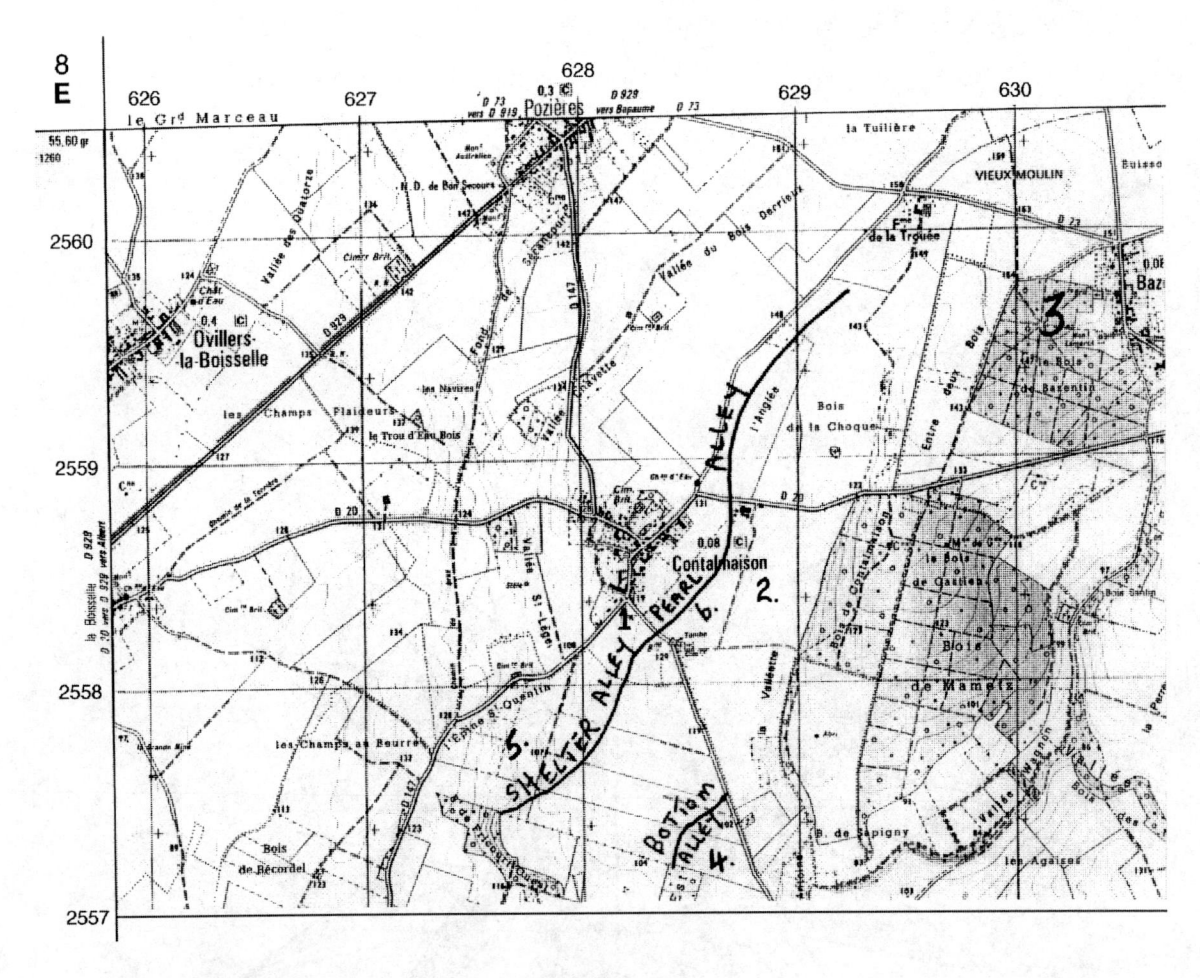

A few examples from the top left corner of Map 8E

1. Bell's Redoubt, a small location situated at the junction of Fricourt/Mametz road, heading south from Contalmaison can have a precise reference number:
    Map 8 E   6282/255834

2. The village of Contalmaison occupies almost a quarter of a kilometer square on Map 8E. The reference is, therefore:
    Map 8 E   628/2558

3. Bazentin le Petit Wood starts in kilometer line 629, but extends east into line 630. Its full reference is:
    Map 8 E   629+/2559

4. Bottom Alley has been drawn on the map. As it is contained in one square, its reference is therefore:
    Map 8 E   628/2557

5. Shelter Alley infringes both east and north and so its reference will be:
    Map 8 E   627+/2557+

6. Similarly, the reference to Pearl Alley will be:
    Map 8 E   628+/2558+

ABOUT THE TRENCHES

Included in the trench list are British and German front line trenches attacked and defended on the 1st July as well as those involved in the remainder of the Somme campaign about which this guide is concerned. The reason being, that apart from trenches in the Mametz and Montauban area, plus a small section of trenches near the Leipzig Salient, any German trenches captured on the 1st July could not be held and were the subject of subsequent attacks. Whereas the allocation to objective numbers of short trenches presented no particular difficulties, it was considered necessary to divide some of the longer ones into north/south or east/west sections. Two such examples are Worcester and Bayonet Trenches. The former runs from High Wood in a south-easterly direction towards Delville Wood and was, therefore, involved in the capture of both woods. Accordingly, Worcester Trench North is allocated to High Wood (Objective 23) and Worcester Trench South to Delville Wood (Objective 21). Similarly, Bayonet Trench East covers the area just north of Gueudecourt (Objective 32) whereas Bayonet Trench West is included in the area north-east of Eaucourt Abbey (Objective 36). The division of some trenches has made it possible to follow the battalions' movements more clearly.

There are over one thousand nine hundred trenches and sites listed showing the main centres of action in the campaign. Reading the diaries, memoires and notes of some of the men who took part in the momentous events between July and November 1916 and on into 1917, a trench had many different meanings for those who had to dig, live in and eventually go over the top. A trench could be:

    A good place to be - clean, deep and dry, giving good protection
    A place of great apprehension while waiting for the unknown
    A place of great elation when the enemy trench was captured
    A place which was the target of enemy artillery
    A place of mud and water sometimes over a foot deep, cold and hostile
    A place so damaged by shell-fire that little cover was left
    A place of filth, death, mutilated bodies, rats and millions of flies
    A place of stress so great that minds were shattered without a physical wound
    A place from which the men had to endure the most sickly smell of putrifying corpses lying a few yards away
    A place which seemed to have no resemblance to a civilised world of art, music, poetry and the simple beauty of England's green fields
    A place in which men feared death, not so much for themselves, but for the future of their families, their sweethearts and friends
    A place in which men realised the improbability of enjoying again the simple pleasures of life, of experiencing love and marriage
    A place in which to wait, followed by intense action
    A place where many died

Whatever they were to the men, they gallantly stuck to their task, usually with a moan in the way soldiers have always done, put up with the dreary life in the trenches, often enlivened by the typical humour of these fighting men whose lot it was to continue the struggle.

TRENCHES, BATTLESITES ETC IN ALPHABETICAL ORDER

Site/trench	Location	Map	MapRef	Obj
10th Street Trench	800m N of North Practice Trenches	7 E	629/2563	29
1st Avenue	W of Y Ravine	7 W	622/2564	35
26th Avenue East	Assem. pos. for attack on Le Sars	7 E	631/2562+	36
26th Avenue West	Runs E from Courcelette	7 E	629/2562	29
2nd Avenue	E off old Beaumont Rd.	7 W	621/2565	35
2nd Graben	400m S of Mouquet Farm	7 E	627/2561	22
3rd Avenue	Runs SW from White City	7 W	621/2565	35
4th Avenue	Just S of 5th Avenue	7 W	621/2565	35
5th Avenue	W of Leipzig Salient	7 W	624/2561	34
5th Avenue	Runs W from White City	7 W	621/2565	35
67th Street	S of Regina Trench East	7 E	629/2563	29
6th Avenue	1200m WNW of Baz-le-Petit	8 E	629/2560	28
70th Avenue	WNW of Bazentin le Petit	8 E	629/2560	28
88th Trench	700m W of Hawthorn Crater	7 W	621/2565	35
Abbey Lane	Just N of Eaucourt Abbey	7 E	632/2563	36
Abbey Road	Heads NW out of Flers	7 E	632/2563	36
Abbey Trench	980m E of Le Sars	7 E	632/2563	36
Abeyne Street	ESE of Chapes Spur	8 W	625/2557	16
Acid Drop Copse	440m ESE of Contalmaison	8 E	6287/255844	18
Adolf Trench	850m SSE of Dublin Redoubt	8 E	632/2554	18
Aeroplane Trench	1300m W of Mametz Halt	8 E	627/2555	18
Aintree Street	Authuille Wood	7 W	624/2559+	17
Albert Street	NW of Hawthorn Crater	7 W	622/2565	35
Alcohol Trenches	East of Delville Wood	8 E	634/2558+	25
Ale Alley/Trench	Runs E from Delville Wood	8 E	634+/2559	25
Allen Avenue	South Mesnil	7 W	622/2562	35
Alt Trench	400m SW of Glatz Redoubt	8 E	631/2555	18
Alte Jäger Stellung	800m E of La Boisselle	8 E	626/2558	16
Ancre Trench	Runs E from Beaucourt	7 W	625/2565	35
Andrews Avenue	Nr. Keats Redan	8 W	624/2558	17
Angle Trench	200m E of Delville Wood	8 E	634/2558	21
Angle Wood	1800m SSE of Guillemont	8 E	635/2555+	33
Anna (pioneer trench)	420m WNW of Rossignol Wood	7 W	623/2570	35
Anstruther Street	SE of Ovillers Post	8 W	624/2558	16
Ant Trench	560m WSW of Rossignol Wood	7 W	623/2570	35
Antelope Trench	800m E of Lesboeufs	7 E	638/2560	31
Appendix, The	500m SE of Butte de Warlencourt	7 E	633/2564	36
Apple Alley	600m W of Mametz Halt	8 E	627/2555	18
Aquaduct Road	NW from Le Sars	7 E	631/2564	36
Aragon Trench	1100M SE of Rossignol Wood	7 W	624/2569	35
Arbroath Street	Just N of Chapes Spur	8 W	625/2557	16
Ardgour Street	Just E of Authuille village	7 W	624/2560	34
Ardnshaig Street	Just E of Authuille village	7 W	624/2560	34
Argyll Alley	1150m E of Baz-le-Petit	8 E	631/2560	23
Argyll Street	700m E of Ovillers Post	8 W	625/2558	17
Arrow Head Copse, site of	150m S of Trônes Wd/Guillemont Rd	8 E	6346/25571	24
Arrow Lane	Just N of Lonely Lane	8 E	627/2556	15
Arthur Street	E of Thiepval Wood	7 W	624/2561+	34
Artillery Alley	1400m NNE of Beaucourt	7 W	625/2565	35
Artillery Lane	NNW of Beaucourt sur Ancre	7 W	624/2565	35
Arun Trench	470m S of Rossignol Wood	7 W	624/2570	35
Ash Grove	S from Oblong Wood	7 W	624/2562	34
Ashtown Street	Just N of Chapes Spur	8 W	625/2557	16
Aston Trench	Bazentin Ridge	8 E	629+/2558+	20
Athall Street	NNW of Chapes Spur	8 W	625/2557	16
Atom Trench	1800m NE of Gueudecourt	7 E	637/2563	31
Austrian Junction	E off S end of Back Alley	8 E	629/2556	18
Austrian Trench	S of Austrian Junction	8 E	629/2556	18
Authuille Bridge	Bridge over Ancre NW Authuille	7 W	623/2560	34
Authuille Quarry	Granatloch/Leipzig Salient	7 W	6247/2561	34
Authuille Wood	Assem. of 70th Brig.1st July attack	8 W	624+/2559+	34

TRENCHES, BATTLESITES ETC

Site/trench	Location	Map	MapRef	Obj
Avoca Valley	Between Tara Hill & Becourt	8 W	624/2556	16
Babylon	N from Observation Wood	7 W	621/2567	35
Back Alley	1300m NNW of Carnoy	8 E	629/2556	18
Back Lane	N of Breslau Alley	8 E	631/2555	18
Back Trench	240m N of Breslau Trench	8 E	631/2555	18
Bacon Trench	1000m N of Gueudecourt	7 E	636/2563	31
Baden Street	1000m NNW of Mouquet Farm	7 E	626/2562	34
Bailiff Wood	600m W of Contalmaison	8 E	6276/25587	18
Baillescourt Farm	540m NNW of Grandcourt	7 E	6267/25656	34
Bainbridge Trench	Contn. of Bulgar Trench	7 E	625+/2562	34
Ball Lane	Joins Lozenge Alley from W	8 E	626/2556	15
Bank Trench	1400m ESE of Butte de Warlencourt	7 E	634/2564	36
Barley Trench	1600m N of Gueudecourt	7 E	636/2564	31
Barlow Street	Just W of Ovillers	8 W	625/2559	17
Barrow Road	Near Ovillers	8 E	626/2559	17
Basin Wood	CCS & mass burial site nr.La Signy Farm	7 W	621/2567	35
Bateman Trench	Southern edge of Oblong Wood	7 W	624/2562	34
Bath Lane	SE from Puisieux	7 W	625/2568	35
Battalion Trench	E of Hawthorn Crater	7 W	622/2565	35
Battery Valley	North from Stuff Redoubt	7 E	626/2563+	34
Battle Street	200m N of Cambridge Copse	8 E	631/2554	18
Bay Lane	750m S of Pommiers Redoubt	8 E	629/2555	18
Bay Point	300m W of Casino Point	8 E	630/2555	18
Bay Trench	NW from Casino Pt. Mine	8 E	630/2556	18
Bay Trench	2000m NW of Gueudecourt	7 E	634/2563	36
Bayonet Trench East	Runs NW from N of Gueudecourt	7 E	635+/2563	32
Bayonet Trench West	1700m ENE of Eaucourt Abbey	7 E	634/2563	36
Baz-le-Pet. Civil Cem.	Just E of village	8 E	6309/25596	20
Bazentin Ridge East	E of Bazentin le Grand	8 E	631+/2558	20
Bazentin Ridge West	SW of Bazentin-le-Petit	8 E	630+/2558	20
Bazentin le Grand Wood	560m W of Baz-le-Grand	8 E	630+/2558+	20
Bazentin le Petit Wood	400m SW of Baz-le-Petit	8 E	629+/2559	20
Bazentin-le-Grand	2000m W of Longueval	8 E	631/25589	20
Bazentin-le-Petit	2400m NE of Contalmaison	8 E	630/2559	20
Beadles Trench	Crucifix Corner towards High Wood	8 E	631/2559	20
Beam Lane	1800m ENE of Gueudecourt	7 E	637+/2563	31
Beau Regard Alley	700m E of Wundt Werk	7 E	626/2567	35
Beaucourt Château (ruins)	E Beaucourt	7 W	624/2564	35
Beaucourt Mill	800m SSW of Beaucourt	7 W	624/2564	35
Beaucourt Redoubt	350m NW of Beaucourt	7 W	624/2565	35
Beaucourt Road	Beaumont to Beaucourt	7 W	623+/2565	35
Beaucourt Trench	W from Beaucourt	7 W	624/2565	35
Beaucourt-Hamel Station	1500m SE of Beaumont Hamel	7 W	624/2564	35
Beaucourt-sur-Ancre	2000m E Beaumont Hamel	7 W	624+/2564+	35
Beaumont Alley	E exit from Beaumont Hamel	7 W	623/2565	35
Beaumont Hamel	2000m E Auchonvillers	7 W	622+/2565	35
Beaumont Hamel Civil Cem.	250m E of village	7 W	623/2565	35
Beaumont Trench	Runs N from Beaumont Hamel	7 W	622/2565	35
Bécourt Avenue	Comm. T. crossing Tara Hill	8 W	624/2557	16
Bedford Street	1300m E of Baz-le-Petit	8 E	631/2560	23
Bedford Street	250m S of Mary Redan	7 W	622+/2563	35
Bedford Trench	E from German's Wood	8 E	632/2555	18
Beef Trench	Just N of Leuze Wood	8 E	637/2558	33
Beer Trench	N from Ginchy/Longueval Rd	8 E	635/2558+	25
Beetle Alley	Just N of Pommiers Redoubt	8 E	630/2556	18
Bell's Redoubt	Start of Mametz Rd from Contalmaison	8 E	6282/255834	18
Below Support Trench	W of Dyke Road, Courcelette	7 E	629/2562	29
Below Trench North	E of Pys	7 E	629/2565+	34
Below Trench South	Pys to Dyke Road	7 E	629/2563+	29
Bennett Trench	Adjacent Lesboeufs	8 E	636+/2559+	30
Berg Graben	SE from Rossignol Wood	7 W	624/2570	35
Bergwerk	Just N of Beaumont Hamel	7 W	623/2565	35
Berkley Street	Just W of Talus Boisé	8 E	631/2555	18
Berkshire Avenue	CT, Albert to Avoca Valley	8 W	623+/2556	16

Site/trench	Location	Map	MapRef	Obj
Berlin Trench	Just NE of Serre	7 W	624/2567	35
Bernafay Wood	1140m ENE of Montauban	8 E	632+/2556+	19
Berwick Avenue	450m E of Carnoy civ. cem.	8 E	630/2554	18
Bess Street	700m S of Matthew Copse	7 W	622/2567	35
Billi Trench	1600m SE of Butte de Warlencourt	7 E	634/2562	36
Birch Tree Avenue	SSW of Contalmaison	8 E	627/2557	18
Birch Tree Trench	NE from Birch Tree Wood	8 E	627/2557	15
Birch Tree Wood, site of	800m NNW of Fricourt Farm	8 E	6273/25574	15
Birtle Post	Authuille Wood	8 W	624/2560	34
Bisset Trench	Just W of Leipzig Salient	7 W	624/2561	34
Bite Trench	At W end of Bayonet Trench W	7 E	634/2563	36
Black Alley	600m SSW of Pommiers Redoubt	8 E	629/2555	18
Black Hedge	900m SSW of Pommiers Redoubt	8 E	629/2555	18
Black Horse Bridge	Bridge SSW of Authuille	7 W	623/2560	34
Black Horse Road	Just SW of Authuille	7 W	623/2560	34
Black Horse Shelters	Just W of Authuille Civ. Cem.	8 W	6237/25604	34
Black Road	Longueval to High Wood	7 E	632/2559	23
Black Trench	W of Black Hedge	8 E	629/2555	18
Black Watch Alley	1800m SE of Pozières	8 E	629/2559	20
Black Watch Trench	Adjacent High Wood	8 E	631+/2560	23
Blackfriars Bridge	Bridge over trenches at Serre	7 W	622/2567	35
Blighty Valley - Nab Valley	NE from Authuille Wood E	8 W	624+/2559+	34
Blind Alley	800m E of Pommiers Redoubt	8 E	630/2556	18
Bloater Trench	500m E of Lochnagar crater	8 E	626/2558	16
Bloomfield Avenue	Runs W from S of Rough Trench	7 W	622/2565	35
Blue Trench	850m NE of Schwaben Redoubt	7 E	625/2563	34
Bluff, The	900m SW of Petit Miraumont	7 E	627/2565	34
Bluff, The	600m N of Authuille	7 W	624/2561	34
Board Street	450m S of Matthew Copse	7 W	622/2567	35
Boche Trench	Just S of The Loop	8 E	630/2555	18
Bodmin Trench	W of Combles	8 E	637/2557	33
Boggart Hole Clough	Just N of Lonsdale Cem.	7 W	624/2560	34
Bois d'Hollande	NE of Beaucourt	7 W	625/2565	35
Bois de Biez	E of Gommecourt Park	7 W	624/2571	35
Bois du Sartel (see Gommecrt. Wd)	SW of Foncquevillers	7 W	622/2571+	35
Boisselle Street	400m E of Ovillers Post	8 W	624/2558	17
Bolt Alley	2300m ENE of Gueudecourt	7 E	638/2563	31
Bond Street	Delville Wood	8 E	634/2559	21
Bond Street	S of Y Ravine	7 W	622/2564	35
Bonte Redoubt	1500m SW of Rose Cottage	8 E	626/2555	15
Boom Ravine	1200m E of Coulee Trench	7 E	629/2564	29
Boritska Trench	1100m E of Lesboeufs	7 E	638+/2560	31
Bottom Alley	NE from Bottom Wood	8 E	628/2557	18
Bottom Road	950m S of Martinpuich	7 W	630/2560	28
Bottom Trench	S of Tangle Trench	7 E	630/2561	28
Bottom Trench East	900m N of Bazentin le Petit	7 E	630/2560	20
Bottom Trench West	1050m NNW of Baz-le-Petit	7 E	630/2560	20
Bottom Wood	860m ENE of Fricourt Farm	8 E	6285/2556+	15
Bouleaux Wood	1100m NW of Combles centre	8 E	637/2557+	33
Bouvat Trench	NNW of Fort Briggs	7 W	623/2568	35
Bovril Trench	Adjacent Morval	8 E	637/2558	30
Bow Trench	Just W of Serre Trench	7 W	623/2567	35
Bow, The	1200m E of Martinpuich	7 E	631/2561	28
Bowdler Redoubt	Western edge of Pys	7 E	629/2565	34
Bowery, The	200m N of New Beaumont Road	7 W	6219/25658	35
Bowl Trench	950m S of Serre	7 W	623/2566	35
Box Alley	S of Box Wood	7 W	624/2569	35
Box Lane	1000m SE of Mouquet Farm	7 E	627/2561	22
Box Trench	600m NNE of Flers	7 E	635/2561	27
Box Wood	1000m W of Puisieux	7 W	624/2569	35
Braemar Street	N of Chapes Spur	8 W	625/2557	16
Brandy Trench	700m W of Lozenge Wood	8 E	626/2557	15
Brasserie Trench	240m E of Foncquevillers	7 W	621/2572	35

TRENCHES, BATTLESITES ETC

Site/trench	Location	Map	MapRef	Obj
Braun Stellung	German 2nd Line	8 E	626+/2555+	17
Brawn Trench	550m SW of Thiepval crossroads	7 W	624+/2561	34
Bray Street	350m WSW of Keats Redan	8 W	624/2558	16
Bread Trench	1500m NNE of Gueudecourt	7 E	637/2563	31
Brecon Sap	W of High Wood	7 E	631/2560	23
Breslau Alley	NE off Mine Trench	8 E	630/2555	18
Breslau Point	Just E of Breslau Trench	8 E	631/2555	18
Breslau Salient	Just S of Breslau Trench	8 E	631/2555	18
Breslau Trench	600m E of The Castle	8 E	630+/2555	18
Brick Lane	S from Alt Trench	8 E	631/2555	18
Brick Point	700m W of German's Wood	8 E	631/2554	18
Brickfield	800m SE of Montauban church	8 E	6326/25563	18
Brickfield, Ginchy	S Ginchy	8 E	635/2558	26
Brickworks Chimney	950m NNE of German's Wood	8 E	632/2555	18
Bridge Head	SW of Ovillers Post	8 W	624/2558	17
Bridgend	Just W of Hawthorn Crater	7 W	622/2565	35
Brigade Headquarters	Just E of Observation Wood	7 W	622/2567	35
Bright Alley	Runs S from Valley Trench	8 E	627/2556	18
Brimstone Trench	W of Ration Trench	7 E	626/2560	22
Brimstone Trench	2500m ENE of Gueudecourt	7 E	638/2562+	31
Brinds Road	650m SE of Mouquet Farm	7 E	627/2561	22
Briqueterie Trench	Immed. E of the Briqueterie	8 E	632/2556	18
Brisour Trench	N of Fort Briggs	7 W	623/2568	35
British Railway (1917)	Puisieux to Colincamps	7 W	619+/2567+	35
Broad Avenue	W of Gommecourt Park	7 W	621/2571	35
Broadway	W of Y Ravine	7 W	622/2564	35
Brock's Benefit	Opposite Mesnil Château	7 W	622/2562	34
Brompton Road	NW Guillemont	8 E	634/2557	24
Brook Street	570m SW of Y Ravine	7 W	622/2564	35
Broomielaw Street	E of Thiepval Wood	7 W	624/2562	34
Brown Trench	N of Delville Wood	8 E	633+/2559+	21
Buchanan Avenue	E of Thiepval Wood	7 W	624/2561+	34
Buchanan Street	Delville Wood	8 E	633+//2558+	21
Bucket Trench	ESE from Danzig Alley Cem.	8 E	629/25599	18
Buckingham Street	S of Y Ravine	7 W	622/2564	35
Bulford Trench	Immed.E of Gueudecourt	7 E	636/2562	32
Bulgar Alley	Runs N 500m E of Mametz	8 E	629/2555	18
Bulgar Point	700m SSE of Mametz	8 E	629/2554	18
Bulgar Trench	E of Bulgar Point	8 E	629/2554	18
Bulgar Trench	NE from N of Thiepval	7 W	625/2562	34
Bull's Eye	At NW Leipzig Salient	7 W	624/2561	34
Bullen Trench	N from W of Waterlot Farm	8 E	634/2558	21
Bullock Road	East Mesnil	7 W	622/2562	34
Bulls Road	Runs E from N Flers	7 E	635+/2561	27
Bully Road	550m ESE of Casino Point	8 E	630/2555	18
Bully Trench	NW of Bouleaux Wood	8 E	636/2558	33
Bulow Weg	800m SW of Rossignol Wood	7 W	623/2569	35
Bund Support	700m SE of Pommiers Redoubt	8 E	630/2555	18
Bund Trench	700m SSE of Pommiers Redoubt	8 E	630/2555	18
Bunny Alley	Just S of Willow Stream & N of Mametz	8 E	629/2556	18
Bunny Wood	1100m ENE of Rose Cottage	8 E	6284/25562	18
Burghead	Just NW of Thiepval Wood	7 W	624/2562	34
Burlington Arcade	600m S of Y Ravine	7 W	622/2564	35
Burnaby Support Trench	E of Lesboeufs	7 E	638/2560	30
Burnaby Trench	E of Lesboeufs	7 E	638/2560	30
Burnt Island	320m SE of Keats Redan	8 W	625/2558	16
Burnwurk	520m E of Jacob's Ladder	7 W	622/2565	35
Burrel Avenue	300m ENE of Hamel	7 W	623/2563	35
Burrow Trench	Adjacent Munich Trench	7 W	623/2565+	35
Burrow's Post	230m WSW of Adanac Cem.	7 E	628/2564	29
Bury Avenue	Just E of Authuille village	7 W	624/2560	34
Butte Alley	Immed. S of Butte de Warlencourt	7 E	632/2564	36
Butte Trench	Just S of Butte Alley	7 E	632/2564	36

Site/trench	Location	Map	MapRef	Obj
Butte de Warlencourt	1040m NE of Le Sars church	7 E	63286/25645	36
Butterworth Trench	1220m WNW of Baz-le-Petit	8 E	629/2560	28
Caber Trench	WNW of John Copse	7 W	623/2567	35
Cake Trench	Immed. left of Waggon Road	7 W	623/2565	35
Calf Alley	850m NNE of Ginchy	8 E	635/2559	27
Cambridge Copse	1300m E of Carnoy	8 E	631/2554	18
Cameron Trench	formerly part of Switch Line West	8 E	630/2560	28
Campbell Avenue	Ovillers Rd from Authuille	7 W	624/2560	34
Campbell Post	Near Leipzig Salient	7 W	624/2560	34
Campbell Street	Delville Wood	8 E	634/2558+	21
Campbell Work	200m E of Hawthorn Crater	7 W	622/2565	35
Campion Trench	Just SW of old Touvent Farm	7 W	622/2567	35
Canal Trench	S from canal near Grandcourt	7 W	625/2563+	34
Candy Trench	SE from Sugar Factory on D929	8 E	627/2560	29
Cape Avenue	1200m NE of Foncquevillers	7 W	621/2572	35
Cardiff Street	500m W of Hawthorn Crater	7 W	622/2565	35
Cardiff Trench	450m ENE of Baz-le-Petit	8 E	630/2560	23
Carlisle Street	550m SW of Y Ravine	7 W	622/2564	35
Carlton Trench	Between High Wood & Delville Wood	7 E	632+/2559+	21
Carr Trench	Runs E 80m N of Thiepval XR	7 W	625/2562	34
Casement Trench	60m N of German's Wood	8 E	632/2555	18
Casino Point Mine	1000m N of Carnoy	8 E	6302/25552	18
Castle, The	800m SSE of Pommiers Redoubt	8 E	630/2555	18
Castor Post	410m SE of Hamel	7 W	623/2564	34
Casualty Corner	200m W of Bailiff Wood	8 E	6274/25587	18
Cat Lane	600m E of Destremont Farm	7 E	631/2563	36
Cat Street	E of Watling Street	7 W	622/2566	35
Cat Street Tunnel	Heidenkopf to Lager Alley	7 W	622/2566	35
Cateau Trench	E from Brigade H.Q.	7 W	622/2567	35
Caterpillar Copse	Just S of Thiepval Wood	7 W	6243/25617	34
Caterpillar Trench	S from Caterpillar Wood	8 E	630/2557	18
Caterpillar Valley	E of Caterpillar Wood	8 E	630+/2557	18
Caterpillar Wood	1500m NW of Montauban	8 E	630/2557+	18
Causeway	Between Lesboeufs & Gueudecourt	8 E	636/2562	30
Causeway Side	Just N of Johnstone Post	7 W	624/2561+	34
Cawdor Trench	W of John Copse	7 W	623/2568	35
Cemetery Circle	750m SW of Le Transloy	7 E	638/2562	31
Cemetery Trench	Just S of Mametz village	8 E	628/2555	18
Central Avenue	W from Mark Copse	7 W	623/2567	35
Central Trench	Trônes Wood	8 E	633/2557	19
Centre Way	500m SW of Gibraltar	7 E	627/2560+	22
Chalk Alley	200m N of Ten Tree Alley	7 W	622/2566	35
Chalk Cliff	Immed. E of Gueudecourt	7 E	636/2562	32
Chalk Pit	900m WNW of Fabrique Farm	7 W	620/25667	35
Chalk Pit	1100m W of Courcelette	7 E	628/2562	29
Chalk Pit	950m S of Gibraltar	8 E	627/2559	22
Chalk Trench	Adjacent Martinpuich	7 E	630+/2561	28
Chapes Spur	600m SSW of Lochnager crater	8 W	625/2557	16
Charing Cross	S of Y Ravine	7 W	622/2564	35
Charles Avenue	CT to Hamel	7 W	622+/2563	35
Chasseur Hedge	Just N of old Touvent Farm	7 W	622/2568	35
Château Keep	N Maricourt	8 E	632/2554	18
Château Redoubt	S Thiepval	7 W	625/2561	34
Château Trench	Thiepval Wood	7 W	624/2561+	34
Chatham Trench	1000m N of Hawthorn Crater	7 W	622/2566	35
Cheapside	Trônes Wood to Delville Wood	8 E	633/2558	21
Cheese Road	500m WNW of Gueudecourt	7 E	635/2563	32
Chequer Bent Street	Just W of Authuille Wood	7 W	624/2561	34
Cheshire Trench	E of Leuze Wood	8 E	636/2557	33
Cheshire Trench	Northern edge of Stuff Redoubt	7 E	626/2563	34
Chester Street	E of Bazentin le Petit	8 E	631/2559	20
Chimney Trench	Runs E from Nord Alley	8 E	632/2556	18
Chimpanzee Trench	SSW off Maltzhorn Trench	8 E	633/2555	24

Site/trench	Location	Map	MapRef	Obj
Chorley Street	Authuille Wood	7 W	624/2559+	17
Chowbent Street	SW of Leipzig Salient	7 W	624/2561	34
Church Square	Guillemont Centre	8 E	635/2557	24
Church Street	400m ESE of Beaucourt	7 W	625/2564	35
Circus Trench	Joins the 2 Bazentin Woods	8 E	630/2558	20
Circus Trench	Goose Alley to The Circus	7 E	633/2563	36
Circus Trench	850m ESE of Y Ravine	7 W	623/2564	35
Circus, The	E off Back Alley	8 E	629/2556	18
Circus, The	350m NE of Eaucourt Abbey	7 E	633/2563	36
Clarges Street	W from Longueval centre	8 E	633/2558	21
Clark Trench	W of High Wood	8 E	631/2560	23
Clay Trench	930m NW of Stuff Redoubt	7 W	625/2563	34
Clieres Trench	Runs E of Pozières	8 E	628/2560	22
Cliff Trench	Mametz Wood	8 E	6291/25569	18
Clifford Avenue	S from Jacob's Ladder	7 W	622/2561+	35
Clive Trench	Runs S off Watling Street	7 W	622/2565	35
Clonmel Avenue	1000m SW of Hawthorn Crater	7 W	622/2564	35
Cloudy Trench	1000m NE of Gueudecourt	7 E	637/2562+	31
Clyde Avenue	E of Thiepval Wood	7 W	624/2561+	34
Cochrane Alley	400m E of Maltzhorn Farm	8 E	634/2556	24
Cochrane Sap	Near Maltzhorn Farm	8 E	634/2556	24
Cockshy Avenue	1640m WSW of Grandcourt Cemetery	7 E	625/2564	34
Cocoa Lane	Just NE of Delville Wood	8 E	634/2559	27
Coffee Lane	1200m N of Longueval	8 E	633/2560	27
Coffee Trench	Just SW of Boom Ravine	7 E	627/2564	29
Coin Lane	1500m ENE of Butte de Warlencourt	7 E	634/2565	36
Coke Avenue	700m NW of Cambridge Copse	8 E	630/2554	18
Combles	2800m ESE of Guillemont	8 E	637+/2556+	33
Combles Ravine	West Combles	8 E	637/2557	33
Combles Station	NNE Combles	8 E	637/2557	33
Combles Trench	W of Combles	8 E	637/2556+	33
Connantry Avenue	230m N of Hamel	7 W	623/2563	35
Conniston Post	1300m N of Ovillers Post	8 W	624/2560	17
Conniston Street	from SE edge of Authuille Wood	8 W	624/2559	17
Constance Trench	WSW of Mouquet Farm	7 E	626/2561	22
Constitution Hill	S of Y Ravine	7 W	622/2564	35
Contalmaison	2500m E of La Boisselle	8 E	628/2558	18
Contalmaison Château	340m WSW of The Cutting	8 E	6282/25587	18
Contalmaison Civil Cemetery	500m SSE of Contalmaison	8 E	6285/25582	18
Contalmaison Villa	1200m NNE of Contalmaison	8 E	628/2559	18
Contalmaison Wood	800m NW of Contalmaison	8 E	6278/25592	18
Coombe Alley	880m SSW of Mametz	8 E	628/2554	18
Copper Trench	SE from Rose Alley	8 E	627/2555	15
Copse Alley	Runs N from D929 Pozières	7 E	628/2560	22
Copse Trench	SE of old Touvent Farm	7 W	622/2568	35
Cornish Alley	Runs E from Trônes Wood	8 E	634/2557	24
Cornwall Trench	S from The Gully	7 E	628/2564	29
Cornwall Trench	W of Combles	8 E	637/2557	33
Cough Drop	Southern end of Drop Alley	7 E	633/2561	36
Coulee Trench	W Miraumont Rd to S Pys	7 E	629/2565	34
Coupe Trench	Runs E from Warlencourt	7 E	632/2565	36
Courcelette	1500m NW of Martinpuich	7 E	629/2562	29
Courcelette Civil Cem.	E Courcelette	7 E	629/2562	29
Courcelette Track	Courcelette to Ovillers	7 E	626+/2560+	29
Courcelette Trench	N from Courcelette	7 E	629/2563	29
Cox Trench	500m NNE of Flers	7 E	634/2561	27
Crater Lane	Joins Beaumont Trench and Lager Alley.	7 W	622+/2565	35
Crawl Boys Lane	1260m NNE of Foncquevillers	7 W	621/2573	35
Crescent Alley	1000m E of Martinpuich	7 E	631/2561	28
Crescent Avenue	Just SW of Eaucourt Abbey	7 E	632/2563	36
Crescent, The	Adjacent Martinpuich	7 E	631/2561	28
Crest Farm	N of Longueval (NZ Mem. site)	8 E	6334/25605	23
Crest Trench	E of High Wood	8 E	632+/2560	23

Site/trench	Location	Map	MapRef	Obj
Crest Trench	400m N of Coulee Trench	7 E	629/2565	34
Cripps Cut	Runs N from Pilk Street	7 W	622/2565	35
Cromerty Avenue	Thiepval Wood	7 W	624/2562	34
Cross Road Support	Immed. E of Le Sars	7 E	632/2563	36
Cross Street	S of Gommecourt Park	7 W	621/2570	35
Cross Trench	Just W of The Ravine	7 E	628/2564	29
Crows Nest	540m NE of Hamel	7 W	623/2563	34
Crucifix Corner	350m N of Baz-le-Grand	8 E	6315/25592	20
Crucifix Corner	700m ENE of Aveluy	7 W	623/2559	34
Crucifix Trench	From Poodles towards Round Wood	8 E	627/2556+	15
Crucifix, Mametz Civil cemetery	M/c gun opposing advance of 9th Devons	8 E	6285/25552	18
Crucifix, The	700m N of Thiepval XR	7 W	6254/25627	34
Crucifix, The	E of Death Valley, nr. Trônes Wood	8 E	6342/25563	19
Crucifix, The	200m N of New Fricourt Farm	8 E	6276/255698	15
Crucifix, The	Near W edge of Gommecourt Park	7 W	621/2571	35
Crucifix, The	500m W of Rossignol Wood	7 W	623/2570	35
Crucifix, The	300m E of Baz. northern XR	8 E	6307/2560	23
Crucifix, The	680m ESE of Thiepval XR	7 W	625/2561	34
Cuesclin Trench	NW of Staff Copse	7 W	621/2568	35
Cutting, The	500m NE of Contalmaison	8 E	6286/25589	18
Dalhousie Street	650m S of Glory Hole	8 W	625/2557	16
Danger Tree	Newfoundland Park	7 W	622/2564	35
Daniell Alley	250m N of Maltzhorn Farm	8 E	634/2556	19
Danube Trench	500m S of Mametz Halt	8 E	628/2554+	18
Danube Trench	1000m SE of Thiepval	7 W	625/2561	22
Danzig Alley	520m ENE of Mametz	8 E	6293/25551	18
Danzig Trench	Just SW of Cemetery Trench, Mametz	8 E	628/2555	18
Dart Lane	Just N of Lonely Lane	8 E	627/2556	15
Davaar	Just E of Authuille Village	7 W	624/2560	34
Death Valley	ENE of Southern tip of Mametz Wood	8 E	6295/25574	18
Death Valley	600m S of Trônes Wood	8 E	6334/25563	19
Death Valley	NE of Courcelette	7 E	629/2562	29
Death Valley	SE from High Wood	8 E	631/2559+	23
Dell, The = Y Ravine	Newfoundland Park	7 W	622+/2564	35
Delville Wood	Immed. E & NE of Longueval	8 E	634/2559	21
Dent Street	N of John Copse	7 W	623/2568	35
Derby Street	NW of Gommecourt Park	7 W	621/2571	35
Desire Support	North of Regina Trench	7 E	628+/2563+	29
Desire Trench	North of Regina Trench	7 E	628+/2563+	29
Destremont Farm	840m SW of XR at Le Sars	7 E	6312/25635	36
Devil's Staircase	510m ENE of Hamel	7 W	623/2563	34
Devil's Trench	Delville Wood	8 E	633/2559	21
Dewdrop Trench	750m NE of Lesboeufs	7 E	638/2561	31
Diagonal Trench	see Snag Trench	7 E	633/2564	36
Diamond Wood, site of	200m W of Thiepval Château	7 W	6249/25619	34
Dingle, The	350m N of Lozenge Wood	8 E	627/2557	15
Dinkum Alley	340m W of Chalk Pit	8 E	627/2559	22
Done's Redoubt	W Maricourt	8 E	632/2553	18
Donnet Post	400m N of Ovillers Post	8 W	62452/255916	17
Donnet Street	S from Donnet Post	8 W	624/2558	17
Dorset Road	SW of Ovillers	8 W	624/2558	17
Dorset Street	600m ESE of Ovillers Post	8 W	624/2558	17
Dorset Trench	Nr. Longueval	8 E	633/2558+	21
Dot Trench	Just NW of Pozières Windmill	7 E	628/2561	22
Dougle Trench	W of Munich Trench South	7 W	623/2566	35
Dover Street	South Longueval	8 E	633/2558	21
Down Street	South East Longueval	8 E	633/2558	21
Down Street	W from Guillemont Centre	8 E	634/2557	24
Dressler Post	900m SW of Y Sap crater	8 W	624/2557	16
Drop Alley	NE from Cough Drop to Goose Alley	7 E	633/2561+	36
Drop Trench	S of High Wood	8 E	632/2559	21
Dublin Alley	S of Bernafay Wood	8 E	632/2556	19
Dublin Redoubt	880m SSE of the Briqueterie	8 E	632/2555	18

TRENCHES, BATTLESITES ETC

Site/trench	Location	Map	MapRef	Obj
Dublin Street	550m SE of Keats Redan	8 W	625/2558	16
Dublin Trench	S of Guillemont	8 E	635/2557	24
Dublin Trench	500m N of German's Wood	8 E	632/2555	18
Dug-out	300m NW of Bailiff Wood	8 E	6275/25589	18
Dugout Lane	800m N of Star Wood	7 W	624/2569	35
Dugout Trench	Just N of Warren Trench	8 E	631/2555	18
Duhollue Street	Glory Hole	8 W	625/2557	16
Duke Street	W from N Longueval	8 E	633/2559	21
Duke Trench	N from Zollern Redoubt	7 E	627/2562	34
Dumbarton Track	Authuille Wood	8 W	624/2560	34
Dummy Trench	N of Bernafay Wood	8 E	633/2557	19
Dump, Martinpuich	E Martinpuich	7 E	630/2561	28
Dundee Street	Just N of Chapes Spur	8 W	625/2557	16
Dunfermline Street	SE of Ovillers Post	8 W	624/2558	16
Dunmow Trench	S from Brigade H.Q.	7 W	622/2567	35
Durham Street	Leipzig Salient	7 W	624/2561	34
Durham Trench	1000m S of Le Sars	7 E	631/2562	36
Dust Road	1000m SE of Mesnil	7 W	623/2561	35
Dyke Road	NE from Courcelette	7 E	629+/2563+	29
Dyke Street	NW of Gommecourt Park	7 W	621/2571	35
Dyke Trench	NE from Courcelette	7 E	630/2563	29
East KOYLI Trench	Just N of Thiepval Wood	7 W	624/2562	34
East Trench	200m E of Caterpillar Trench	8 E	630/2557	18
Eaucourt Abbey	1300m SE of Le Sars	7 E	632+/2563	36
Eclipse Trench	1700m E of Gueudecourt	7 E	637/2562	31
Eczema Trench	E of Brigade H.Q.	7 W	622/2567	35
Eden Trench	SE of Gommecourt Park	7 W	622/2571	35
Eden Trench	W of Staff Copse	7 W	621/2568	35
Edge Trench	On E edge of Delville Wood	8 E	634/2559	21
Edward Avenue	400m SSW of Casino Point	8 E	630/2554	18
Edward Trench	SE from Arrow Head Copse	8 E	634/2557	24
Eighth Street	W of Pozières	7 E	626/2560+	22
Elbe Trench	SE of Gommecourt Park	7 W	622/2571	35
Elbow, The	390m NNW of Pozières Windmill	7 E	628/2561	22
Elbow, The	800m E of Munster Alley	8 E	629/2560	22
Elgin Avenue	Assem. pos.for attack on Schwaben Rdt.	7 W	624/2562	34
Elie Street	250m WSW of Keats Redan	8 W	624/2558	16
Emden Trench	700m SSW of Pommiers Redoubt	8 E	629/2556	18
Empress Trench	N of Konig Trench	8 E	626/2556	15
Engine Trench	Just W of Beaucourt sur Ancre	7 W	624/2564	35
Epte Trench	E of Gommecourt Park	7 W	622/2571	35
Erin Trench	Just S of The Maze	7 W	621/2571	35
Erstwaite Street	1600m NE of Ovillers Post	8 W	625/2559	17
Esau Alley	Fort Jackson to Hamel	7 W	622+/2563	35
Esau's Way	Contn. of Tenderloin Trench	7 W	622/2565	35
Essex Street	500m SW of Hawthorn Crater	7 W	622/2565	35
Essex Trench	Hawthorn Ridge	7 W	622/2565	35
Essex Trench	300m E of Thiepval	7 W	625/2562	34
Etch Trench	In Gommecourt Park	7 W	622/2571	35
Euston Dump	on Colincourt Road	7 W	620/2567	35
Eva Street	400m SE of Mesnil	7 W	622/2561	35
Evacuation Trench	SSW from Johnstone Post	7 W	624/2561	34
Eve Alley	Immed. E of Gueudecourt	7 E	636/2562	32
Exe Trench	Just S of The Maze	7 W	621+/2571	35
Eye Trench	Between High Wood and Martinpuich	7 E	630+/2560+	28
Fabeck Graben	see Fabeck Trench	7 E	627+/2561+	22
Fabeck Trench	Mouquet Farm towards Courcelette	7 E	627+/2561+	22
Fable Trench	920m W of Rossignol Wood	7 W	623/2570	35
Face Trench	1850m SSE of The Maze	7 W	621/2571	35
Fact Trench	900m SE of Rossignol Wood	7 W	623/2570	35
Factory Corner	At Fabrique Farm, nr. Gueudecourt	7 E	6351/2563	32
Factory Lane	Just W of Martinpuich	7 E	629+/2561	28
Fag Trench	Just W of Flers	7 E	634/2561	27

Site/trench	Location	Map	MapRef	Obj
Fagan Trench	200m E of Trônes Wood	8 E	634/2557	24
Fair Trench	1730m SSE of The Maze	7 W	621/2571	35
Fairmad Street	Immed. S of Glory Hole	8 W	625/2558	16
Falfemont (Faffement) Farm	1400m SW of Combles monument	8 E	6369/25565	33
Falfemont (Faffemont) Wood	850m W of Falfemont Farm	8 E	636/2556	33
Fall Trench	Near "Met." Trenches	7 E	636+/2561+	31
Fall Trench	1120m W of Rossignol Wood	7 W	622/2570	35
Fame Trench	840m W of Rossignol Wood	7 W	623/2570	35
Fancy Trench	980m SW of Rossignol Wood	7 W	623/2570	35
Farce Trench	Just E of Gommecourt Wood	7 W	622/2572	35
Farlock Trench	Runs SE from Ovillers	8 E	626/2559	17
Farmer Trench	920m SSE of The Maze	7 W	621/2571	35
Farmers Road	NW of Destremont Farm	7 E	630/2563	36
Farmyard Trench	980m SSE of The Maze	7 W	621/2571	35
Fat Trench	980m SW of Rossignol Wood	7 W	623/2570	35
Fate Trench	1260m SSE of The Maze	7 W	621/2571	35
Fatigue Alley	750m NW of Gueudecourt	7 E	635/2563	32
Fatt Trench	W of Flers	7 E	634/2561	27
Favière Trench	400m S of German's Wood	8 E	632/2555	18
Feed Trench	640m SSE of The Maze	7 W	622/2571	35
Feint Trench	840m SSE of The Maze	7 W	622/2571	35
Fell Trench	E of Gommecourt Park	7 W	621/2571	35
Fellon Trench	E edge of Gommecourt Park	7 W	622/2571	35
Fellow Trench	E of Gommecourt Park	7 W	621/2571	35
Felt Trench	1060m SSE of The Maze	7 W	622/2571	35
Female Trench	500m SSE of The Maze	7 W	622/2571	35
Fen Trench	530m SSW of The Maze	7 W	621/2571	35
Ferden Trench	1050m NNE of Schwaben Redoubt	7 W	625/2563+	34
Ferguson Trench	1070m NW of Stuff Redoubt	7 W	625/2563	34
Ferme, la Grande	800m N of Thiepval	7 W	6256/25629	34
Fern Trench	640m SSE of the Maze	7 W	622/2571	35
Ferret Trench	600m W of S Flers	7 E	634/2561	27
Ferret Trench	560m S of The Maze	7 W	621/2571	35
Feste Schwaben	1000m NNE of Thiepval XR	7 W	6256/2563	34
Feste Soden	SSW of Serre	7 W	624/2567	35
Feste Zollern (Goat Redoubt)	1000m N of Mouquet Farm	7 E	6269/25627	34
Fethard Street	W of Y Ravine	7 W	622/2564	35
Fetlock Trench	W of Wold Redoubt	8 E	626+/2557+	16
Fettor Trench	1120m SSE of The Maze	7 W	621/2571	35
Feud Trench	E of Gommecourt Park	7 W	621/2571	35
Fever Trench	890m SSE of the Maze	7 W	622/2570	35
Field Trench	560m W of The Maze	7 W	621/2571	35
Field Trench	N from Zollern Redoubt	7 E	627/2562+	34
Fiennes Street	Just W of Schwaben Redoubt	7 W	624/2562	34
Fifth Avenue	W of Gommecourt Park	7 W	621/2571	35
Fifth Avenue	see Ration Trench	7 E	627/2561	34
Fifth Street	900m WSW of Gibraltar	7 E	627/2559	22
Fig Trench	800m SW of The Maze	7 W	621/2571	35
Fight Trench	In Gommecourt Park	7 W	621/2571	35
Fillet Trench	In Gommecourt Park	7 W	621/2571	35
Film Trench	Just E of The Maze	7 W	622/2571	35
Fin Trench	470m WSW of The Maze	7 W	621/2571	35
Find Trench	890m WSW of The Maze	7 W	621/2571	35
Fine Trench	In Gommecourt Park	7 W	621/2571	35
Fir Trench	800m W of S Flers	7 E	634/2561	27
Fire Trench	2000m NE of Gueudecourt	7 E	637/2563	31
Firm Trench	550m SW of The Maze	7 W	621/2571	35
First Aid Trench	220m N of Chalk Pit	8 E	628/2559	22
Fish Alley	800m SW of Flers	7 E	633/2560	27
Fish Trench	600m SW of The Maze	7 W	621/2571	35
Fist Trench	In Gommecourt Park	7 W	621/2571	35
Fitzpatrick Trench	Western edge of Zollern Redoubt	7 E	626/2562	34
Fix Trench	550m W of The Maze	7 W	621/2571	35
Flag Alley	W from Serre	7 W	623/2567	35

Site/trench	Location	Map	MapRef	Obj
Flag Avenue Trench	SW from Matthew Copse	7 W	622/2567	35
Flag Lane	1200m NE of crater in High Wood	7 E	633/2561	36
Flag Switch	350m S of Matthew Copse	7 W	622/2567	35
Flank Street	400m SE of Thiepval	7 W	625/2561	34
Flank Trench	Runs E from just N of Serre	7 W	624/2567	35
Flare Alley	From NE Flers to W Gueudecourt	7 E	635/2562	32
Flat Iron Copse	200m E of Eastern edge of Mametz Wood	8 E	63034/25583	20
Flat Iron Valley	NE from Mametz Wood	8 E	630/2558	18
Flatiron Trench	Just S of Baz. le Pet. Wood	8 E	630/2558	20
Flea Trench	600m NNE of Flers	7 E	635/2562	27
Fleche, The	Northern tip of Talus Boisé	8 E	631/2555	18
Fleet Street	SE from Delville Wood	8 E	634/2558	25
Fleet Street	500m W of Guillemont	8 E	634/2557	24
Flers	3200m WNW Lesboeufs	7 E	634/2561	27
Flers Line	Between Le Sars & Flers	7 E	632+/2561+	36
Flers Reserve 1	Just S of Flers Trench West	7 E	631+/2562+	36
Flers Reserve 2	Just S of Flers Trench West	7 E	631+/2562+	36
Flers Support East	NW of Flers	7 E	634/2562	27
Flers Support West	SE from S Le Sars	7 E	631+/2562+	36
Flers Switch	1800m WNW of Flers	7 E	633/2562	36
Flers Trench East	Eaucourt Abbey towards S Flers	7 E	633+/2561+	27
Flers Trench West	SE from Le Sars towards Eaucourt Abb.	7 E	631+/2562+	36
Flig Lane	1400m W of Flers	7 E	633/2561	27
Flinders Support	Just N of Gueudecourt	7 E	636/2563	32
Fob Trench	1960m WNW of Rettemoy Farm	7 W	622/2572	35
Focus Trench	550m WNW of The Maze	7 W	621/2571	35
Foe Trench	50m W of Gommecourt Wood	7 W	621/2571	35
Folk Trench	Just W of Gommecourt Wood	7 W	622/2572	35
Folly Trench	W of Sixteen Road	7 E	627/2564	29
Folly Trench	W of Gommecourt Wood	7 W	622/2572	35
Folly Trench	NW from northern tip of Trônes Wood	8 E	633/2558	19
Font Trench	W of Gommecourt Wood	7 W	621/2571	35
Food Trench	W of Gommecourt Wood	7 W	622/2572	35
Fool Trench	W of Gommecourt Wood	7 W	622/2572	35
Foolery Trench	140m W of Gommecourt Wood	7 W	621/2571	35
Foot Trench	890m W of Pigeon Wood	7 W	622/2572	35
Forage Alley	900m NW of Gueudecourt	7 E	635/2563	32
Forage Trench	E of Gommecourt Wood	7 W	622/2572	35
Ford Trench	40m W of Gommecourt Wood	7 W	621/2571	35
Forehead Trench	Just E of Gommecourt Wood	7 W	622/2572	35
Foreign Trench	Just E of Gommecourt Wood	7 W	622/2572	35
Fores Avenue	Thiepval Wood	7 W	624/2561+	34
Foresight Trench	Runs NE from top of Gommecourt Wood	7 W	621/2571	35
Forest Trench	Through Baz. le Petit Wood	8 E	630/2559	20
Fork Trench	Runs NW from Gommecourt Wood	7 W	621/2571	35
Fork Trench	N from Puisieux	7 W	625/2569	35
Fork, The	1400m N of Ginchy	8 E	635/2560	27
Form Trench	Just E of Gommecourt Wood	7 W	622/2572	35
Fort Anley	1600m SW of Y Ravine	7 W	621/2563	35
Fort Briggs	W of John Copse	7 W	623/2568	35
Fort Grosvenor	1920m WNW of Sheffield Park	7 W	621/2568	35
Fort Hindenburg	Bet. Leipzip Sal. & Wonderwork	7 W	624/2561	34
Fort Jackson	Newfoundland Park	7 W	622/2564	35
Fort Lemberg	N of Fort Hindenburg	7 W	623/2560	34
Fort Moulin	1800m SW of Y Ravine	7 W	621/2563	35
Fort Prowse	2200m SW of Y Ravine	7 W	621/2563	35
Fort Southdown	1600m W of old Touvent Farm	7 W	621/2568	35
Fort Sussex	N of Red Cottage	7 W	621/2567	35
Fort Trench	Just E of Gommecourt Wood	7 W	622/2572	35
Fortified Road	Runs S from Guillemont	8 E	634/2557	24
Fortune Trench	W of Pigeon Wood	7 W	623/2572	35
Fount Trench	170m W of Gommecourt Wood	7 W	621/2571	35
Fourth Avenue	1000m W of Pozières	7 E	626/2560	22

Site/trench	Location	Map	MapRef	Obj
Fourth Street	Ovillers to Bailiff Wood	8 E	6267/2559	18
Fowl Trench	700m W of Pigeon Wood	7 W	622/2572	35
Fox Street	1050m NNW of Hawthorn Crater	7 W	622/2566	35
Fox Trench	640m W of Pigeon Wood	7 W	622/2572	35
Foxbar Street	E of Johnstone Post	7 W	624/2561	34
Frankfort Trench	200m E of Munich Trench South	7 W	623+/2565+	35
Franz Alley	750m ESE of German's Wood	8 E	632/2555	18
French Street	S of Johnstone Post	7 W	624/2561+	34
Fricourt	1600m W of Mametz	8 E	627/2555	15
Fricourt Château	160m NNW of Rose Cottage	8 E	6274/2556	15
Fricourt Farm (new)	840m NNE of Fricourt	8 E	6275/255676	15
Fricourt Farm (old site of)	300m E of Fricourt farm (new)	8 E	62766/255664	15
Fricourt Spur	NE from E of Bécourt	8 E	626+/2556+	15
Fricourt Station	200m SSW of German Tambour	8 E	6266/25556	15
Fricourt Trench	SSE from Tambour Mines	8 E	626+/2555	15
Fricourt Wood	Immediately NE of Fricourt	8 E	627/2556	15
Fritz Avenue	800m N of Serre	7 W	624/2568	35
Fritz Folly	Just N of Gueudecourt	7 E	636/2563	32
Fritz Trench	600m W of Pommiers Redoubt	8 E	629/2556	18
Frontier Lane	N from Beaumont Hamel	7 W	622/2566	35
Frosty Trench	750m E of Lesboeufs	7 E	638/2560	31
Fuel Trench	1570m NNE of Gommecourt Park	7 W	623/2573	35
Fume Trench	NE of Gommecourt Wood	7 W	622/2572	35
Fun Trench	1120m NE of Gommecourt Park	7 W	623/2572	35
Fungus Trench	500m NNE of Pigeon Wood	7 W	623/2573	35
Funk Trench	1820m NNE of Gommecourt Park	7 W	623/2573	35
Fur Trench	1600m NNE of Gommecourt Park	7 W	623/2573	35
Furness Street	Just W of Ovillers	8 W	625/2559	17
Furrier Trench	1450m NNE of Gommecourt Park	7 W	623/2573	35
Fury Trench	1400m NNE of Gommecourt Park	7 W	622/2573	35
Fuse Trench	NE of Gommecourt Wood	7 W	622/2572	35
Fusilier Alley	450m N of Bazentin le Petit	8 E	630/2560	20
Fusilier Trench	S of Leuze Wood	8 E	636/2556+	33
Fuss Trench	1820m NNE of Gommecourt Park	7 W	623/2573	35
Galgen	see Keats Redan	8 W	625/2558	17
Gallwitz Support Trench	W of Dyke Road, Courcelette	7 E	630/2563	29
Ganter Weg	NW of Pozières	7 E	627/2561	22
Gap Alley	1000m SSE of Flers	7 E	635/2560	27
Gap Trench	800m SE of Flers	7 E	635/2561	27
Gas Alley	1000m E of S Flers	7 E	635+/2561	32
Gate Lane	800m S of Flers	7 E	634/2560	27
Gate Trench	700m NW of Star Wood	7 W	623/2569	35
Gaul Avenue	SE from Mesnil	7 W	622+/2561	35
Gemmel Trench	E of Thiepval Wood	7 W	624/2561+	34
George Street	Thiepval Wood	7 W	624/2561+	34
George Street	900m ESE of Casino Point	8 E	631/2554	18
German Lane	900m NNW of Star Wood	7 W	623/2569	35
German Tambour	Just S of Triple Tambour Mines	8 E	626/2555	15
German's Wood (Schrapnell Way)	1600m SSE of Montauban	8 E	631+/2556	18
Giants Causeway	Between River Ancre & Beaucourt	7 W	624/2564	35
Gibraltar/Panzeaturm	German strongpoint at S Pozières	8 E	6277/25602	22
Gierich Weg (Ration Trench)	NW of Pozières	7 E	626+/2561	22
Gin Alley	Just WNW of Lozenge Wood	8 E	627/2556	15
Ginchy	1000m NNE Guillemont	8 E	635/2558	26
Ginchy Avenue	E from Waterlot Farm	8 E	634/2558	25
Ginchy Civil Cemetery	N of village	8 E	635/2558	26
Ginchy Farm	SW Ginchy	8 E	635/2558	26
Ginchy Telegraph, site of	450m E of Ginchy	8 E	6361/255876	26
Gypsy Hill	900m S of Y Ravine	7 W	622/2563	35
Gird Support East	W & S of Gueudecourt	7 E	634++/2562+	32
Gird Support West	E of Le Sars	7 E	632++/2563+	36
Gird Trench East	W & S of Gueudecourt	7 E	634++/2562+	32
Gird Trench West	E of Le Sars	7 E	632++/2563+	36
Glasgow Trench	SW of Desire Trench	7 E	627/2563	29

TRENCHES, BATTLESITES ETC

Site/trench	Location	Map	MapRef	Obj
Glatz Alley	Runs S from Glatz Redoubt	8 E	632/2555+	18
Glatz Redoubt	800m SSE of Montauban	8 E	632/2556	18
Glebe Street	N out of Flers	7 E	634/2562	32
Glory Hole, The	250m E of Mem. Seat, La Bois.	8 W	625/2557	16
Glory Lane	E of Munich Trench South	7 W	623/2565	35
Gloster Alley	1500m E of Pozières	8 E	629/2560	20
Gloster Deviation	1000m NW of Baz-le-Petit	8 E	629/2560	28
Goat Redoubt (Feste Zollern)	1000m N of Mouquet Farm	7 E	6269/25627	34
Goat Trench	E of Flers	7 E	635/2561	32
Gommecourt Park	SW of Foncquevillers	7 W	621+/2571	35
Gommecourt Wood (Bois du Sartel)	SW of Foncquevillers	7 W	622/2571+	35
Gommecourt village	1300m SE Foncquevillers	7 W	621+/2571	35
Gooch Street	W of Gommecourt Park	7 W	621/2571	35
Good Street	SW of Gueudecourt	7 E	635/2562	32
Goodwin Post	700m N of Gueudecourt	7 E	636/2563	32
Goose Alley	Connects Flers Support & Gird Trenches	7 E	633+/2561+	36
Gordon Alley	1200m WNW of Baz-le-Petit	8 E	629/2560	28
Gordon Castle	Thiepval Wood	7 W	624/2562	34
Gordon Dump	E of La Boisselle	8 E	626/2558	16
Gordon Post	700m NE of Lochnagar crater	8 E	626/2558	16
Gordon Trench	W of Serre	7 W	622/2567	35
Gordon Trench	350m SE of Mary Redan	7 W	623/2564	35
Gounod Trench	NNW of Fort Briggs	7 W	623/2568	35
Gourlay Trench	700m E of Contalmaison Villa	8 E	629/2559	20
Gouroch Street	E of Thiepval Wood	7 W	624/2561+	34
Govan Street	E of Thiepval Wood	7 W	624/2561+	34
Gowrie Street	400m SSW of Glory Hole	8 W	625/2558	16
Granatloch	see Leipzig Salient	7 W	624/2561	34
Grandcourt Line	S from W Grandcourt	7 E	626/2563+	34
Grandcourt Rd. Courcelette	From Courcelette to Grandcourt	7 E	627+/2562+	29
Grandcourt Trench	E from S Grandcourt	7 E	627/2564	29
Grass Lane	N off Bulls Road, Flers	7 E	635/2561+	32
Grease Trench	600m N of Gueudecourt	7 E	636/2563	32
Great North Road	Just S of Baz.-le-Petit	8 E	630/2559	20
Great Northern Railway	La Signy to Touvent Farm	7 W	621+/2567+	35
Green Line	W of Railway Road	7 W	624/2564	35
Green Street	Adjacent Guillemont	8 E	634+/2557	24
Green Trench	Just W of Fricourt	8 E	626/2555	15
Greenock Avenue	Just NE of Authuille	7 W	624/2560	34
Grevillers Trench	1400m NW of Warlencourt	7 E	631/2556	36
Gropi Trench	Just W of Bouleaux Wood	8 E	636/2557	33
Grossherzgraben	NE from E of the Cutting	8 E	628/2558+	20
Grove Alley	1000m NNW of Flers	7 E	634/2562	32
Grundy Road	1500m W of Warlencourt	7 E	631/2565	36
Gudgeon Trench	Just S of Puisieux	7 W	625/2568	35
Gueudecourt	2000m NE Flers	7 E	636/2562	32
Guillemont	1100m SSW of Ginchy	8 E	634+/2557	24
Guillemont Alley	Trônes Wood to Guillemont	8 E	634/2557	24
Guillemont Civil Cemetery	E Guillemont	8 E	6352/255762	24
Guillemont Ridge	SSW of Guillemont	8 E	6347/2557	24
Guillemont Station	500m NNW of Guillemont	8 E	6347/25579	24
Guillemont Track	Runs E from Trônes Wood	8 E	634/2557	24
Gully, The	W end of Desiree Trench	7 E	629/2564	29
Gun Alley	700m SE of Flers	7 E	635/2561	27
Gun Pits	Just E of S Practice Trenches	7 E	630/2563	29
Gun Pits, The	German pos. 700m E of Lesboeufs	7 E	638/2560	31
Gunpit Lane	from Courcelette to Martinpuich	7 E	629+/2561+	29
Gunpit Trench	SE from Courcelette	7 E	629+/2561+	29
Gusty Trench	E of Lesboeufs	7 E	637/2562	31
H.L.I. Alley	750m NW of Bazle-Petit	8 E	629/2560	28
Hail Trench	800m SW of le Transloy	7 E	638/2561	31
Hair Alley	1200m NW of Star Wood	7 W	622/2570	35
Halt, The	Railway halt S of Mametz	8 E	628/2555	18
Ham Trench	750m N of Gueudecourt	7 E	636/2563	32

Site/trench	Location	Map	MapRef	Obj
Hamel	1500m NE of Mesnil Martinsart	7 W	623/2563	35
Hamel Bridge	Bridge over Ancre ESE of Hamel	7 W	623/2562	34
Hamilton Alley	Trônes Wood	8 E	633/2557	19
Hamilton Avenue	E of Thiepval Wood	7 W	624/2561+	34
Hamilton Work	400m N of Y Ravine	7 W	623/2565	35
Hammerhead	E side of Mametz Wood	8 E	630/25579	18
Hammerhead Sap, site of	120m E of Thiepval Wood	7 W	624/2562	34
Handel Road	1000m E of Petit Miraumont	7 E	628/2565	34
Hans Crescent	W Ginchy	8 E	635/2558	26
Hansa Line	N of Schwaben Rdbt. to R. Ancre	7 W	625/2563+	34
Hansa Road	Contn. of Mill Trench to Grandcourt	7 E	625+/2564	35
Happy Valley	W of Hawthorn Crater	7 W	622/2565	35
Happy Valley	see Flat Iron Valley	8 E	630/2558	18
Hardwick Trench	800m E of Beaumont-Hamel	7 W	623/2565	35
Hare Lane	E of Triple Tambour Mines	8 E	626/2556	15
Hat Trench	300m NW of Star Wood	7 W	623/2569	35
Hawthorn Ridge Redoubt Mine	500m W of Beaumont Hamel	7 W	6225/25653	35
Hawthorn Road	W of Beaumont Hamel	7 W	622/2565	35
Haymarket	Delville Wood East	8 E	634/2559	21
Haymarket	W of Y Ravine	7 W	622/2564	35
Hazy Trench	950m E of Lesboeufs	7 E	638/2560+	31
Heaton Road	S of old Beaumont Road	7 W	622/2565	35
Hedgerow Trench	150m N of Hamel	7 W	623/2563	35
Heiden Kopf	Site of Serre Rd. No. 2 Cem.	7 W	622/2566	35
Heligoland - Sausage Rdt.	600m ESE of Lochnagar crater	8 E	6265/25576	18
Hell Lane	450m SE of Stuff Redoubt	7 E	626/2562	34
Hennslow Road	S of old Beaumont Road	7 W	622/2565	35
Hessian Trench West	E from N of Bulgar Trench	7 E	626+/2562+	34
Hessian Trench East	W from N of Courcelette	7 E	628+/2563	29
Hexham Road	NE of Eaucourt Abbey	7 E	633/2563	36
Hidden Lane	SSW from Hidden Wood	8 E	628/2554	18
Hidden Valley	S of High Wood	8 E	632/2559	23
Hidden Wood	1100m SE Fricourt church	8 E	628/25552	18
High Alley	Crucifix Corner to High Wd (E of Lane)	8 E	631/2559+	23
High Holborn	Delville Wood twds. Guillemont	8 E	634/2557+	25
High Trench	200m NE of Mouquet Farm	7 E	627/2561	22
High Wood	1800m SE of Martinpuich (on 7 E & 8 E)	7 E	631+/2560+	23
Highland Trench	1100m NW of Baz-le-Petit	8 E	629/2560	28
Hill Trench	Joins Zollern & Bulgar Trenches	7 W	625/2562	34
Hilt Trench	500m N of Gueudecourt	7 E	636/2563	32
Hindenberg Trench	1000m NE of Baillescourt Farm	7 E	627/2566	35
Hindenburg Trench	Just N of Leipzig Salient	7 W	624/2561	34
Hock Trench	400m SE of Star Wood	7 W	623/2567	35
Hodder Street	N of Ovillers Post	8 W	624/2558	17
Hogg's Back Trench	Adjacent Flers	7 E	635/2561	27
Hogshead	400m E of N Flers	7 E	635/2561	27
Hohenzollern Trench	SW from Wonder Work	7 W	625/2561	34
Holmes Redoubt	400m NE of Contalmaison	8 E	628/2558	18
Hooge Alley	300m S of Trônes Wood	8 E	633/2556	19
Hook Sap	Contn. of Butte Trench	7 E	633/2564	36
Hook Trench	1000m SE of Martinpuich	7 E	631/2560	23
Hoop Trench	Just W of Guillemont Quarry	8 E	634/2557	24
Hoop, The	1350m WNW of Baz-le-Petit	8 E	629/2560	28
Hop Alley/Trench	E from Delville Wood	8 E	634/2559	25
Hope Trench	Just N of Y Ravine	7 W	622/2564	35
Horn Alley	700m ESE of German's Wood	8 E	632/2555	18
Hornby Trench	N from Bainbridge Trench	7 E	626/2562	34
Horseshoe Trench	NE of Scots Redoubt	8 E	626/2557+	16
Hosky Trench	Fricourt Wood	8 E	627/2556	15
Hospital Trench	300m ESE of Hawthorn Crater	7 W	622/2565	35
Houghton Street	300m E of Ovillers Post	8 W	624/2558	17
Hounslow Road	Just E of old Beaumont Road	7 W	622/2565	35
Howitzer Avenue	600m NE of Chalk Pit	8 E	627/2560	22
Howson Road	850m SE of Mesnil	7 W	623/2561	35

Site/trench	Location	Map	MapRef	Obj
Hun Alley	400m W of Guillemont	8 E	634/2557	24
Hunters Trench	S of Hawthorn Mine	7 W	622/2564	35
Hyde Park Corner	S of Y Ravine	7 W	622/2564	35
Hyde Road East	500m NNW of Carnoy	8 E	629/2554	18
Iban Trench	400m ENE of Rettemoy Farm	7 W	624/2572	35
Ilot, The	Immed. W of the Glory Hole	8 W	625/2558	16
Inch Street	Adjacent to Glory Hole	8 W	625/2558	16
Indre Trench	550m W of Rettemoy Farm	7 W	623/2572	35
Indus Trench	W of Bois du Biez	7 W	622/2571	35
Inner Trench	Delville Wood	8 E	634+/2559	21
Intermediate Trench	From N of Baz. east towards High Wood	7 E	631/2560	23
International Trench	1000m S of Martinpuich	7 E	630+/2560	20
Inverness Avenue	Thiepval Wood	7 W	624/2562	34
Inverrary Street	Just W of Leipzig Salient	7 W	624/2561	34
Invicta Alley	250m E of Trônes Wood	8 E	634/2557	24
Iona Street	Just W of Leipzig Salient	7 W	624/2561	34
Irish Alley	Just S of Trônes Wood	8 E	633/2556	19
Ironside Trench	Just W of Adanac Cem.	7 E	629/2564	29
Irving Post	1080m WSW of Adanac Cem.	7 E	628/2563	29
Irwin Trench	800m E of St. Pierre Divion	7 W	625/2563	34
Islay Street	Just W of Leipzig Salient	7 W	624/2561	34
Itchin Trench	670m SW of Rettemoy Farm	7 W	623/2571	35
Jackson's Trench	400m N of Maltzhorn Farm	8 E	634/2556	24
Jacob's Ladder	CT Mesnil to Hamel	7 W	622+/2563	35
Jacob's Ladder Trench	450m NW of Hawthorn Crater	7 W	622/2565	35
Jäger-Höhe	1700m E of La Boisselle	8 E	627/2558	16
Jam Street	Immed. E of John Copse	7 W	623/2568	35
Jardine Bridge	Bridge over Ancre S of Authuille	8 W	623/2559	34
Jean Bart Trench	Nr. new Touvent Farm	7 W	622/2568	35
Jenburg Street	Just N of Leipzig Salient	7 W	624/2561	34
John Alley	NW from Serre	7 W	623/2567+	35
John Bull Trench	Adjacent Lesboeufs	7 E	637/2561	30
John Copse	Assem. point for Serre attack	7 W	623/2568	35
John O'Gaunt Street	Just E of Donnet Post	8 W	625/2559	17
John Street	South Hamel	7 W	623/2563	35
Johnstone Post	SE corner of Thiepval Wood	7 W	624/2562	34
Johnstone Street	Just SW of Johnstone Post	7 W	624/2561	34
Jones Trench	N of John Copse	7 W	623/2568	35
Joseph Trench	600m SE of Thiepval	7 W	625/2561	34
Junction Street	Links The Ravine to Boom Ravine	7 E	628/2564	29
Junction Trench	1150m ENE of Baz-le-Petit	8 E	631/2560	23
Junk Trench	650m E of Mary Redan	7 W	623/2564	35
Jura Street	Just W of Leipzig Salient	7 W	624/2561	34
Jutland Alley	600m NE of Baz-le-Petit	8 E	630/2560	23
K Trench (see Western Trench)	Runs N from just W of Gibraltar	7 E	627/2560	22
KOYLI Redoubt	520m W of Bazentin le Petit	8 E	629/2560	20
KOYLI Trench	720m W of Bazentin le Petit	8 E	629/2560	20
Kabelgraben	NE from D929 Ovillers/Pozières road	8 E	627/2560	22
Kaiser Lane	540m N of Serre	7 W	624/2569	35
Kaiser Str.	700m E of La Boisselle	8 E	626/2558	16
Kaiser's Oak Tree	SW corner of Gommecourt Wood	7 W	6214/25712	35
Kaisergraben Quadrangle Trench	Contalmaison to Mametz Wood	8 E	628/2557+	18
Kaufmann Gr.	400m N of Lochnagar crater	8 E	626/2558	16
Keats Redan	1000m NW of Lochnagar crater	8 W	625/2558	16
Keilson Street	430m WNW of Thiepval Crossroads	7 W	624/2562	34
Keep, The	Just S of Gommecourt Park	7 W	621/2571	35
Keep, The	Authuille village	7 W	624/2560	34
Kemp Trench	1000m W of Stuff Redoubt	7 W	625/2563	34
Kendal Sap	1100m N of Sugar Factory	7 E	629/2562	29
Kendri Trench	1000m NW of Courcelette	7 E	628/2563	29
Kenora Trench	Runs NW from Sudbury T to Regina T	7 E	628/2563	29
Kern Redoubt	E edge of Gommecourt Park	7 W	622/2571	35
Kerrera Street	Just W of Leipzig Salient	7 W	624/2561	34
Kersall Street	400m SW of Leipzig Salient	7 W	624/2560	34

Site/trench	Location	Map	MapRef	Obj
Kiel Lane	300m SW of Mametz Halt	8 E	628/2555	18
Kilberry	Just E of Authuille village	7 W	624/2560	34
Kilmun Street	Just W of Leipzig Salient	7 W	624/2561	34
Kilometer Lane	NE from Auchonvillers to Serre Rd.	7 W	621/2566	35
Kinfauns Street	770m S of Glory Hole	8 W	625/2557	16
King Street	Delville Wood	8 E	634/2558+	21
King Street	E from Tenderloin Trench	7 W	622/2565	35
King's Walk	Delville Wood	8 E	633+/2558+	21
Kingsgate Street	Just N of Chapes Spur	8 W	625/2557	16
Kintyre Trench	E of Authuille village	7 W	624/2560	34
Kipper Trench	700m W of Round Wood	8 E	626/2557	16
Kirkaldy Street	800m ESE of Ovillers Post	8 W	625/2558	16
Kitchen Road	East Mesnil	7 W	622/2561	35
Kitchen Trench	SSE off Rose Alley	8 E	627/2555	18
Kite Copse	E of Pigeon Wood	7 W	623/2572	35
Knife Trench	N from Puisieux	7 W	625/2569	35
Knot Trench	600m W of Guillemont	8 E	634/2557	24
Knox Street	NW of John Copse	7 W	623/2568	35
Kollmann Trench	Contn. of Fabeck Trench Eastward	7 E	628/2561+	22
König Lane	Just N of Fricourt	8 E	627/2556	15
König Support	Immed. E of König Trench	8 E	626/2555	15
König Trench	N-S through Triple Tambour Mines	8 E	626/2555+	15
La Boisselle	1200m SSW Ovillers	8 W	625/2558	16
La Brayelle Farm	N of Pigeon Wood	7 W	623/2572	35
La Fabrique Farm	800m S of Courcelette	7 E	629/2561	29
La Louvière Farm	400m N of Star Wood	7 W	624/2569	35
Lager Alley	E from Munich Tr to join Ten Tree Alley	7 W	623+/2566	35
Lager Lane	800m E of Delville Wood	8 E	635/2559	25
Lamb Trench	Near Arrow Head Copse	8 E	634/2557	24
Lancashire Lane - Zollern Trench	ENE from Thiepval twds Goat Rdt.	7 W	625+/2562	34
Lancashire Post	570m NE of Hamel	7 W	623/2563	34
Lancashire Sap	350m N of Bazentin le Petit	8 E	630/2560	20
Lancashire Trench	1000m WNW of Baz-le-Petit	8 E	629/2560	28
Lancaster Avenue	400m N of Ovillers Post	8 W	624/2559	17
Lanwick/Lerwick Trench	N from Jacob's Ladder	7 W	622/2565	35
Larkhill Trench	Near "Met." trenches	7 E	636+/2561+	31
Lattorf Graben	Protects S side of Pozières	8 E	627/2560	22
Le Barque Switch	From Le Barque twds Butte de Warlencourt	7 E	633+/2564	36
Le Sars	2200m NE of Courcelette	7 E	631+/2563+	36
Le Sars Civil Cemetery	N Le Sars	7 E	632/2564	36
Le Sars Line	NW from Le Sars	7 E	631/2564	36
Le Transloy	4000m NE of Lesboeufs	IGN	Fold 4	30
Lean Alley	700m NW of Baz-le-Petit	8 E	629/2560	28
Leave Avenue	E from Beaumont Hamel	7 W	623+/2565	35
Leeds Avenue	460m E of Carnoy civ. cem.	8 E	630/2554	18
Leek Way	400m S of Beaumont Hamel Cem.	7 W	623/2564	35
Legend Trench	Adj. Munich Trench North	7 W	623/2566	35
Lehmgruben Höhe	1000m ESE of Lochnagar crater	8 E	626/2557	16
Leicester Street	NW of Gommecourt Park	7 W	621/2571	35
Leiling-Schlucht	see Y Ravine	7 W	622+/2564	35
Leinster Alley	C.T. to Trônes Wood	8 E	633/2556	19
Leipzig Salient	1200m SSW of Thiepval	7 W	62475/2561	34
Leith Walk	200m SW of W edge of High Wood	8 E	631/2560	23
Lemberg Trench	Just N of Leipzig Salient	7 W	624/2561	34
Lesboeufs	3200m ESE of Flers	7 E	637+/2560	30
Leuze Wood	1600m SE of Ginchy	8 E	636/2557	33
Leuzemake Trench	S edge of Leuze Wood	8 E	637/2557	33
Ligny Road	N from Flers	7 E	634+/2562+	27
Lime Street	CT to Boggart Hole Clough	7 W	624/2560	34
Limerick Junction	W of Y Ravine	7 W	622/2564	35
Lincoln Lane	NW of Gommecourt Park	7 W	621/2571	35
Lincoln Redoubt	500m N of Scots Redoubt	8 E	626/2558	16
Little Wood	750m WSW of Warlencourt	7 E	631/2565	36

Site/trench	Location	Map	MapRef	Obj
Little Z	810m WNW of Pigeon Wood	7 W	622/2572	35
Liverpool Street	250m S of Casino Point	8 E	630/2555	18
Loch Aire Street	Just W of Leipzig Salient	7 W	624/2561	34
Lochnagar Mine Crater	600m SSE of La Boisselle	8 E	6259/255776	16
Lochnagar Street	Just N of Chapes Spur	8 W	625/2557	16
Lonely Copse	400m N of Fricourt	8 E	6272/25564	15
Lonely Lane	Runs N from Red Cottage	8 E	626/2556	15
Lonely Lane	900m N of Pendant Copse	7 W	624/2567	35
Lonely Support	Immed. E of Lonely Trench	8 E	627/2556	15
Lonely Trench	Just N of Fricourt	8 E	627/2556	15
Lonely Trench (later Shropshire Tr)	800m S of Guillemont	8 E	635/2556	24
Long Acre	700m S of Y Ravine	7 W	622/2564	35
Long Drive	600m SE of Pozières	8 E	628/2559	22
Long Sap	300m SW of Mary Redan	7 W	622+/2563	35
Longueval	2000m E of Baz-le-Grand	8 E	633/2558+	21
Longueval Alley	Bernafay Wd to NW side Trônes Wood	8 E	633/2557	19
Loop Trench	E of Leuze Wood	8 E	636+/2557	33
Loop Trench	650m E of Pommiers Redoubt	8 E	630/2556	18
Loop, The	600m SE of Pommiers Redoubt	8 E	630/2555	18
Lorna Pass	Just W of Leipzig Salient	7 W	624/2561	34
Louvercy Street	600m SSE of Mary Redan	7 W	623/2563	35
Louvière Alley	500m NW of Star Wood	7 W	623/2569	35
Lower Horwich Street	Just E of Conniston Post	8 W	624/2559	17
Lower Wood	1200m ENE of Contalmaison	8 E	62942/25591	20
Lozenge Alley	Runs W of Fricourt Farm	8 E	627/2556	15
Lozenge Trench	Runs S from Lozenge Alley	8 E	627/2556	15
Lozenge Wood	400m W of Fricourt Farm	8 E	6273/25568	15
Lozenge, The	see Wundt Werk	7 W	625/2567	35
Lucky Way	Just E of Schwaben Redoubt	7 W	625/2562	34
Luisen Trench	1700m NW of Gueudecourt	7 E	634/2563	32
Luisenhof Farm	1700m NNW of Gueudecourt	7 E	635/2564	36
Luke Alley	NW from Serre	7 W	623/2567	35
Luke Copse	Assem. point for Serre attack	7 W	623/2568	35
Lunar Trench	Runs S from the Bowery	7 W	621/2565	35
Lupton Lane	NW out of Mesnil	7 W	622/2562	35
Luxton Trench	E of Pozières	7 E	629/2560	29
MacDonnell Trench	SW of Courcelette	7 E	628/2561+	29
Machine Gun Alley	500m SE of Waterlot Farm	8 E	634/2558	25
Machine Gun House	500m SE of Waterlot Farm	8 E	634/2558	25
Machine Gun Wood	600m NW of Château Keep	8 E	631/2554	18
Maclaren Lane	W of Beaumont Hamel	7 W	622/2565	35
Madam Trench	E corner of Rossignol Wood	7 W	624/2570	35
Maid Trench	Just SE of Rossignol Wood	7 W	624/2570	35
Maidstone Avenue	400m N of Carnoy	8 E	629/2554	18
Main Trench	SW of Rossignol Wood	7 W	624/2570	35
Maisie Lane	SSW from Strasburg Line	7 W	624/2563	34
Maison grise sap	E of Thiepval Wood	7 W	624/2561+	34
Male Trench	NE of Rossignol Wood	7 W	624/2570	35
Malt Trench	1400m E of Warlencourt	7 E	633/2565	36
Maltzhorn Farm	640m SSE of S tip of Trônes Wood	8 E	6342/25563	19
Maltzhorn Trench	S from E side of Trônes Wood	8 E	633+/2555+	19
Mama Trench	450m SE of RossignolWood	7 W	624/2570	35
Mametz	1600m E of Fricourt	8 E	628/2555	18
Mametz Trench	W of Bulgar Trench	8 E	628/2555	18
Mametz Wood	1000m E of Contalmaison	8 E	628+/2557+	18
Man Trench	NE edge of Rossignol Wood	7 W	624/2570	35
Manchester Alley	N from Silesia Trench	8 E	631/2555	18
Manchester Avenue	550m E of Carnoy civ. cem.	8 E	630/2554	18
Mansel Copse	400m SE of Mametz Halt	8 E	6286/25549	18
Map Trench	Just SW of Rossignol Wood	7 W	623/2570	35
Maple Trench	380m E of Pommiers Redoubt	8 E	629/2556	18
Mark Copse	Assem. point for attack on Serre	7 W	623/2568	35
Market Cross	Just SW of Dressler Post	8 W	624/2557	16
Market Trench	650m N of Thiepval	7 W	625/2562	34

Site/trench	Location	Map	MapRef	Obj
Marlborough Trench	N from Marlborough Wood	8 E	630/2558	18
Marlborough Wood	1250m NW of Montauban	8 E	6309/25578	18
Marsh Street	E from South Hamel	7 W	623/2563	35
Marsh Trench	Just S of Rossignol Wood	7 W	624/2570	35
Martin Alley	N from Tangle, Martinpuich	7 E	631/2561	28
Martin Trench	300m E of S Martinpuich	7 E	631/2561	28
Martin's Lane	N from Thiepval to Schwaben Rdt.	7 W	625/2562	34
Martinpuich	1800m NW of High Wood	7 E	630/2561	28
Mary Redan	500m SSE from S tip of Y Ravine	7 W	622/2564	35
Mary Redan Trench	S of Beaumont Hamel	7 W	622/2564	35
Mary Street	300m W of Talus Boisé	8 E	631/2555	18
Mash Valley	N of D929 bet.La Boiselle & Ovillers	8 W	625/2558	17
Marshal Trench	740m E of Mary Redan	7 W	623/2563+	35
Mat Trench	Just SW of Rossignol Wood	7 W	623/2570	35
Mat's Post	300m ESE of Adanac Cem.	7 E	629/2564	29
Match Trench	750m SE of Rossignol Wood	7 W	624/2570	35
Matthew Alley	W from Serre	7 W	623/2567	35
Matthew Copse (now agric. land)	Assem. point for attack on Serre	7 W	622/2567	35
Maurepas Ridge	Bet. Maurepas & Oakhanger Wood	8 E	636/2555	33
Maxim Trench	SE from Serre	7 W	624/2567	35
Maxwell Support	1200m E of Le Sars	7 E	633/2564	36
Maxwell Trench	Butte de Warlencourt	7 E	632/2564	36
Maxwell Trench	ESE from Thiepval XR	7 W	625/2562	34
Maze, The	1500m NE of Eaucourt Abbey	7 E	634/2564	36
Meadow Trench	360m NNW of Rossignol Wood	7 W	623/2571	35
Medway Trench	Adjacent Schwaben Redoubt	7 W	624/2563	34
Meed Trench	300m NW of Rossignol Wood	7 W	623/2571	35
Meet Trench	Just NW of Rossignol Wood	7 W	623/2570	35
Mend Trench	420m N of Rossignol Wood	7 W	623/2571	35
Mere Trench	750m NW of Rossignol Wood	7 W	623/2571	35
Mersey Street	Authuille Wood	7 W	624/2560	34
Mersey Trench	Adjacent Leipzig Redoubt	7 W	625/2561	34
Mesnil Château	East Mesnil Martinsart	7 W	6227/2562	35
Mess Trench	980m NW of Rossignol Wood	7 W	623/2571	35
Metal Trench	Just E of Bois du Biez	7 W	624/2571	35
Meteor Trench	1500m NW of Le Transloy	7 E	638/2563	31
Meteorological Trenches	Between Lesboeufs & Gueudecourt	7 E	636+/2561+	31
Methuen Street	300m S of Glory Hole	8 W	625/2557	16
Middle Alley	Joins Mametz Wd & Baz-le-Petit Wd.	8 E	630/2558	20
Middle Avenue	Runs N off Back Trench	8 E	631/2555	18
Middle Copse	Bet. Quadrilaterial & Bouleaux Wood	8 E	6368/25582	33
Middle Street	NW of Hawthorn Crater	7 W	622/2565	35
Middle Trench	Behind Foncquevillers	7 W	621/2573	35
Middle Wood	1150m W of Baz-le-Petit	8 E	6294/25594	20
Midland Trench	NW of Gommecourt Park	7 W	621/2571	35
Midway Line	Schwaben Rdt to Mouquet Farm	7 W	625+/2561+	34
Might Trench	550m WNW of Bois du Biez	7 W	623/2571	35
Mild Alley	1400m NE of Gueudecourt	7 E	637/2563	32
Mild Trench	900m NE of Gueudecourt	7 E	637/2563	31
Mill Bridge	River bridge at Authuille Mill	7 W	623/2560	34
Mill Road	SE from N of Hamel	7 W	624/2563	34
Mill Street	NNE from Morval	8 E	638/2560	30
Mill Street	600m ENE of Baz-le-Petit	8 E	631/2559	20
Mill Trench	Just N of St. Pierre Divion	7 W	624/2563	34
Mill Trench	Just SW of Montauban	8 E	631/2556	18
Mill Trench	430m NE of Thiepval	7 W	625/2562	34
Mill, site of	800m WNW of Ulster Tower	7 W	623/2563	34
Mill, site of	800m SW of Beaucourt	7 W	624/2564	35
Mill, site of	400m NW of Eaucourt Abbey	7 E	632/2563	36
Mill, site of	By River Ancre, Miraumont	7 E	627/2565	35
Mill, site of	N Authuille	7 W	623/2560	34
Mill, site of	800m NE of Pozières	7 E	628/2560	22
Mill, site of	500m W of Longueval	8 E	633/2558	21
Mill, site of	N Martinpuich	7 E	630/2561	28

TRENCHES, BATTLESITES ETC

Site/trench	Location	Map	MapRef	Obj
Mill, site of	800m NW of Lesboeufs	7 E	637/2561	31
Millner Alley	W from Serre	7 W	623/2567	35
Mince Trench	Adjacent Morval	8 E	637/2558	30
Mince Trench	550m E of Bois du Biez	7 W	623/2571	35
Mine Alley	Runs NE from The Castle	8 E	631/2556	18
Mine Trench	E from Casino Point Mine	8 E	630/2555	18
Mining tunnel entry (one of)	La Boisselle civil cem.	8 W	625/2558	16
Mint Trench	280m NNE of Rettemoy Farm	7 W	624/2572	35
Minx Trench	330m S of Pigeon Wood	7 W	623/2572	35
Mirage Trench	700m E of Lesboeufs	7 E	638/2560	31
Miraumont Road East	N from Courcelette to Miraumont	7 E	628+/2562+	29
Miraumont Road West	NNW from Courcelette to Miraumont	7 E	628+/2562+	29
Miraumont Trench	600m SE of Petit Miraumont	7 E	628/2565	34
Mist Trench	Just N of E edge of Bois du Biez	7 W	624/2571	35
Misty Trench	1400m E of Gueudecourt	7 E	637/2562	31
Mole Trench	420m ENE of Pigeon Wood	7 W	623/2572	35
Moltke Graben	800m W of Rossignol Wood	7 W	623/2570	35
Monk Trench	420m NW of Rettemoy Farm	7 W	624/2572	35
Monkey Trench	430m NW of Rettemoy Farm	7 W	624/2572	35
Montauban	2200m S of Baz-le-Grand	8 E	631+/2556	18
Montauban Alley	ENE from Pommiers R. to Bernafay Wd	8 E	630+/2557	18
Montauban Avenue	450m SSE of Casino Point	8 E	630/2554	18
Moon Trench	Just S of Le Transloy	7 E	639/2561	31
Moon Trench	470m NNE of Rettemoy Farm	7 W	624/2572	35
Mop Trench	Just E of Pigeon Wood	7 W	623/2572	35
Morval	3000m ENE of Ginchy	8 E	638/2559	30
Morval Mill	W of Morval	8 E	638/2559	30
Moss Side	450m ENE of Hamel	7 W	623/2563	34
Moss Trench	640m E of Pigeon Wood	7 W	623/2572	35
Mound Keep	NE Maricourt	8 E	632/2554	18
Mound Keep	Just NW of Authuille	7 W	623/2561	34
Mound, The	N of Stuff Redoubt	7 W	625/2563	34
Mound, The	550m W of St. Pierre Division	7 W	624/2563	35
Mound, The	250m E of Bulgar Point	8 E	6291/25542	18
Mount Street	Guillemont centre	8 E	634+/2557	24
Mountain Alley	1000m N of Pendant Copse	7 W	624/2568	35
Mountjoy Street	1070m NNW of Hawthorn Crater	7 W	622/2566	35
Mouquet Farm	1600m NW of Pozières	7 E	627/25617	22
Mouquet Farm Track	Park Lane to Mouquet Farm	7 E	627/2561	22
Mouquet Quarry	200m S of Mouquet Farm	7 E	6269/25615	22
Mouquet Road	From Courcelette to Mouquet Farm	7 E	627+/2561+	29
Mouquet Switch	Mouquet Farm towards Schwaben Rdbt	7 E	626/2562	34
Mouse Trench	550m E of Pigeon Wood	7 W	623/2572	35
Moy Avenue	NE of Ovillers	7 E	626/2560	22
Muck Trench	Towards Beaucourt from Leave Av.	7 W	624/2565	35
Mud Trench	Just N of Pigeon Wood	7 W	623/2572	35
Mug Trench	615m E of la Brayelle Farm	7 W	623/2572	35
Munich Trench North	N from Ten Tree Alley to Serre	7 W	623/2566+	35
Munich Trench South	S from Ten Tree Alley to Beaumont Alley	7 W	623/2565+	35
Munster Alley, East	S of the Sugar Factory	7 E	629/2561	29
Munster Alley, West	1200m E of Pozières	7 E	628/2560	22
Music Trench	Just N of Pigeon Wood	7 W	623/2572	35
Mute Trench	300m E of la Brayelle Farm	7 W	623/2572	35
Mutton Trench	S of Morval	8 E	637+/2558	30
Muzzle Trench	550m E of la Brayelle Farm	7 W	623/2572	35
Nab Valley (later Blightly Valley)	NE from Authuille Wood E	7 W	624+/2559+	34
Nab Wood	1600m NE of Aveluy XR	8 W	624/2559	34
Nab, The	550m SE of Leipzig Redoubt	7 W	6253/25603	34
Nairne Street	Just E of Old Touvent Farm	7 W	622/2568	35
Nairne Trench	N of Serre	7 W	624/2568	35
Nameless Farm (Bock Farm)	400m SE of Gommecourt Park	7 W	622/2570	35
Nameless Trench	NW of Star Wood	7 W	623/2569	35
Napier's Redoubt	SW Maricourt	8 E	632/2553	18

Site/trench	Location	Map	MapRef	Obj
Naze, The	Just W of Leipzig Salient	7 W	624/2560	34
Needle Trench	800m E of E Flers	7 E	635/2561	32
Neilstone Street	E of Thiepval Wood	7 W	624/2561+	34
Nestler Höhlen	Just S of Bottom Wood	8 E	627/2556	15
New Beaumont Road	Auchonvillers to Beaumont Hamel	7 W	621+/2565	35
New Cut	780m ESE of Mary Redan	7 W	623+/2563+	35
New Munich Trench	S from Waggon Rd to Leave Avenue	7 W	623/2565+	35
New Street	300m NE of Adanac Cem.	7 E	629/2564	29
New Street	W of Gommecourt Park	7 W	621/2571	35
New Trench	Just SW of Gueudecourt	7 E	635/2562	32
New Trench	300m E of Trônes Wood	8 E	634/2557	24
New Trench	680m S of Y Ravine	7 W	622/2564	35
Newham Road	NE of Eaucourt Abbey	7 E	633/2563	36
Niemeyer Weg	see German Lane	7 W	623/2569	35
Nimble, the	S of Butte de Warlencourt	7 E	632/2564	36
Ninth Alley	750m N of Eaucourt Abbey	7 E	633/2563	36
Nord Alley	Montauban to Glatz Redoubt	8 E	632/2556	18
Nordwerk	900m NNE of Ovillers	7 E	626/25607	22
Norfolk Crescent	Guillemont Centre	8 E	635/2557	24
North Causeway	Just SW of Thiepval Wood	7 W	623/2561	34
North Practice Trench	NNE of Courcelette	7 E	630/2563	29
North Street	N out of Longueval	8 E	633/2559	21
North Street	N from Mount St. Guillemont	8 E	635/2557	24
North Street	Just S of Watling Street	7 W	622/2566	35
North Tangle	SE Martinpuich	7 E	630/2561	28
Northern Avenue	W from Luke Copse	7 W	623/2567	35
Nose, The	600m E of Adanac Cem.	7 E	629/2564	29
Nose, The	700m NNW of Eaucourt Abbey	7 E	632/2564	36
Nose, The	300m SSE of Lochnagar crater	8 E	626/2557	16
Nottingham Street	NW of Gommecourt Park	7 W	621/2571	35
OG1 - German trench	NE of Poz. to Baz-le-Petit Wood	8 E	628+/2559+	22
OG2 - German trench	From NE of Poz.to Baz. Ridge	8 E	628+/2559+	22
Oakham Trench	W of Gommecourt Wood	7 W	621/2571	35
Oakhanger Wood	1600m SW of Combles church	8 E	6367/25559	33
Oban Avenue	Just E of Authuille	7 W	624/2560	34
Oban Trench	100m S of Granatloch, Authuille Rd.	7 W	624/2560	34
Oblong Wood, site of	270m WNW of Thiepval Château	7 W	6249/25621	34
Observation Wood	W of Matthew Copse	7 W	623/2567	35
Odiham Trench	Just W of Gommecourt Park	7 W	622/2572	35
Ohio Trench	Just N of Gommecourt Wood	7 W	622/2572	35
Oily Lane	Just N of Gueudecourt	7 E	636/2563	32
Old Beaumont Road	NE from Auchonvillers	7 W	621/2565	35
Old German Alley	NE from N of Trônes Wood	8 E	633+/2558	19
One Tree Hill	800m N of New Beaumont Road	7 W	621/2565	35
Orange Trench	550m W of Pigeon Wood	7 W	622/2572	35
Orchard Alley	800m W of Mametz Halt	8 E	627/2555	18
Orchard Alley	E from Puisieux	7 W	625/2568	35
Orchard Copse	600m W of Montauban	8 E	630/2556	18
Orchard Trench	NW of Tea Trench	7 E	627/2564	29
Orchard Trench	Immed. W of Delville Wood	8 E	632+/2559	21
Orchard Trench	300m W of Mametz	8 E	628/2555	18
Orchard, Ginchy	N Ginchy	8 E	635/2558	26
Orchard, Pozières	375m N on right of Thiepval Road	7 E	627/2560	22
Orchard, The	N Combles	8 E	6376/25575	33
Orchard, The	Junction of Orchard & Willow Trenches	8 E	628/2556	18
Orchard, site of	Immed. E of Thiepval Mem. Cem.	7 W	6254/25614	34
Orion Trench	1250m NE of Lesboeufs	7 E	638/2562	31
Orionoco Trench	Just N of Gommecourt Wood	7 W	622/2572	35
Orléans Trench	Just W of Gommecourt Wood	7 W	622/2572	35
Oundle Trench	Just W of Gommecourt Wood	7 W	622/2572	35
Ouse Trench	ENE of Gommecourt Wood	7 W	622/2572	35
Ovillers	1200m NNE of La Boisselle	8 E	625+/2559	17
Ovillers Post	1100m WNW of Y Sap crater	8 W	624/2558	17
Ovillers Spur	N of Ovillers	8 E	625/2559	17

TRENCHES, BATTLESITES ETC

Site/trench	Location	Map	MapRef	Obj
Ovillers Tramway	La Boisselle to Ulverston	8 W	625/2558+	17
Ox Trench	Adjacent Lesboeufs	7 E	637/2561	30
Oxford Circus	1140m W of Adanac Cem.	7 E	628/2564	29
Oxford Copse	1200m ESE of Carnoy	8 E	631/2554	18
Oxus Trench	NE of Gommecourt Park	7 W	623/2572	35
Padre Street	E of Ovillers Post	8 W	624/2558	17
Paisley Avenue	E from Thiepval Wood	7 W	624/2561	34
Paisley Dump	Just E of Thiepval Wood	7 W	624/2562	34
Palestine Trench	Communication Trench Serre	7 W	623/2567	35
Pall Mall	South from Longueval	8 E	633/2558	21
Papen Trench	800m WNW of Mametz Halt	8 E	627/2555	18
Paradise Alley	S from Hessian Trench	7 E	626/2561+	34
Parallelogram	see Schwaben Redoubt	7 W	624+/2562+	34
Park Corner	680m SSW of Y Ravine	7 W	622/2563	35
Park Lane	Joins Ration Trench to OG1	7 E	627/2561	22
Park Lane	W Guillemont	8 E	634/2557	24
Passerelle de Magenta	Just SW of Thiepval Wood	7 W	623/2561	34
Patch Alley	250m W of The Dingle	8 E	627/2557	15
Pea Trench	750m E of Pozières	7 E	628/2560	22
Peake Woods	800m SW of Contalmaison	8 E	6277/2558	18
Pear Street	Just W of Delville Wood	8 E	633/2559	21
Pear Trench	Just W of Apple Alley	8 E	627/2555	18
Pearl Alley	Runs N bet.Contalmaison & Mametz Wd.	8 E	628+/2558+	18
Pearl Wood	1000m ENE of Contalmaison	8 E	6292/2559	20
Peel Trench	830m W of Stuff Redoubt	7 W	625/2562	34
Peg Trench	E of Pozières	7 E	629/2560	29
Pendant Alley East	Contn. of Pendant Alley West	7 W	624/2567	35
Pendant Alley West	ENE from Munich Trench North	7 W	624/2567	35
Pendant Copse	940m SE of Serre centre	7 W	6247/25671	35
Pendant Trench	N of Pendant Copse	7 W	624/2567	35
Pendlehill Street	1300m NE of Ovillers Post	8 W	625/2559	17
Peterhead Sap	Just NW of Thiepval Wood	7 W	624/2562	34
Petrol Lane	Just N of Gueudecourt	7 E	636/2563	32
Piccadilly	S of Y Ravine	7 W	622/2564	35
Piccadilly	West Longueval	8 E	633/2559	21
Pig Trench	1000m NE from old Touvent Farm	7 W	623/2568	35
Pigeon Wood	1600m NE of Gommecourt	7 W	623/2572	35
Pilgrim's Way	NE Flers to S Gueudecourt	7 E	635/2562	32
Pilk Street	W of Hawthorn Crater	7 W	622/2565	35
Pill Trench	NW of Gommecourt Park	7 W	621/2571	35
Pilsen Lane	E from Delville Wood S	8 E	634/2558	25
Pimple Alley	E from The Pimple	7 E	632/2564	36
Pimple, The	730m NNW of Eaucourt Abbey	7 E	632/2563	36
Pint Trench	Runs N from Ginchy	8 E	635/2559	25
Pioneer Alley	400m NNW of Baz-le-Petit	8 E	630/2560	28
Pioneer Support	750m NNE of Eaucourt Abbey	7 E	633/2563	36
Pioneer Trench	850m E of E Flers	7 E	635/2561	32
Pip Street	1500m S of Thiepval	7 W	625/2561	34
Pipe Trench	70m E of Western edge of Stuff Redoubt	7 E	626/2563	34
Pit Alley	500m E of Trônes Wood	8 E	634/2557	24
Pithie Post	1350m WNW of Baz-le-Petit	8 E	629/2560	28
Pitlochry Street	NNW of Chapes Spur	8 W	625/2557	16
Pitt Street	W of Watling Street	7 W	622/2565	35
Plug Street	S from Rossignol Wood	7 W	624/2569+	35
Point 110	S of Fricourt on 110m contour line	8 E	627/2554	15
Point 48	North of Oakhanger Wood	8 E	637/2556	33
Point 60	240m E of S tip of Y Ravine	7 W	622/2564	35
Point 66	Strong point in Desire Trench	7 E	627/2563	35
Point 71 N	1400m S of Tambour Mines	8 E	626/2554	15
Point 89	80m from W end of Y Ravine	7 W	622/2564	35
Point, The	N of John Copse	7 W	623/2568	35
Pole Trench	1250m SW of Mouquet Farm	7 E	626/2560	22
Pole Trench	650m NE of Gueudecourt	7 E	636/2563	32

Site/trench	Location	Map	MapRef	Obj
Polish Trench	800m ENE of Gueudecourt	7 E	637/2563	32
Polo Road	550m NE of Pozières	8 E	627/2560	22
Pom-Pom Alley	E out of N Serre	7 W	624/2567	35
Pommiers Lane	SE from Pommiers Redoubt	8 E	630/2556	18
Pommiers Redoubt	E of Danzig Alley	8 E	630/2556	18
Pommiers Trench	S of Pommiers Redoubt	8 E	630/2555	18
Pond Street	280m SW of Mary Redan	7 W	622+/2563	35
Pont Street	Longueval windmill to High Wood	8 E	632/2559	21
Poodles, site of	Just N of Fricourt Farm	8 E	6276/25568	15
Pope's Nose	1200m NNW of Thiepval	7 W	6244/2562	34
Popoff Lane	500m S of Pommiers Redoubt	8 E	630/2555	18
Pork Trench	1250m NW of Gueudecourt	7 E	635/2563	32
Port Louis	300m SW of Ovillers Post	8 W	624/2558	17
Porter Trench	600m W of Ginchy	8 E	635/2558	25
Possum Reserve	800m SSW of Gueudecourt	7 E	635/2561	32
Post Trench	formerly Sanderson & Cameron Trenches	8 E	629+/2560	28
Potsdam Trench	450m E of Munich Trench South	7 W	623/2565	35
Pottage Street	South Hamel	7 W	623/2563	35
Pottage Trench	SW from Hamel	7 W	622+/2563	35
Pozières	2000m N of Contalmaison	8 E	627/2560	22
Pozières Civil Cemetery	At N end of village	7 E	62752/25609	22
Pozières Trench	Just S of Pozières village	8 E	627+/2560	22
Price Street	650m N of Thiepval XR	7 W	625/2562	34
Price Trench	700m NW of Gueudecourt	7 E	635/2563	32
Prince Street	NNE from Leipzig Salient	7 W	624/2561	34
Prince's Street	Delville Wood	8 E	634+/2559	21
Princes Street	750m ESE of Casino Point	8 E	630/2555	18
Princess Street	North Hamel	7 W	623/2563	35
Prindlehill Street	Just W of Ovillers	8 W	625/2559	17
Pritchard Trench	Immed. E of New Munich Trench	7 W	622/2566	35
Prospect Row	440m SSW of Hamel	7 W	622+/2562	35
Prue Trench	Just E of Martinpuich	7 E	631/2561	28
Pub Street	1200m NE of old Touvent Farm	7 W	623/2568	35
Puisieux Alley	SW from Wundt Werk	7 W	625/2566	35
Puisieux Trench	N from Grandcourt	7 E	626/2565+	35
Punch Trench	1600m SE of Flers	8 E	635+/2560	27
Purfleet	Just W of König Lane	8 E	626/2555	15
Push Alley	Adjacent Martinpuich	7 E	630+/2560+	28
Push Alley	Adjacent Pendant Copse	7 E	626/2565	35
Pys	NW of Le Sars	7 E	629/2565	35
Quadrangle Alley	N off Quadrangle Trench	8 E	628/2557	18
Quadrangle Support	Joins Pearl Alley to Quadrangle Alley	8 E	628/2557+	18
Quadrangle Trench - Kaiser Graben	Twds. Contalmaison from N of Quad.Wood	8 E	628/2557	18
Quadrangle Wood	1400m ENE of Fricourt Farm	8 E	6287/25573	18
Quadrangle, The	Group of trenches E of Mametz Wood	8 E	6288/25574	18
Quadrilateral	1000m E of Ginchy	8 E	6365/25586	26
Quadrilateral	Adjacent Thiepval	7 W	625/2562	34
Quadrilateral	1100m NW of Le Sars	7 E	631/2564	36
Quadrilateral	1820m NE of Courcelette	7 E	630/2564	29
Quadrilateral (Heindenkopf)	80m S of Serre Brit. Cem. No. 2	7 W	6227/25666	35
Quadrilateral - Gommecourt	835m E of Kern Redoubt	7 W	622/2571	35
Quadrilaterial	N of Leuze Wood	8 E	636/2558	33
Quarry	800m N of Montauban	8 E	6321/25576	18
Quarry	Just NE of Falfemont Farm	8 E	637/2556	33
Quarry	Just S of Mouquet Farm	7 E	6269/25615	22
Quarry	East Courcellete	7 E	629/2562	29
Quarry	260m W of Guillemont	8 E	6347/25576	24
Quarry	NNE of Quadrilateral, nr. Ginchy	8 E	637/2559	26
Quarry	N of Bernafay Wood	8 E	632/2557	19
Quarry	800m NW of Le Sars	7 E	631/2564	36
Quarry	Immed. W of Butte de Warlencourt	7 E	632/2564	36
Quarry	Immed. S of Baz-le-Pet. cemetery	8 E	6309/25596	23
Quarry Branch	Railway N twds. Authuille Wood	8 W	624/2558+	17
Quarry Post	Immed. E of Authuille Wood	8 W	6248/2560	34

Site/trench	Location	Map	MapRef	Obj
Quarry Trench	E - W from S of Mouquet Farm	7 E	626+/2561	22
Queen Victoria Street	300m E of Talus Boisé	8 E	631/2555	18
Queen's Cross Street	E of Thiepval Wood	7 W	624/2561+	34
Queen's Nullah	Just SW of Mametz Wood	8 E	629/2556	18
Queen's Redoubt	1000m W of Triple Tambour Mines	8 W	6268/25561	15
Queen's Street	600m ESE of Casino Point	8 E	630/2555	18
Queen's Trench	runs through High Wood	8 E	632/2560	23
Quemart Street	Immed. S of Glory Hole	8 W	625/2558	16
Quergraben II	E of D929 NE of La Boisselle	8 E	627/2558+	16
Quergraben III	Runs SE from Ovillers twds. Scots Redbt.	8 E	626/2557+	16
Rabbit Lane	E of Triple Tambour Mines	8 E	626/2555	15
Radcliffe Street	Just E of Authuille	7 W	624/2560	34
Rail Avenue	800m E of Carnoy	8 E	630/2554	18
Railland Trench	600m N of Authuille	7 W	624/2561	34
Railway Alley	Fricourt Farm to Bottom Wood	8 E	627+/2556	18
Railway Alley	Joins Station Road & Beaucourt Trench	7 W	624/2564	35
Railway Avenue Trench	Nr. Euston Dump	7 W	620/2567	35
Railway Copse	450m ENE of Fricourt Farm	8 E	6282/25569	15
Railway Road	Beaucourt-Miraumont road	7 E	626/2565	35
Railway Trench	700m NNW of Beaucourt	7 W	623/2564	35
Railway View	700m E of Mesnil	7 W	622/2562	35
Rainbow Trench	1000m E of Gueudecourt	7 E	636+/2562	31
Rainy Trench	600m NE of Lesboeufs	7 E	638/2561	31
Ranger Trench	Just W of Bouleaux Wood	8 E	636/2557	33
Ransome Trench	Immed. S of Schwaben Rdt.	7 W	625/2562	34
Ration Trench (Gierich Weg)	NW of Pozières	7 E	626+/2561	22
Ravine, The	900m N of Regina Trench East	7 E	629/2564	29
Raymond Avenue	NW of Gommecourt Park	7 W	621/2571	35
Rectangle, The	700m SSW of Rose Cottage	8 E	627/2555	15
Red Château - site of	Just N of Courcelette church	7 E	629/2562	29
Red Cottage	N Fricourt	8 E	62714/25561	15
Red Cottage	400m N of Basin Wood	7 W	621/2567	35
Red Lane	W of N Fricourt	8 E	627/2556	15
Red Trench	S from Red Lane to Rabbit Lane	8 E	626/2555	15
Red Trench	E off K Trench	7 E	627/2560	22
Redan Ridge	N of Beaumont Hamel	7 W	622/2566	35
Redan Ridge Redoubt	400m N of Bergwerk	7 W	622/2566	35
Redoubt Alley	Joins Station Road to Beaucourt Trench	7 W	624/2564+	35
Regent Street	Delville Wood	8 E	634/2559	21
Regent Street	S of Y Ravine	7 W	622/2564	35
Regent Street	NW of Gommecourt Park	7 W	621/2571	35
Regina Trench West	Contn. of Stuff Trench	7 E	626+/2562+	34
Regina Trench East	Protects Courcelette from the north east	7 E	628+/2563+	29
Reids Alley	770m NW of Baz-le-Petit	8 E	629/2560	28
Rettemoy Farm	2300m ENE of Gommecourt	7 W	624/2572	35
Rex Trench	1400m NW of Courcelette	7 E	627/2563	29
Ribble Street	S from Donnet Post	8 W	624/2558	17
Riddell Trench	Northern edge of Schwaben Redoubt	7 W	625/2563	34
Ridge Redoubt	SW of Serre	7 W	623/2567	35
Ridge Trench	700m W of Hamel	7 W	622/2563	35
Riebel Stellung	N of Pozières	7 E	628/2561	22
Rifle Trench	N from bet. Mouquet Fm. & Courcelette	7 E	627/2562+	29
Rim Trench	From quarry at Guillemont	8 E	634/2557	24
Ripley Trench	240m N of Thiepval XR	7 W	625/2562	34
River Trench	700m NE of Bois d'Hollande	7 E	626/2565	35
Rivington Street	100m NE of Donnet Post	8 W	625/2559	17
Road Trench	Leipzig Salient towards Thiepval Mem.	7 W	624/2561	34
Rob Roy Trench	S of Matthew Copse	7 W	622/2567	35
Robert's Trench	E of Englebelmer	7 W	620/2562	35
Roberts Avenue	NW of Gommecourt Park	7 W	621/2571	35
Roberts Trench	Near Beaucourt	7 W	624+/2564+	35
Rodley Road	Just E of Talus Boisé	8 E	631/2555	18
Roedergraben	S of D929 SW of Pozières	8 E	627/2559	22

Site/trench	Location	Map	MapRef	Obj
Rolland Trench	N from Great Northern Trench	7 W	622/2567	35
Roman Road Trench	Commun. Trench Sucrerie	7 W	620/2566	35
Rooney's Sap	300m SW of Hawthorn Crater	7 W	622/2565	35
Rose Alley	SE Fricourt	8 E	627/2555	15
Rose Cottage	E Fricourt	8 E	6275/25559	15
Rose Reserve	500m SE of Gueudecourt	7 E	636/2562	32
Rose Trench	SE from Fricourt Wood	8 E	627/2555	15
Ross Castle	Thiepval Wood	7 W	624/2561+	34
Ross Street	S of Regina Trench East	7 E	629/2563	29
Ross Street	Thiepval Wood	7 W	624/2561+	34
Rossignol Farm	Near Rossignol Wood	7 W	624/2570	35
Rossignol Trench	Rossignol Wood to Puisieux	7 W	624+/2569+	35
Rossignol Wood	2600m E of Hébuterne	7 W	623+/2570	35
Rothes Street	Thiepval Wood	7 W	624/2562	34
Rotten Row	Delville Wood	8 E	634+/2559	21
Rotten Row	NW of Gommecourt Park	7 W	621/2571	35
Rough Trench	250m SW of Hawthorn Crater	7 W	622/2565	35
Round Wood	740m NW of Fricourt Farm	8 E	6271/25574	15
Round Wood Alley	S edge of Scots Redoubt	8 E	627/2556	16
Royal Avenue	670m WNW of Baz-le-Petit	8 E	629/2560	20
Rum Trench	1200m SE of Baillescourt Farm	7 E	626/2564	29
Rump Trench	N of Beaumont Hamel	7 W	623/2565	35
Rutherford Alley	1000m SE of Le Sars	7 E	632/2563	36
Rycroft Alley	E from Munich Trench S	7 W	623/2566	35
Rycroft Street	1000m NE of Ovillers Post	8 W	625/2559	17
S1, S2 & S3	German strongpoints in Trônes Wood	8 E	633/2557	19
Sabot Copse	Just NE of Flat Iron Copse	8 E	6306/25585	20
Sackville Street	Just N of Redan Ridge	7 W	622/2567	35
Safety Trench	1600m NE of Bois d'Hollande	7 E	626/2566	35
Salford Trench	Runs W off Munich Trench South	7 W	623/2566	35
Sanda Street	Just W of Leipzig Salient	7 W	624/2561	34
Sanderson Trench	700m NW of Baz-le-Petit	8 E	629/2560	20
Sapper Trench	400m W of Guillemont	8 E	634/2557	24
Sauchiehall Street	E of Thiepval Wood	7 W	624/2561+	34
Sausage Rdt. - Heligoland	600m ESE of Lochnagar crater	8 E	6265/25576	16
Sausage Valley	1000m S of La Boisselle	8 W	625/2557	16
Savernake Wood	1050m SSW of Combles	8 E	637/2556	33
Savoy Trench	Longueval	8 E	633/2558+	21
Scabbard Trench	1000m N of Gueudecourt	7 E	636/2563	32
Schrapnell Way	see German's Wood	8 E	632/2555	18
Schwaben Höhe	200m N of Lochnagar crater	8 E	626/2557	16
Schwaben Nest	Just N of Gommecourt Wood	7 W	622/2572	35
Schwaben Redoubt	950m NNE of Thiepval XR	7 W	625/2562+	34
Schwaben Trench	ESE of Thiepval	7 W	625+/2561	34
Schwarzwald Graben	NW of Pozières	7 E	627/2561	22
Schweickert Graben	900m WNW of Rossignol Wood	7 W	622/2570	35
Score Street	200m SW of Glory Hole	8 W	625/2558	16
Scotch Alley	550m W of Baz-le-Petit	8 E	629/2560	20
Scots Redoubt	400m NW of Birch Tree Wood	8 E	62686/25576	16
Scottish Alley	Just SW of Arrow Head Copse	8 E	634/2556	24
Scottish Trench	Just W of Arrow Head Copse	8 E	634/2556	24
Seaforth Trench	Just W of High Wood	7 E	631/2560	23
Seaforth Trench	300m SW of Hawthorn Ridge	7 W	622/2565	35
Serb Switch	700m E of St. Pierre Divion	7 W	624+/2563	34
Serheb Road	NW out of Serre	7 W	623/2568	35
Serpentine Trench	NW from the Triangle	8 E	635+/2559	26
Serre	2000m SW Puisieux	7 W	623+/2567	35
Serre Alley	Serre to Puisieux	7 W	624+/2567+	35
Serre Trench	Runs SSW from Serre	7 W	623/2566+	35
Seven Dials	800m W of N Gueudecourt	7 E	635/2562	32
Seven Elms	SSW of Eaucourt Abbey	7 E	632/2562	32
Seventh Street	500m NW of Gibraltar	7 E	627/2560	22
Shaftesbury Avenue	S of Y Ravine	7 W	622/2564	35
Shelter Alley	Runs NE from Shelter Wood	8 E	627+/2557+	18

Site/trench	Location	Map	MapRef	Obj
Shelter Wood	700m N of Fricourt Farm	8 E	627/2557	15
Sherwood Trench	200m E of Trônes Wood	8 E	634/2557	24
Shetland Alley	1050m NW of Baz-le-Petit	8 E	629/2560	28
Shin Trench	W of Le Transloy	IG4	Fold 5	30
Shine Trench	ENE of Lesboeufs	7 E	638/2561	31
Short Street	300m NE of Hamel	7 W	623/2563	35
Shrapnel Street	600m W of Mouquet Farm	7 E	626/2561	34
Shrapnel Trench	Adjacent Ovillers	8 E	626/2559	17
Shrapnel Trench	570m NW of Mouquet Farm	7 E	626/2561+	34
Shrine Alley	Mametz to Mametz Civil Cem.	8 E	628/2555	18
Shrine, The	German m/c gun post Mametz Civ. Cem	8 E	62855/25554	18
Shropshire Tr (formerly Lonely Tr)	800m S of Guillemont	8 E	635/2556	24
Shute Trench	400m E of Maltzhorn Farm	8 E	634/2556	24
Sickle Trench	N of Ovillers	8 E	626/2560	17
Sigel Graben	1300m ESE of Lochnagar crater	8 E	627/2557	16
Sight Trench	680m N of Stuff Redoubt	7 E	626/2563	34
Signal Alley	Just E of Trônes Wood	8 E	634/2557	24
Silesia Alley	N & S from Silesia Trench	8 E	631/2555	18
Silesia Trench	1200m S of Montauban	8 E	631/2555	18
Sinclair Road	Authuille village	7 W	624/2560	34
Sixteen Loop	Folly Trench to Sixteen Trench	7 E	627/2564	29
Sixteen Road	800m SE of Grandcourt Cemetery	7 E	627/2564	29
Sixteen Trench	Just W of Sixteen Road	7 E	627/2564	29
Sixth Avenue	see Skyline Trench	7 E	626+/2561	34
Sixth Street	W of Gommecourt Park	7 W	621/2571	35
Sixth Street	500m W of Gibraltar	7 E	627/2560	22
Skyline Trench	SW of Mouquet Farm	7 E	626/2561	22
Sleet Trench	1200m ENE of Lesboeufs	7 E	638/2561	31
Slight Valley	300m E of Ovillers	8 E	626/2559	17
Sloane Street	West Longueval	8 E	633/2558	21
Sloane Street	E of Englebelmer	7 W	620/2562	35
Smoke Trench	900m NNE of Flers	7 E	635/2562	32
Smoke Trench	700m N of Beaumont Hamel	7 W	623/2566	35
Smyth Valley	Runs N from Chalk Pit	8 E	627/2559+	22
Snag Trench	800m N of Eaucourt Abbey	7 E	633/2564	36
Snagg Support	N of Eaucourt Abbey	7 E	632/2563	36
Snout, The	Just S of Baz-le-Grand Wood	8 E	630/2558	20
Snow Trench	Adjacent Lesboeufs	7 E	637/2560	30
Snuff Alley	1300m NE of old Touvent Farm	7 W	623/2568	35
Soap Alley	SE from Bath Lane	7 W	625/2568	35
Soda Trench	550m W of The Dingle	8 E	626/2557	15
Somme Alley	500m NNE of Baz-le-Petit	8 E	630/2560	23
South Causeway	Just SW of Thiepval Wood	7 W	623/2561	34
South Miraumont Trench	400m S of Pet. Miraumont twds. Pys	7 E	628/2565	34
South Practice Trench	NE of Courcelette	7 E	630/2563	29
South Sausage	800m WSW of Round Wood	8 E	626/2557	15
South Street	Southern edge of Delville Wood	8 E	634/2558	21
South Street	Runs S from Mount St, Guillemont	8 E	635/2557	24
South Street	NW of Hawthorn Crater	7 w	622/2565	35
South Tangle	400m SE of Martinpuich	7 E	631/2561	28
South Trench	S of Montauban	8 E	631/2556	18
Southern Avenue	Just S of Matthew Copse	7 W	622/2567	35
Southern Trench	Bet. Maltzhorn Farm & Trônes Wood	8 E	634/2556	19
Southern Trench	Just S of Montauban	8 E	631/2556	18
Spectrum Trench	1100m NNE of Lesboeufs	7 E	638/2561	31
Spence Trench	S of Le Sars	7 E	631/2562	36
Spider Trench	Near "Met" trenches	7 E	636+/2561+	31
Splutter Trench	East from Schwaben Redoubt	7 W	625/2563	34
Spring Gardens	500m E of Ovillers	8 E	626/2559	22
Square Wood	Just SW of Bois de Biez	7 W	624/2571	35
St. Andrew's Avenue	CT west from Usna Hill	8 W	624/2558	16
St. Helen's Road	N from New Beaumont Road	7 W	622/2565	35
St. James's Street	S of Y Ravine	7 W	622/2564	35

Site/trench	Location	Map	MapRef	Obj
St. John's Road	Auchonvillers towards Hamel	7 W	621+/2564	35
St. John's Road Trench	Assem. point for att. on Y Ravine	7 W	621+/2564	35
St. Martin's Lane	W of Gommecourt Park	7 W	621/2571	35
St. Monans Street	500m SE of Keats Redan	8 W	625/2558	16
St. Pierre Divion	1600m NNW Thiepval	7 W	624/2563	34
St. Vincent Street	Just W of Ovillers	8 W	625/2559	17
Stabgraben	200m E of D929 NE of La Boisselle	8 E	627/2559	16
Staff Copse	300m W of Old Touvent Farm	7 W	622/2568	35
Staffa Street	Just W of Leipzig Salient	7 W	624/2561	34
Stafford Avenue	NW of Gommecourt Park	7 W	621/2571	35
Standish Street	700m E of Ovillers Post	8 W	625/2558	17
Star Alley	W from Star Wood	7 W	622/2567	35
Star Wood	1200m N of Serre	7 W	624/2569	35
Starfish Line	East from Martinpuich	7 E	631+/2561	28
Starfish Trench	See Starfish Line	7 E	631/2561	28
Starfish, The	Junc. of Starfish Line & Flig Lane	7 E	633/2561	28
Station Alley	Joins Station Road to Beaucourt Trench	7 W	623/2564+	35
Station Road	SE from Beaumont Hamel	7 W	623+/2564	35
Station Road	S from Guillemont Station	8 E	634/2557	24
Station Trench	Just S of Station Road, Beaumont	7 W	623/2564	35
Staufen Riegel	see Regina Trench East	7 E	629+/2564	29
Steinbruch Graben	Bet. La Bois. & Pozières on D929	8 E	626/2559	16
Steingraben	E from D929 NE of La Boisselle	8 E	627/2559	16
Stephenson Avenue	South Mesnil	7 W	622/2561	35
Stewart Trench	Just West of Courcelette	7 E	628/2562	29
Stewart Work	550m NE of Beaumont Hamel	7 W	624/2565	35
Stocker Avenue	Just W of Maltzhorn Farm	8 E	634/2556	19
Stoney Trench	S from St. Pierre Divion	7 W	624/2563	34
Stormy Trench	1050m NE of Gueudecourt	7 E	636+/2563	31
Stout Trench	650m W of Ginchy	8 E	634/2558	25
Straight Trench	Triangle to Quadrilateral nr. Ginchy	8 E	6365/2558+	26
Strand	Delville Wood	8 E	633/2559	21
Strasburg Line	St. Pierre Div. to Schwaben Rdt.	7 W	624+/2563	34
Strip Trench	On W edge of Mametz Wd. S	8 E	6293/2557	18
Stuff Redbout	1600m NNW of Mouquet Farm	7 E	6267/2562+	34
Stuff Trench	N of Stuff Redoubt	7 W	625+/2563	34
Stump Alley	SE from Rossignol Wood	7 W	623/2570	35
Stump Road	North from Stuff Redoubt	7 E	626/2563+	34
Stuttgart Lane	400m N of Ten Tree Alley	7 W	623/2567	35
Sudbury Trench	800m NW of Courcelette	7 E	628/2563	29
Sugar Factory	900m S of Courcelette	7 E	6293/25616	29
Sugar Refinery	SE of Longueval	8 E	634/2558	24
Sugar Trench	600m SW of Courcelette	7 E	628+/2561+	29
Sulphur Avenue	2700m ENE of Gueudecourt	7 E	638+/2563	31
Sun Alley	950m SE of Flers	7 E	635/2560	27
Sun Trench	Met. trench W of Le Transloy	7 E	638/2562	31
Sunken Lane	Mansel Copse	8 E	628/2554	18
Sunken Lane	400m N of Hawthorn Crater	7 W	622/2565	35
Sunken Road	N from Courcelette	7 E	628/2564	29
Sunken Road	S from Crucifix Trench, Fricourt	8 E	627/2556	15
Sunken Road	On Ginchy-Gueudecourt road	7 E	635/2561	32
Sunken Road	Bet. Lesboeufs & Gueudecourt	7 E	637/2561+	31
Sunken Road	Bet. Flers and Gueudecourt	7 E	635/2562	32
Sunken Road	Ginchy to NW corner Leuze Wood	8 E	635+/2557+	33
Sunken Road	see Abbey Road	7 E	632/2563	36
Sunken Road	Between Ginchy & Leuze Wood	8 E	635+/2558	26
Sunken Road	Immed. S of Trônes Wood	8 E	634/2556	19
Sunken Road	400m S of Flers	7 E	634/2560	27
Sunken Road	Auchonvillers to Fabrique Farm	7 W	621/2565	35
Sunken Road Trench	Runs SSE from Fricourt	8 E	627/2555	15
Sunray Trench	1500m NE of Gueudecourt	7 E	637/2563	31
Sunshine Alley	Southern boundary of Fricourt Wood	8 E	627+/2556	15
Sutherland Avenue	Immed. W of Thiepval Wood	7 W	624/2562	34

Site/trench	Location	Map	MapRef	Obj
Sutherland Trench	1150m E of Baz-le-Petit	8 E	631/2560	23
Suvla Lane	1200m E of Beaumont-Hamel	7 E	626/2564	35
Suvla Trench	700m NE of Beaucourt	7 W	625/2565	35
Swan Trench	1500m N of Grandcourt	7 E	626/2566	35
Swansea Trench	1000m W of W side of High Wood	8 E	631/2560	28
Swiss Trench	SW of Gueudecourt	7 E	635/2562	32
Switch Elbow	see Sanderson Trench	8 E	629/2560	20
Switch Line Centre	Runs through N Corner of High Wood	7 E	631+/2560	23
Switch Line East	East-West from South of Flers	8 E	633+/2560	27
Switch Line West	S of Martinpuich to High Wood	7 E	630+/2560+	28
Switch Trench	Bet. High Wood and Delville Wood	8 E	632+/2559+	21
Switch Trench	Leipzig Salient	7 W	624/2561	34
Sydney Street	600m W of Pozières Windmill	7 E	627/2560	22
Taffy Trench	Starts at left fork N from Pozières	7 E	629/2561	29
Tail, The	S of Butte de Warlencourt	7 E	632/2564	36
Talus Boisé	1500m ENE of Carnoy	8 E	631/2554	18
Tangle Alley	Just E of Le Sars	7 E	632/2563	36
Tangle South	530m S of Martinpuich	7 E	630/2561	28
Tangle Trench	Just S of Martinpuich	7 E	630/2561	28
Tangle, North	Just E of Martinpuich	7 E	630/2561	28
Tangle, The	400m SE of Le Sars	7 E	632/2563	36
Tank Alley	1150m ESE of Beaumont-Hamel	7 W	624/2564	35
Tank Trench	E of Leuze Wood	8 E	637/2557	33
Tara Hill	Just S of Bapaume Rd. before La Bois.	8 W	624/2557	16
Tara Redoubt	1500m SW of Y Sap crater	8 W	624/2557	16
Tarbert Street	Leipzig Salient	7 W	624/2561	34
Tatler Trench	NE of Ginchy	8 E	635+/2558+	26
Tay Street	130m S of Glory Hole	8 W	625/2558	16
Taylor's Post	350m NE of Adanac Cemetery	7 E	629/2564	29
Tea Lane	Runs NE from Tea Trench	8 E	634/2560	27
Tea Support Trench	N of Delville Wood	8 E	634/2560	27
Tea Trench	N of Delville Wood	8 E	632+/2559+	21
Tea Trench	900m SE of Baillescourt Farm	7 E	626/2564	29
Teale Trench	SE from Guillemont Alley	8 E	634/2557	24
Ten Tree Alley	Runs West-East 800m S of Serre	7 W	623+/2566+	35
Tenderloin Street	SSE from White City	7 W	622/2565	35
Thiepval	3000m NW of Pozières	7 W	625/2561+	34
Thiepval Avenue	Authuille to Thiepval	7 W	624/2561	34
Thiepval Château, site of	200m SW of Thiepval XR	7 W	6251/25619	34
Thiepval Civil Cemetery	420m N of Thiepval XR	7 W	6254/25625	34
Thiepval Fort	Just W of Thiepval Château	7 W	625/2562	34
Thiepval Point North	150m SW of Hammerhead Sap XR	7 W	624/2562	34
Thiepval Point South	400m SSW of Hammerhead Sap XR	7 W	624/2561	34
Thiepval Post	Just S of Thiepval Wood	7 W	624/2561	34
Thiepval Ridge	N & SE of Thiepval	7 W	625/2561+	34
Thiepval Spur	see Thiepval Ridge	7 W	625/2561+	34
Thiepval Wood	Just W of Thiepval	7 W	624/2562	34
Thiepval plateau	see Thiepval ridge	7 W	625/2561+	34
Thistle Dump	On left of Orchard Trench	8 E	632/2559	20
Thorn Lane	900m WNW of Mametz Halt	8 E	627/2555	15
Thorpe Street	W of Gommecourt Park	7 W	621/2571	35
Thorsby Street	1700m NNE of Ovillers Post	8 E	625/2560	17
Thunder Trench	E of Bouleaux Wood	8 E	639/2558	30
Thurles Dump	W of Y Ravine	7 W	622/2564	35
Thurso Road	Authuille village	7 W	624/2560	34
Thurso Street	Thiepval Wood	7 W	624/2562	34
Tipperary Avenue	W of Y Ravine	7 W	622/2564	35
Tirpitz Trench	900m SSW of Mametz	8 E	628/2554	18
Tithebarn Street	N edge of Authuille Wood	7 W	625/2561	34
Tobermory Street	Just E of Authuille	7 W	624/2560	34
Tollcross Street	Just E of Authuille	7 W	624/2560	34
Tom's Cut	N of Pozières Cemetery	7 E	627/2561	22
Torr Trench	WNW from Munster Alley	8 E	628/2560	22

Site/trench	Location	Map	MapRef	Obj
Toten Wald	see Round Wood	8 E	6271/25574	18
Tournai Trench	SW from Watling Street	7 W	622/2566	35
Touvent Farm (new)	2400m W of Puisieux	7 W	623/2568	35
Touvent Farm (old, destroyed)	380m NW of Railway Hollow Cem.	7 W	622/2568	35
Towser Post	120m W of Adanac Cem.	7 E	629/2564	29
Train Alley	700m S of Montauban	8 E	631/2556	18
Tramway Trench	E of K Trench, Pozières, to join D929	7 E	627+/2560	22
Transport Road	Between Sailly au Bois and Hébuterne	7 W	619+/2569	35
Treacle Trench	S Le Transloy	7 E	639/2562	31
Triangle Point	500m N of Montauban	8 E	631/2557	18
Triangle, The	1200m NNW of Thiepval	7 W	6244/25629	34
Triangle, The	1300m NE of Ginchy	8 E	6363/25595	26
Triangle, The	ESE of Thiepval	7 W	625/2562	34
Triangle, The	Immed. S of Leave Avenue	7 W	623/2655	35
Triangle, The	700m S of Pommiers Redoubt	8 E	629/2555	18
Triangle, The	800m NE of Scots Redoubt	8 E	627/2554	18
Triangle, The	E from Horseshoe Trench	8 E	626/2558	16
Triangle Trench	E from Lincoln Redoubt	8 E	626+/2558	16
Triple Tambour Mines	500m W of Fricourt	8 E	6268/255+	15
Trônes Alley	Links Trônes & Bernafay woods at S end	8 E	633/2556+	19
Trônes Trench	Just E of Trônes Wood	8 E	634/2557	24
Trônes Wood	2000m ENE of Montauban	8 E	633/2556+	19
Trongate	E of Thiepval Wood	7 W	624/2562	34
Tulip Road	Adjacent Morval	8 E	638/2559	30
Tummel Street	160m S of Glory Hole	8 W	625/2558	16
Turk Lane	N of Delville Wood	8 E	633/2559+	21
Turk Street	Adjacent Wonder Work	7 W	625/2561	34
Twenty Road	NE of Courcelette twds Destrement Fm.	7 E	630/2562	29
Twenty Sixth Avenue	E of S. Courcelette	7 E	629+/2562	29
Twenty Three Road	Runs N from W of Courcelette	7 E	628/2562	29
Twins, The	940m WSW of Montauban	8 E	6309/25565	18
Tyler's Redoubt	Assem. pos. for attack on La Boisselle	8 E	626/2557	16
Tyndrum Street	Just W of Leipzig Salient	7 W	624/2561	34
Tyndrum Trench	CT towards Leipzig Salient	7 W	624/2561	34
Tyne Trench	1400m N of Ovillers	7 W	625/2561	34
Uhlenfeld Graben	SE from N of Rossignol Wood	7 W	624/2570	35
Ullmer Graben	WSW of Mouquet Farm	7 E	626/2561	34
Ulverstone Street	Just W of Ovillers	8 W	625/2559	17
Union Street	Thiepval Wood	7 W	624/2561+	34
Union Street	800m W of Pozières Windmill	7 E	627/2560	29
Union Trench	1360m SE of Mouquet Farm	7 E	627/2561	29
Upper Horwich Street	Just E of Conniston Post	8 W	624/2559	17
Upper Road	Lane from Baz-le-Petit to High Wood	8 E	630+/2559+	23
Ur Trench	560m N of Pigeon Wood	7 W	623/2573	35
Ural Trench	780m N of Pigeon Wood	7 W	623/2573	35
Usk Trench	550m N of Pigeon Wood	7 W	623/2572	35
Usna Hill	Just N. of Bapaume Rd. before La Bois.	8 W	624/2557	16
Usna Redoubt	1100m WSW of Y Sap crater	8 W	624/2558	16
Valley Avenue	W of Gommecourt Park	7 W	621/2571	35
Valley Support	350m E of The Warren	8 E	631/2555	18
Valley Trench	1600m ESE of Pommiers Rdt.	8 E	631/2555	18
Valley Trench	NNE from Wedge Wood	8 E	636/2556+	33
Vancouver Trench	400m N of N Practice Trenches	7 E	629/2563	29
Vat Alley	Just E of Delville Wood	8 E	634/2559	25
Vauban Trench	NNW of Fort Briggs	7 W	623/2568	35
Vaughan Redoubt	300m ESE of Contalmaison Château	8 E	628/2558	18
Vercingetorix Trench	Contn. of Sackville Street	7 W	621/2568	35
Vernon Street	Just E of Talus Boisé	8 E	631/2555	18
Victoria Street	800m NW of Hamel	7 W	622/2563+	35
Victoria Trench	Just N of Leuze Wood	8 E	637/2558	33
Villa Trench	Bazentin Ridge	8 E	629+/2559	20
Villa Wood	600m ESE of Contalmaison Villa	8 E	6296/25595	20
Village Trench	1100m NNW of Wundt Werk	7 W	625/2568	35
Wade Lane	350m W of Thiepval Château	7 W	624/2561+	34

TRENCHES, BATTLESITES ETC

Site/trench	Location	Map	MapRef	Obj
Waggon (Wagon) Road	NNE from B.Hamel towards Serre	7 W	623/2565	35
Wagram Trench	W from Fort Sussex	7 W	621/2568	35
Walker Trench	1600m S of Courcelette	7 E	629/2561	29
Walnut Trench	600m SE of Serre	7 W	624/2567	35
Walter Trench	Runs N-S from W of Serre	7 W	623/2567	35
Waltney Street	1650m NNE of Ovillers Post	8 W	625/2560	17
Warlencourt-Eaucourt	1500m NE of Le Sars	7 E	632/2565	36
Warren Trench	1300m ESE of Pommiers Rdt.	8 E	631/2555	18
Warren Trench	Just N of Courcelette	7 E	629/2562	29
Warrier Street	S of Gommecourt Park	7 W	621/2570	35
Waterloo Bridge	400m S of La Signy Farm	7 W	621/2567	35
Waterlot Farm, site of	1100m NW of Guillemont	8 E	6342/25584	25
Waters Trench	750m ENE of Baz-le-petit	8 E	631/2560	23
Watling Street	Sunken Rd, S of Gueudecourt	7 E	635+/2561+	32
Watling Street	NW from B. Hamel to Serre Rd.	7 W	622/2566	35
Wedge Wood	1100m SE of Guillemont	8 E	6357/25567	33
Welcome Street	S of Gommecourt Park	7 W	621/2570	35
Wellington Trench	300m S of Y Ravine	7 W	622/2564	35
Well Lane	North Fricourt	8 E	627/25561	15
Welsh Alley	750m W of Baz-le-Petit	8 E	629+/2559+	20
Wenning Street	Just S of Authuille Wood	8 W	624/2559	17
West KOYLI Trench	Just N of Thiepval Wood	7 W	624/2562	34
West Keep	250m W of Mound Keep	8 E	632/2554	18
Western Trench (see K Trench)	Runs N from just W of Gibraltar	7 E	627/2560	22
Whale Trench	Just N of Gueudecourt	7 E	636/2563	32
Whalley Street	350m ENE of Ovillers Post	8 W	624/2558	17
Wheat Trench	600m ENE of Butte de Warlencourt	7 E	633/2564	36
Whiskey Street	S of Gommecourt Park	7 W	621/2570	35
Whisky Trench	500m W of The Dingle	8 E	626/2557	15
Whitchurch Street	Thiepval Wood	7 W	624/2561+	34
White City	1300m NE of Auchonvillers	7 W	621+/2565	35
White Horse Trench	Runs S from Trônes Wood	8 E	634/2556	19
White Trench	Just S of Mametz Wood	8 E	629/2557	18
White Trench	W off Munich Trench North	7 W	623/2567	35
Whizzbang Avenue	NW of Gommecourt Park	7 W	621/2571	35
Wick Road	Authuille village	7 W	624/2560	34
Wicket Corner	200m W of SW Fricourt	8 E	626/2555	15
Wigan Lane	1240m NW of Courcelette	7 E	628/2562+	29
William Avenue	E off Spence Trench	7 E	632/2563	36
William Redan	1400m SSE of Y Ravine	7 W	623/2563	35
Willow Avenue	Just E of Fricourt	8 E	627/2555	18
Willow Patch	500m SW of Round Wood	8 E	6267/25571	15
Willow Stream	S Fricourt to Mametz Wood	8 E	627+/2555+	18
Willow Trench	SE from Fricourt Wd. to Willow Av.	8 E	627+/2555+	18
Winchester Street	530m S of Mary Redan	7 W	623/2563	35
Windmill	N side of D929 out of Pozières	7 E	628/2561	22
Windmill	Just N of northern exit of Martinpuich	7 E	63105/256212	28
Windmill Trench	W of Le Transloy	IG4	Fold 5	30
Windmill Trench	S of Gueudecourt	7 E	635/2561	32
Windmill, The	300m E of Baz-le-Petit Cem.	8 E	63124/25596	23
Windmill, The	1200m NE of Sugar Factory, Courcelette	7 E	629/2561	29
Wing Corner	400m E of D147 & D938 junction	8 E	6272/25554	15
Wing Trench	1000m E of Serre	7 W	624/2567	35
Wire Road	550m SE of Mesnil	7 W	622/2561	35
Wold Redoubt	500m W of Bailiff Wood	8 E	627/2558	18
Wolf Trench	Between B. Hamel & Serre	7 W	623/2565+	35
Woman Street	S of Gommecourt Park	7 W	621/2570	35
Wonder Work (Wundt Werk)	800m S of Thiepval	7 W	6252/25613	34
Wood Alley	Immed. S of Scots Redoubt	8 E	627/2557	16
Wood Lane North	SE from High Wood	8 E	632/2560	23
Wood Lane South	Runs NW from Delville Wood	8 E	633/2559	21
Wood Post	1970m S of Thiepval	8 W	624/2559	34
Wood Street	S of Gommecourt Park	7 W	621/2570	35
Wood Support Trench	Runs W off Strip Trench	8 E	628+/2557	18

Site/trench	Location	Map	MapRef	Obj
Wood Trench	Runs W off Strip Trench	8 E	628+/25577	18
Wood Trench	NW of Longueval towards High Wood	8 E	632/2559	21
Worcester Trench	1170m W of Pozières	8 E	626/2560	22
Worcester Trench North	SE from High Wood	8 E	632/2559+	23
Worcester Trench South	1000m NW of Longueval	8 E	632/2559	21
Word Work	Between Pozières & Leipzig Salient	7 E	626/2560	34
Worley Trench	S of Matthew Copse	7 W	622/2567	35
Wrangle Avenue	W of John Copse	7 W	623/2568	35
Wretched Tea Trench	Just S of Grandcourt	7 E	626/2564	34
Wundt Werk	1000m E of Pendant Copse	7 W	625/2567	35
Wundt Werk (Wonder Work)	800m S of Thiepval	7 W	6252/25613	34
Wurzel Street	S of Gommecourt Park	7 W	621/2570	35
Y Ravine	Newfoundland Park	7 W	622+/2564	35
Y Sap Mine (now filled in)	N side of main road at La Boisselle	8 W	62535/25583	16
Y Street	400m SW of Hawthorn Crater	7 W	622/2565	35
Yankee Street	S of Gommecourt Park	7 W	621/2570	35
Yankee Trench	700m S of Gommecourt Park	7 W	621/2570	35
Yatman, Bridge	Bridge over Ancre N of Authuille	7 W	624/2560	34
Yellow Line	Brit. defences after 2nd att. on Serre	7 W	622/2567+	35
Yellow Street	S of Gommecourt Park	7 W	621/2570	35
Yiddish Street	S of Commecourt Park	7 W	621/2570	35
York Alley	Just SE of Delville Wood	8 E	634/2558	21
York Road	300m SSW of Casino Point	8 E	630/2555	18
Yorkshire Trench	700m NNE of Beaucourt	7 W	625+2565	35
York Street	400m SE of Casino Point	8 E	630/2554	18
Young Street	S of Gommecourt Park	7 W	621/2570	35
Ypres Street	250m N of Hamel	7 W	623/2563	35
Yussuf Street	S of Gommecourt Park	7 W	621/2570	35
Z Salient	N of Gommecourt Wood	7 W	622/2572	35
Z, The	Just N of Gommecourt Wood	7 W	621/2571	35
ZZ Trench	E of Waterlot Farm	8 E	634/2558	25
Zähringer Graben	500m E of Lochnagar crater	8 E	626/2557	16
Zenith Trench	1400m N of Lesboeufs	7 E	637+/2562	31
Zero Avenue	W of Pozières	7 E	627/2560	22
Zigzag Trench	Just SW of Mouquet Farm	7 E	626/2561	22
Zinc Trench	900m W of Mametz Halt	8 E	627/2555	18
Zitzewitz Graben	NE from Rossignol Wood	7 W	624/2571	35
Zollern Graben East	see Zollern Trench East	7 E	672+/6562	29
Zollern Graben West	see Zollern Trench West	7 W	625+/2562	34
Zollern Redoubt	see Goat Redoubt	7 E	6269/25627	34
Zollern Trench East	W of Courcelette	7 E	627+/2562	29
Zollern Trench West	NE from Thiepval	7 W	625+/2562	34

SITES SITUATED BEHIND THE BRITISH FRONT LINE

The following list gives details of sites behind the British Front Line. To some extent, the word 'behind' is a misnomer, many of these sites being within range of the German artillery and subjected to regular barrages, primarily to prevent reinforcements and supplies moving forward. The hostile barrages caused many casualties among the sick, wounded and other men carrying out their normal duties. At Sailly-Laurette, nestled on the banks of the Somme some eleven kilometres south-south-west of Albert, German artillery shelled the canal adjacent to the River Somme with the aim of hindering the transportation of supplies via the canal. The enemy barrages were such that extensive repairs had to be carried out to the canal wall and today a plaque can be seen to mark the completion of the repairs. Sailly was also a rest centre, Wilfred Owen spending some time in this normally peaceful hamlet. He writes in one of his poems of the barge disappearing slowly round the bend as it gently makes its way westwards. It is a tranquil sight and it is particularly sad to hear of the death of wounded soldiers who, after having been carried to the rear by stretcher-bearers, often under fire, met death in one of the many sites behind the front lines. In order to be in what could be considered a safe zone, it was necessary to travel many kilometres to the west. For troops withdrawn from the front line for a short rest, it was not usually possible to either transport them well to the rear or, reasonably ask them to walk. The list will serve more as a guide to the geographical location of the villages rather than a definitive statement that there was little risk to the occupant.

ALPHABETICAL LIST OF SITES/PLACES SITUATED GENERALLY BEHIND THE BRITISH FRONT LINE

Site/Place	Location	Map	MapRef
Acheux-en-Amiénois	8.8km W of Beaumont Hamel	7 W	613+/2563+
Albert	3600m S of Aveluy Wood	8 W	621+/2555+
Auchonvillers	1.9km W Beaumont Hamel	7 W	620+/2564+
Authuille	4.5km S of Beaumont Hamel	7 W	623/2560
Aveluy	2.4km N of Albert	8 W	622+/2558
Bailleulmont	8km NNW of Foncquevillers	7 W	619/2579+
Baizieux	9km WSW of Albert	8 W	613/2555
Bayencourt	4.4km WSW of Foncquevillers	7 W	616+/2570
Bayonvillers	16km S of Albert	8 W	620/2540
Beauquesne GHQ	20km NW of Albert	52	Fold 8
Beaussart	5.3km W of Beaumont Hamel	7 W	617/2565
Bécordel-Bécourt	2.8km ESE of Albert	8 W	625/2554
Bécourt	2.8km E of Albert	8 W	625/2556
Bécourt Wood (Fr. Bois Planté)	3km E of Albert	8 W	625/2555
Bellevue Farm	Southern tip of Albert	8 W	622/2554
Berles-au-Bois	5200m N of Foncquevillers	7 W	620+/2577+
Bertrancourt	7km WNW of Beaumont Hamel	7 W	616/2566
Bienvillers au Bois	4.3km NNW of Gommecourt	7 W	619+/2575
Bois Brulé	900 WSW of Hardecourt aux Bois	8 E	633/2554
Bois Caffet	200m SW of Carnoy	8 E	629/2554
Bois Carré	900m SSW of Hardecourt aux Bois	8 E	634/2554
Bois Choque	1.7km SSE of Carnoy	8 E	630/2552
Bois Dauvillers	SW of Mailly-Maillet	7 W	617+/2563+
Bois Favier	900m W of Hardecourt aux Bois	8 E	633/2554+
Bois Futaie	NW of Englebelmer	7 W	618/2563+
Bois Hattecourt	7.5km NW of Foncquevillers	7 W	614+/2576+
Bois Hédauville	SE of Hédauville	7 W	617/2560
Bois St. Gauchy	600m SE of Carnoy	8 E	630/2553
Bois Vaquette	4.5km SW of Auchonvillers	7 W	617/2562
Bois d'Auchonvillers	1km NNW of Auchonvillers	7 W	620/2565+
Bois d'Aveluy	S of Mesnil-Martinsart	7 W	622+/2560+
Bois d'Engrement	1km S of Fricourt	8 E	627/2554+
Bois d'en Haut	600m S of Hardecourt aux Bois	8 E	634/2554

Site/Place	Location	Map	MapRef
Bois de Billon	1.5km S of Carnoy	8 E	629+/2552+
Bois de Coigneux	7km NW of Maillet Mailly	7 W	614+/2569+
Bois de Sailly	1000m W of Sailly au Bois	7 W	617/2569
Bois des Annelles	8km NW of Foncquevillers	7 W	615/2577+
Bois des Capelles	1.5km SSW of Mailly Maillet	7 W	618/2563
Bois du Bus	Just N of Bus les Artois	7 W	614/2567+
Bois du Fay	6.5km NW of Maillet Mailly	7 W	613/2567
Bosquet des Hérissons	7km NNW of Foncquevillers	7 W	618/2578
Bosquet, le	Just S of Colincamps	7 W	618/2566
Bouzincourt	3.4km NW of Albert	8 W	619/2558+
Bray-sur-Somme	6.4km S Fricourt	8 E	627/2549
Bresle	6.8km WSW of Albert	8 W	615+/2554
Buire-sur-Ancre	6.1km SW of Albert	8 W	618/2552
Bus-les-Artois	8.5km WNW of Beaumont Hamel	7 W	614/2567
Caffet Wood	500m SW of Carnoy	8 E	629/2554
Cappy village	SE of Bray sur Somme	8 E	630/2547+
Carnoy	SE of Fricourt	8 E	630+/2554
Cerisy	10.5km S of Albert	8 W	621/2545
Chipilly village	1000m E of Cerisy	8 W	622/2545
Coigneux	6km WSW of Foncquevillers	7 W	615/2570
Colincamps	4.3km NW of Beaumont Hamel	7 W	618+/2567
Couin	7.3km WSW of Foncquevillers	7 W	613+/2570+
Courcelles au Bois	5.7km NW of Beaumont Hamel	7 W	617/2567
Dernancourt	3.4km SSW of Albert	8 W	621/2553
Englebelmer	4.5km SW of Beaumont Hamel	7 W	619/2562
Etinehem	8.8km SSE of Albert	8 W	625/2547+
Farm, Bronfay	2km SSW of Carnoy	8 E	629/2552
Farm, Pierrard	2km SW of Puisieux	7 W	623/2568
Farm, de la Haie	2.9km WSW of Foncquevillers	7 W	618/2571
Farm, du Billon	1.4km S of Carnoy	8 E	629/2552
Farm, du Bois de Branlé	600m E of Colincamps	7 W	619/2567
Farm, du Bois de Quesnoy	3.5km E of Hannescamps	7 W	625/2574
Farm, la Fabrique	2km ESE of Colincamps	7 W	620/2566
Farm, la Signy	2.3km W of Serre	7 W	621/2567
Foncquevillers	1.3km NW of Gommecourt	7 W	620+/2572+
Forceville	7.5km WSW of Beaumont Hamel	7 W	615+/2562
Fort Anley	S of Auchonvillers	7 W	621/2564
Fort Briggs	NW of Serre	7 W	622/2568
Fort Southdown (Wagram)	NW of Serre	7 W	621/2568
Fort Sussex	NW of Serre	7 W	621/2568
Garenne Blanche	1.6 km SSW of Maricourt	8 E	631/2552
Garenne de Maricourt	2600m SSW of Maricourt	8 E	631/2551
Garenne de la Grosse Tête	2200m SW of Maricourt	8 E	631/2552
Garenne des Malvaux	1.6 km SW of Maricourt	8 E	631/2552
Garenne des Muterlets	1.6 km WSW of Maricourt	8 E	631/2553
Garenne du Moulin	1200m WSW of Maricourt	8 E	631/2553
Garenne du Vicaire	1 km SW of Maricourt	8 E	631/2553
Gaudiempré	8km WNW of Foncquevillers	7 W	613+/2575+
Hamelet	13.8km SW of Albert	8 W	613/2545
Hannescamps	3km N of Gommecourt	7 W	621/2574
Happy Valley	Opp. Peronne Rd. Cem. at Maricourt	8 E	631+/2553
Hardecourt-aux-Bois	3000m SE of Montauban (Fr. Sector)	8 E	634/2554+
Hébuterne	3.2km NW of Serre	7 W	621/2569+
Hédauville	7.5km SW of Beaumont Hamel	7 W	616/2560+
Heilly	9.55km SW of Albert	8 W	614/2550
Hénu	7.3km W of Foncquevillers	7 W	613/2573
Hénencourt	6km W of Albert	8 W	616/2556
Hubercamps	7.2km NW of Gommecourt	7 W	616+/2576
L'Arbret	8.6km NW of Foncquevillers	7 W	615/2579
La Cauchie	7km NW of Foncquevillers	7 W	617/2578
La Guerre Wood	500m S of Carnoy	8 E	630/2553
La Herlière	8km NW of Foncquevillers	7 W	615+/2579
La Prée Wood	600m SE of Carnoy	8 E	630/2553

Site/Place	Location	Map	MapRef
Lamotte-Warfusée	5000m E of Villers-Bretonneux	8 W	618+/2541
Lancashire Dump	On site of Aveluy Wood Brit. Cem.	7 W	623/2561
Laviéville	5.4km WSW of Albert	8 W	617/2554+
Le Hamel	13km SW of Albert	8 W	616+/2544+
Ludgate Circus	1840m W of Carnoy	8 E	628/2554
Mailly-Maillet	3.8km W of Beaumont Hamel	7 W	618+/2564+
Maple Redoubt	1080m WSW of Mansel Copse	8 E	627/2554
Maricourt village	E of Carnoy	8 E	631+/2553+
Martinsart	5km SSW of Beaumont Hamel	7 W	621/2560
Méaulte	2.6km S of Albert	8 W	622+/2553+
Méricourt l'Abbé	8.4km SW of Albert	8 W	616/2550
Méricourt sur Somme	10.5km SSE of Albert	8 W	624/2545
Mesnil Ridge	N of Mesnil towards Hamel	7 W	622/2562
Mesnil-Martinsart	3.4km S of Beaumont Hamel	7 W	622/2561+
Millencourt	4.5km W of Albert	8 W	617+/2556
Minden Post	800m SW of Carnoy	8 E	629/2553
Monchy-au-Bois	4.6km NNE of Gommecourt	7 W	622+/2575+
Montreuil	General GHQ before Somme offensive	Mich	52 fold8
Morcourt	13km S of Albert	8 W	622/2543+
Morlancourt	6km S of Albert	8 W	620+/2550
Pommier	6km NW of Gommecourt	7 W	618+/2576
Querrieu, Château de, GHQ	17km towards Amiens from Albert	Mich	52 Pli 8
Ribemont sur Ancre	7.8km SW of Albert	8 W	616/2551
Rostrum Trench	N of Auchonvillers	7 W	621/2565
Sailly-Laurette	10.6km SSW of Albert	8 W	619/2546
Sailly-au-Bois	6.1km NW of Beaumont Hamel	7 W	618/2569
Sailly-le-Sec	11km SW of Albert	8 W	617/2547
Saulty	10km NW of Foncquevillers	7 W	613/2579+
Senlis-le-Sec	5.8km NW Albert	8 W	617/2558+
Souastre	4.7km W of Foncquevillers	7 W	615+/2572+
St. Amand	6.8km WNW of Gommecourt	7 W	615+/2574
Sugar Factory	1600m N of Auchonvillers	7 W	620/2566
Treux	6.7km SW of Albert	8 W	618/2551
Vaire-sous-Corbie	12.5km SW of Albert	8 W	614+/2546
Varennes	9.5km WSW of Beaumont Hamel	7 W	613+/2561
Vaux sur Somme	12km SW of Albert	8 W	615/2547
Ville sur Ancre	5.5km SSW of Albert	8 W	619/2551
Villers-Bretonneux	19km SW of Albert	8 W	613/2540+
Vitermont	3.8km SW of Beaumont Hamel	7 W	619/2563
Warloy-Baillon	9km WNW of Albert	8 W	613/2556+

THE ATTACKS

It may be that the soldier or battalion being researched may have taken part in several attacks. Indeed a number of battalions were engaged at seven different objectives and the risk to the soldier of being wounded or meeting death was high. There now follows a description of the action of the infantry attacks presented in numerical order for each objective. Some have been amalgamated to achieve better continuity. Maps showing the main trenches will help guide the reader through the action. Limitations of space prevented a full and thorough description, the aim of this guide being to present a summary of the main events and to refer to as many units as is possible. Even in summarised form, many pages are necessary to describe units engaged in the protracted struggles to capture certain objectives such as Pozières, Mouquet Farm, Delville Wood, High Wood and the November attack north of the Ancre.

OBJECTIVE 15
FRICOURT

On the 1st July the 21st Division and the 22nd Brigade from 7th Division undertook a pincer movement to pinch out the village of Fricourt. The 50th Brigade (17th Div.), attached to the 21st Division, attacked the village from the west. The cost to the 21st Division and the 50th Brigade was a staggering 5,411 casualties including 710 casualties in the 10/West Yorks, this Battalion virtually ceased to exist. The 22nd Brigade's losses were low. In these circumstances it is not surprising there was some confusion to the west and north of Fricourt. Relief was urgent. In the late afternoon the 10/Sherwood Foresters and 7/Border (51st Brig.) moved north to form the right flank of 21st Division in preparation for the planned attack on the 2nd July. The remaining two battalions of 51st Brigade, the 7/Lincoln and 8/South Staffs, moved up a few hours later to take over the positions of the depleted West Yorks of 50th Brigade.

An artillery barrage was planned for 11am and zero hour set for 12.15pm. In the early hours of 2nd July, patrols by elements of the 6/Dorsets and S. Staffs reported little activity in the village, the Staffords bringing in a hundred or so prisoners. These reports were supported by the 2/Royal Irish (22nd Brig. 7th Div.) which had also entered Fricourt unopposed around midnight. Maj.-Gen. Pilcher,

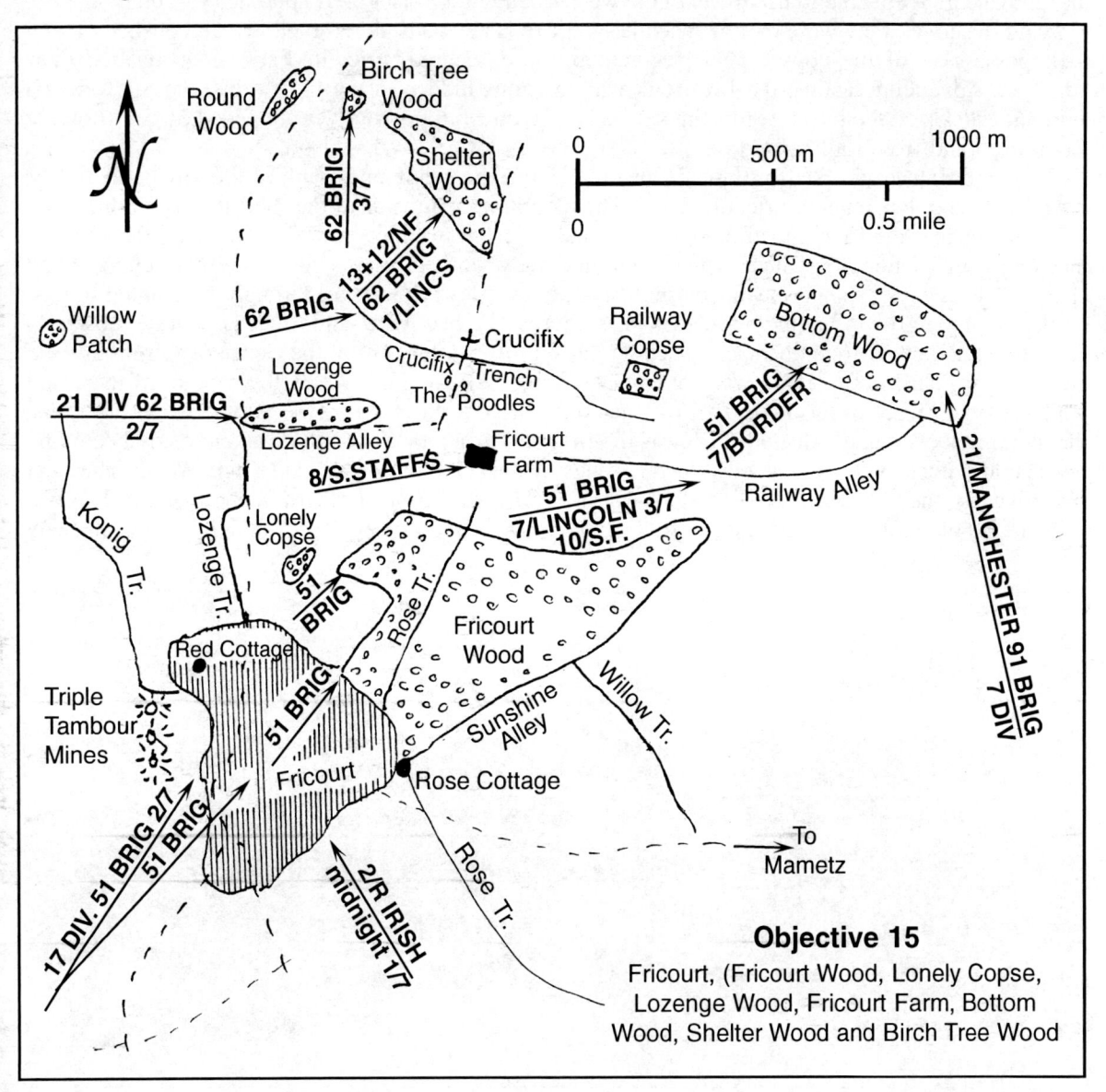

Objective 15
Fricourt, (Fricourt Wood, Lonely Copse, Lozenge Wood, Fricourt Farm, Bottom Wood, Shelter Wood and Birch Tree Wood

commander of 17th Division ordered his 51st Brigade to occupy Fricourt before the scheduled bombardment at 11am but the chain of command and communication did not lend itself to a quick change of plan and Fricourt was not entered until 12.15pm as originally planned.

Lieutenant-General Horne, commander of XV Corps was unaware that the pincer movement on Fricourt on the 1st July had been successful and had indeed, induced General Von Stein's 111th RIR to begin its tactical withdrawal to the next line of defence from midnight onwards.

Before the barrage, stretcher bearers were able to bring in many of the wounded from the previous day. More patrols were sent forward to reconnoitre the terrain. Some Germans, who had not received orders to withdraw, were taken prisoner. To cover the withdrawal, Stein had left a rear-guard which was to be the cause of some small but bitter engagements. Following the barrage, the 8/South Staffs, in the path of the heros of the previous day, approached Fricourt from the west, clearing its way forward by bombing German dugouts and trenches and eventually arriving at the northern end of the village. Fricourt had been well fortified, having many strongholds in the deep cellars. Clearing a village which has been the subject of a long barrage is not an easy task. The heaps of rubble and ruins make good emplacements for machine-guns. But methodically the Staffords cleared the north of the village and then turned north towards Lonely Copse, their next objective. Leaving the village, the Staffords found themselves in open ground with the still untaken Fricourt Wood to the East. However, the Midlanders were able to use the extreme western edge to protect their approach to Lonely Copse. Leaving the cover, they were met by machine-gun fire. They paused, returned fire and pushed on and took possession of the copse. Their second objective secured, Fricourt Farm now awaited their attention. Advancing against the fire from machine-guns hidden in the farm ruins, the Staffords (i) stormed and seized the farm. At this time any advance on Bottom Wood was impeded by fire from the wood itself and from Railway Copse.

On the right of the Stafford, the 7/Lincoln cleared the centre and south of the village including the château and then entered Fricourt Wood. The natural impediments of the shell-torn wood hindered its clearance and the Lincoln suffered some casualties from machine-gun fire. Silencing the German machine-guns by bombing attacks, the Battalion slowly worked its way to the northern edge. Wood clearance is always a dangerous operation. It is difficult to move forward through the tangled trunks, branches, undergrowth including nettles and thorns, barbed wire and trenches, these obstacles providing good cover for snipers and machine-guns. Fortunately, most of the Germans were retreating and the rear guard in the wood was not overly strong, just enough to delay the capture of the wood while the main force of the 111th RIR evacuated the area. Reinforced by the 10/Sherwood Foresters, the wood was eventually cleared. This was the first of a group of woods to be taken, many becoming household names such as Mametz Wood, Trônes Wood, Delville Wood and High Wood. The next objective for the Lincolns and Foresters was Bottom Wood (the northern section of the Bois de Fricourt Est on your IGN Map) but fire from Crucifix Trench, Railway Copse and Alley prevented any

Fricourt Farm before its total destruction. [B. Maes]

immediate approach to the wood.

To the north or left of the 51st Brigade was the 21st Division whose objectives included Lozenge Wood (Bois de la Ferme), Lozenge Alley, The Poodles just north of Fricourt Farm and Shelter Wood (Bois de Fricourt Ouest). Brig.-Gen. Rawling's 62nd Brigade replaced the 64th. While waiting for zero hour, many of the wounded were brought in. Patrols were sent out towards Fricourt Farm (Ferme du Bois) passing behind Lozenge Wood. The 10/Green Howards passed to the north of Fricourt Farm as far as The Poodles, a small clump of trees, and made contact with the 51st Brigade. The centre and western part of Crucifix Trench was secured. After a somewhat lethargic start, the 2nd July had ended well - all objectives had been taken. Casualties in the 62nd Brigade were less than a thousand, about one third of the total German casualties.

The next morning the whole of 51st Brigade led by the 7/Border and closely supported by the previously engaged Lincolns, Staffords and Foresters attacked Railway Alley. Under constant fire and against determined German resistance, a protracted struggle of over two hours was necessary before Railway Alley could be taken. During this period a few Borders managed to get to the western edge of Bottom Wood to find they were all but surrounded. A potentially desperate situation was saved when the 21st Division captured Shelter Wood, 700 meters to the west. This allowed the remainder of the Borders to secure the western corner of Bottom Wood. The opposite corner was taken by the 21/Manchester (91st Brig. 7th Div.) without any resistance.

The capture of Shelter Wood by the 62nd Brigade was not an easy task. Although supported by the Brigade's machine-gunners, the leading battalion, the 1/Lincoln, had much difficulty in approaching the wood. Fire was coming directly from Shelter Wood and also from the smaller triangular-shaped Birch Tree Wood, a few meters to the north-west. The 12/Northumberland Fusiliers (NF) in support were quickly sent forward to reinforce the attack. Against stubborn resistance Birch Tree Wood was taken but Shelter Wood was still inviolate. The 13/NF of the same brigade, supported by Stokes mortars, pinched out the wood allowing the three battalions to take possession with relatively little further loss. Good work by Lewis gun teams repulsed a counter attack on Shelter Wood and Bottom Wood. About 800 Germans surrendered.

The British line was looking much better now. Units from the 7th, 17th and 21st Divisions had formed a continuous line in an advanced position. There was much activity consolidating the new line. It was also necessary to make a defensive flank on the extreme left, that is to say, between the 21st Division at Birch Tree and Shelter Woods and the 34th Division at Round Wood. La Boisselle had yet to fall. The awful losses on the west of Fricourt on the 1st July had been avenged.

(i) In this action, Major R.G. Raper of the 8/South Staffordshire Regiment was killed and buried in a private grave across from the Bray Road British Cemetery and remained undisturbed until 1965. His family then agreed to the reinternment of Major Raper in the adjacent cemetery. A street in Fricourt now bears his name along with that of the town of Ipswich which adopted Fricourt after the war. The Raper family also made a donation towards the rebuilding of the church where there are some interesting memorials.

OBJECTIVE 16
LA BOISSELLE, SAUSAGE VALLEY

Major-General Ingouville-Williams, commanding the 34th Division, was shortly to learn the full extent of his Division's casualties during its heroic but abortive attack on the village of La Boisselle on the 1st July 1916. The first reports were alarming but the final total amounted to a staggering 6,380(i) By the late afternoon on the 1st July General Rawlinson was receiving conflicting reports from the whole front. Well aware that the fall of La Boisselle, lying on the main Albert to Bapaume road, would open the way to Pozières and beyond, Rawlinson ordered the 19th Division (Maj.-Gen. Bridges) to attack the village at 10.30pm the same evening.

However, the 57th and 58th Brigades could not get forward, the ground being littered with dead and wounded. All communication trenches were crowded and blocked with walking wounded, stretcher-bearers and units endeavouring to move supplies up to the front line. The whole area was in a state of disorder and confusion. During the night, the 58th Brigade managed to get forward and the 9/Cheshire relieved units of the 34th Division at the Lochnagar Crater and a few hours later the

6/Wiltshire and 9/RWF made a dawn attack on the village. German resistance was still strong and little progress was made. The 57th Brigade had much difficulty in getting forward and the attack was re-scheduled for 4pm. Inactivity on the German front for a short time allowed supplies to be brought up. A diversionary barrage was placed on Ovillers where there was to be no attack on the 2nd July, successfully diverting German attention away from La Boisselle, at least for the time being. Following the thirty-minute barrage, the 58th Brigade moved forward, the Wiltshires and Welsh Fusiliers attacking the village for the second time. To the right the 9/Cheshire left the crater fighting their way north, first clearing the Kaufmann Graben and then, 200 meters further on, the Alte Jäger Stellung close to the La Boisselle/Contalmaison road. The village had been reduced to a pile of rubble by the long bombardment during the last week of June but these same ruins provided excellent niches for the placement of machine guns and many cellars were still intact. Clearance had to be house by house, street by street. When the supply of Mills bombs had dried up, it was then the work of the rifle and bayonet. By late evening the southern and western parts of the village had been secured and consolidated. Meanwhile, the 7/East Lancs of 56th Brigade made a determined bombing attack on Sausage Redoubt, a few hundred meters to the east of the crater. The redoubt and trenches beyond were secured and some prisoners sent back. The day had ended well, Brig.-Gen. Dowell's 58th Brigade now in control of well over half the village.

The 57th Brigade had been brought forward in the early hours of the 3rd July and the 8/North Staffs, 5/SWB, the divisional pioneers, with the 10/Worcester on the left flank, attacked with bomb, rifle and bayonet from the north, methodically clearing the north-western part of the village. While the Worcesters were advancing north-east towards the German third line, the 10/Royal Warwick and 8/Gloucester joined the fray. Linking up with 58th Brigade which had attacked from the south, both brigades continued the systematic clearance of the village, which involved further close-quarter fighting. It was slow and dangerous work. The men stuck to their task, the village eventually taken and flares sent up to signal the good news. Seeing the flares, the Germans laid a heavy barrage over the area followed by a strong counter attack. Coming from the east, they re-entered the village from the Contalmaison road forcing the brigade back towards the centre of the village. During this action the 57th Brigade was to earn the first of two Victoria Crosses. The first, awarded to Private G.T. Turrall of the 10/Worcester, for refusing to leave his severely-wounded officer, Lt. Jennings. Turrall dragged him to a shell crater, dressed his wounds and fought off numerous German bombing attacks. During the hours of darkness, he carried his officer back to the British front line and then immediately rejoined

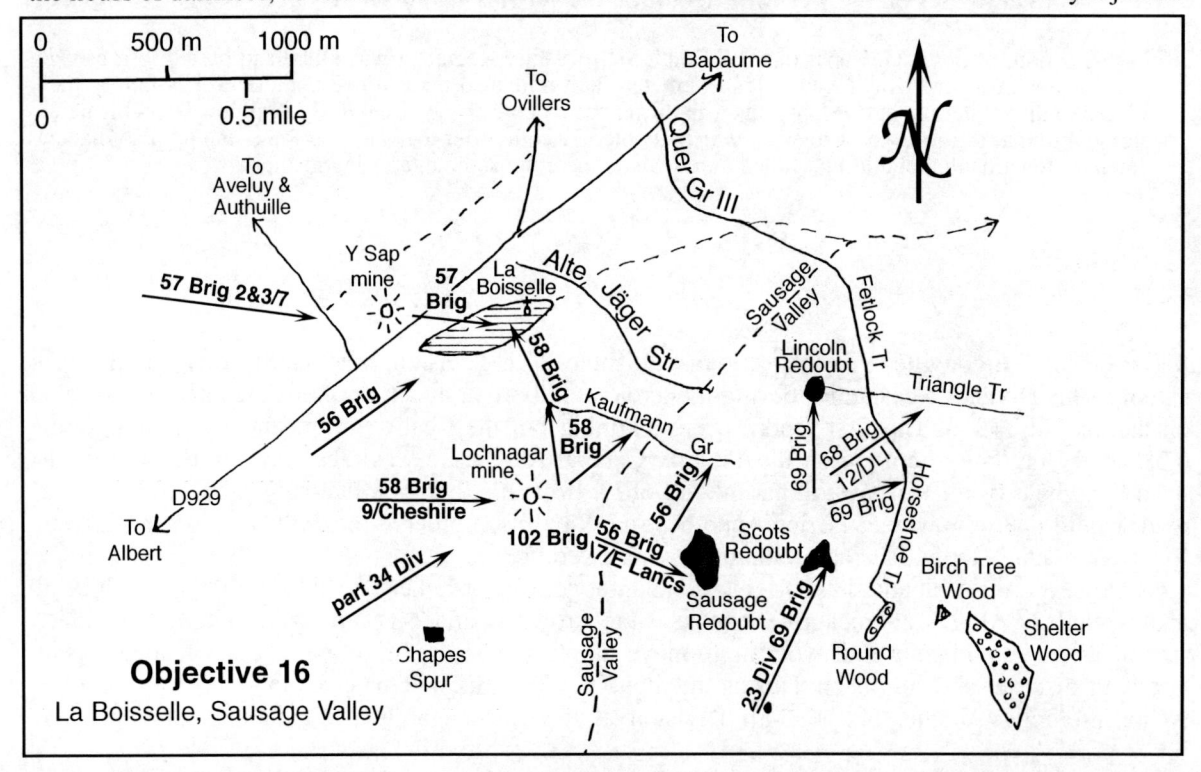

Objective 16
La Boisselle, Sausage Valley

The soldier standing at the bottom adds scale to the size of the Lochnagar Crater [Michelin]

his unit. Unfortunately, Lt. Jennings died a little later but not before giving details of Turrall's courageous actions. The second Cross was awarded to Lt. Col.A. Carton de Wiart, commander of the 8/Gloucester for conspicuous courage and initiative over many hours. He was like a character straight from the pages of a boy's adventure magazine, having lost an eye in Somaliland and a hand at Ypres and it is reported he withdrew the pin from grenades with his teeth. Bridges, commander of the 19th Division, had put de Wiart in charge of the defence of La Boisselle following the death of Lt. Col. Royston-Piggott (10/Worcester) and Major C. Wedgewood (8/North Stafford). Lt.-Col. R.M. Heath (10/Royal Warwicks) was wounded and out of the fight. De Wiart took charge and led the fight in person.

The Germans pushed the 58th Brigade back to their front line. They could not afford to let go now - the fighting became desperate. The 9/RWF, with feelings running high at the loss of the ground so desperately gained the day before, threw the Germans back and the enemy had to content itself with control of its line running by the church at the north-eastern extremity of the village.

The remnants of the 34th Division endeavoured to bomb their way forward to join up with the 19th Division but were unable to do so and the sorely-tried 34th Division was relieved by Maj.-Gen. Babington's 23rd Division under difficult conditions during the night of the 3/4th July when rain turned the white chalk dust into a creamy sticky substance which clung to boots, uniforms and everything with which it came into contact.(ii)

It was evident the Germans did not intend to let go of their hold in the village. The 56th Brigade, (19th Div.) led by the 7/King's Own Royal Lancaster Regiment, aided by machine-guns and mortars advanced against a most stubborn defence. By just after noon, only the northern extremity of the village remained in enemy hands.

Having taken over from 34th Division, the 23rd Division's 69th Brigade attacked Horseshoe Trench which had caused so many problems on the 1st July. The 9/Green Howards and the 11/West Yorks on the left, the 10/Duke of Wellington's on the right, advanced before 6.00am and a most bitter struggle continued for over three hours, during which time some ground was gained only to be lost following a resolute counter-attack. The enemy pushed forward, the struggle was desperate. The 8/Green Howards and remainder of the Wellington's were sent forward to

British tombs in Lochnagar Crater. After the War all human remains were removed from the huge crater and reinterred in nearby British Cemeteries but there are still many hundreds of missing soldiers in this small area [Michelin]

assist and eventually, in the evening, most of Horseshoe Trench was secured along with part of Lincoln Redoubt. 2/Lt. D.S. Bell of the 9/Green Howards accompanied by Corporal Colwill and Private Batey, crawled forward up a communication trench, advanced under fire over open ground to attack an enemy machine-gun crew. Bell shot the gunner and the post was destroyed by grenades. For this action, Bell received the Victoria Cross only to be killed a few days later in front of Contalmaison. (iii)

Meanwhile 19th Division's 56th Brigade, plus units from the 1/Sherwood Foresters (24th Brig. 8th Div.) were still trying to clear the re-entrant near the Bapaume road but their efforts were thwarted at each attempt, the 7/East Lancs having to withdraw to its starting position. The 7/Loyal North Lancs was ordered to retake the lost ground and also pick up a machine-gun which had been left in a forward position. The battalion's commander, Lt. T.O.L. Wilkinson leapt forward with two men and on reaching the gun, put it to good use, holding the position until a bayonet charge by C Coy. secured the position. Later, Wilkinson again rose above the call of duty during a German bombing attack. His day's work not yet finished, he was killed while trying to bring in a wounded man. He was awarded the Victoria Cross, the fourth between the 2nd and 5th July during operations at La Boisselle.

During the night, the 57th Brigade was relieved by Maj.-Gen. Scott's 12th Division. The 19th Division continued its work clearing Horseshoe Trench, still occupied at the northern end. An attack across open ground by the 7/East Lancs (56th Brig.) succeeded where the previous bombing raid had failed. Their objective achieved, they beat off three separate counter attacks.

The 69th Brigade was relieved by the 68th and the 12/DLI secured, without opposition, Triangle Trench which runs east from Horseshoe Trench. La Boisselle, Sausage Valley and its neighbouring network of trenches were now secured.

(i) O.H. Vol. 1, page 391
(ii) Visitors beware! Even after the passing of hundreds of thousands of visitors' feet around the compacted lip of the Lochnagar crater, the slightest fall of rain will turn the normally safe tour of the crater into a skating rink! If you fall, the removal of the white stains on your clothes will demand much time and patience. Spare a thought for those who fought in these conditions.
(iii) At the time of publication of this book, a special memorial is being erected to honour Lt. Bell near the place where he was originally buried.

OBJECTIVE 17
OVILLERS, MASH VALLEY AND TRENCHES NORTH TOWARDS AUTHUILLE WOOD

Ovillers-la-Boisselle, referred to simply as Ovillers in this guide, is situated just to the north of the Albert/Bapaume road and a little over two kilometers south-west of Pozières. The village had been the objective of Maj.-Gen. Hudson's 8th Division during the general attack on the 1st July. Having sustained casualties in excess of five thousand without seriously breaching the German defence system, the division was in no condition to continue the attack and was relieved during the night of 1/2nd July by Maj.-Gen. Scott's 12th Division, the artillery of the 8th Division remaining for a further three days. It is difficult to imagine the thoughts of the relieving units as they witnessed the tragic scene of the evacuation of the wounded. What sense of foreboding as they heard the horrific reports of the hell of no man's land, knowing full well they had to take up the attack and fate alone would decide if they lived or died. The logistical difficulties of the relief were numerous - congested trenches making the normal up/down system of movement impossible. Thousands of dead and wounded lying everywhere and so many more trying to crawl back to their own front line. Exhausted stretcher-bearers worked to the limits of human endurance. Carrying parties had toiled through the night to bring up supplies of food, water and munitions needed to continue the struggle. By dawn, the huge task was completed, a remarkable feat. Orders had been received from Corps headquarters for a dawn attack on Ovillers, but the confusion and congestion made it impossible for the men to get forward. The attack was postponed for twenty-four hours, the remainder of the day being spent clearing the wounded and digging new assembly and communication trenches. This was essential work, the 800 meter width of no man's land being largely responsible for the heavy casualties on the 1st July. Under the cover of night, work continued and assembly trenches were completed to within about 500 meters of the German front line.

The barrage commenced on time at 2.15am. The 5/Royal Berkshire and 7/Suffolk of Brig.-Gen. Solly-Flood's 35th Brigade were on the right flank not far from the 19th Division's left wing at La Boisselle. The 6/Queen's and 6/Royal West Kent, both of 37th Brigade (Brig.-Gen. Cator) made up the left wing of the Ovillers attack. One battalion from the 36th Brigade was detailed to cover the extreme left flank with a smoke screen. The barrage attracted return fire from the Germans who had also taken advantage of the respite on the 2nd July to bring up reinforcements and generally prepare to receive the next attack. Their bombardment caused some casualties in the communication and assembly trenches. Ten minutes before zero at 3.15am, the men left the advanced trenches and crept forward into the unknown. At least, these men were in front of the enemy barrage. At 3.15am they dashed forward, the British artillery now lifting to a strict timetable to destroy the German defences in the centre of the village and around the church. Half an hour later it would extend to shell the eastern end of the village. A further 2,000 meters to the north-east, across open fields, lay the village of Pozières.

Since the first day of the Somme, from Gommecourt in the north down to Ovillers, the men had advanced uphill and eastwards facing the rising sun. Here today, the attack was taking place across the slighest of rises well before sunrise. The initial advance from the assembly trenches before zero hour had given the leading battalions the advantage of surprise. Although hindered by innumerable shell holes, the men found the German wire reasonably well cut, allowing good progress towards the German front line. As soon as the Germans realised the British barrage was increasing its range, they opened fire not only from the western edge of Ovillers, but from La Boisselle which had not yet been taken and from German strongpoints south of Thiepval. The X Corps, which had planned a simultaneous attack at Thiepval, had been obliged to postpone its assault due to assembly problems of a similar nature to those of the 12th Division on the 2nd July. The result of this was that the German defences south of Thiepval were free to concentrate their fire on the left flank of the attack on Ovillers.

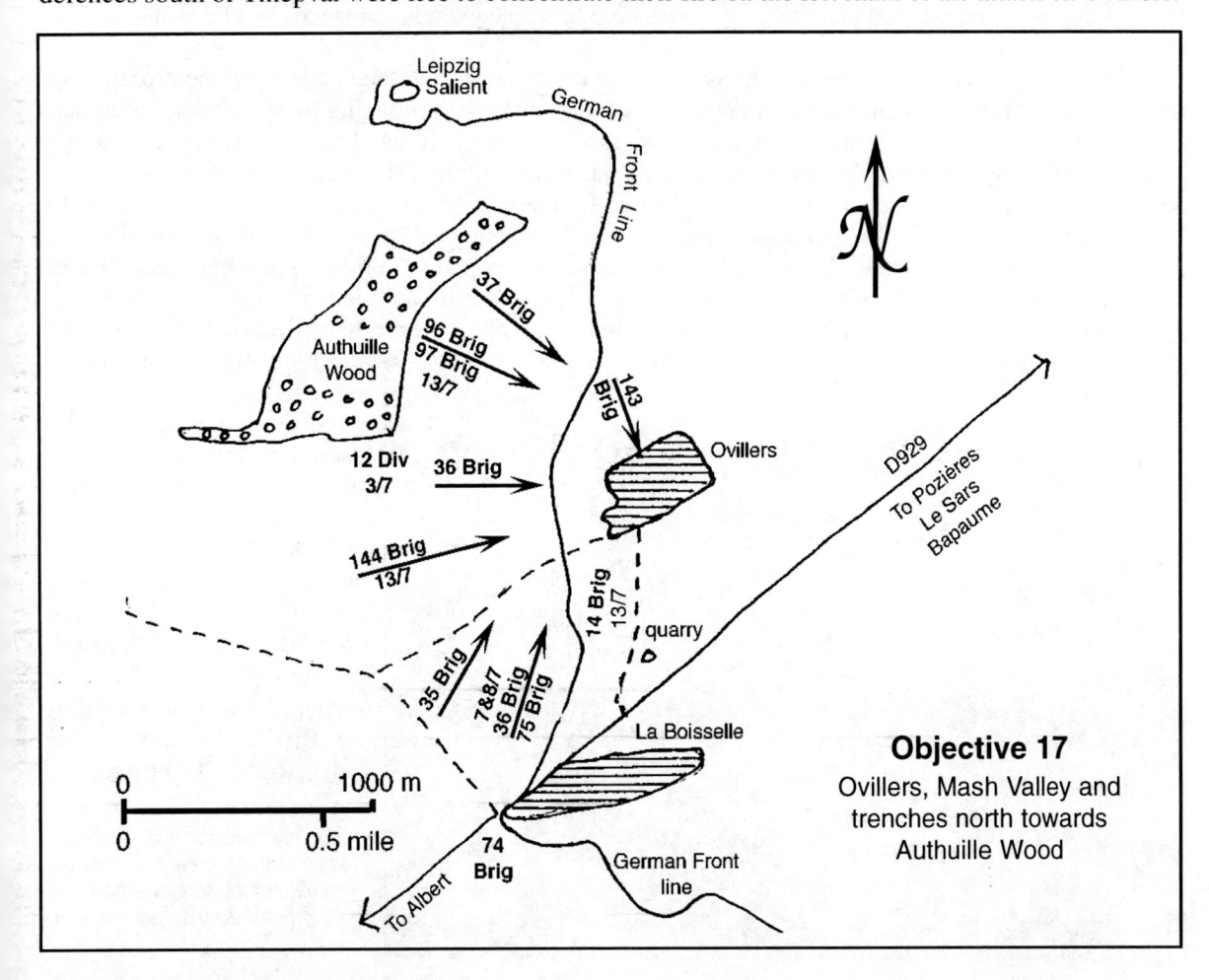

Objective 17
Ovillers, Mash Valley and trenches north towards Authuille Wood

In spite of this, the leading waves of the 6/Queen's arrived at the first German trench. Here, many were cut down by machine-gun fire. With the enemy now on the alert the following waves could not get across no man's land and were wounded long before they could reach the enemy line. The Kents, being closest to the fire from Thiepval, were suffering huge casualties, mostly at the wire, which at their end of the line, was not well cut. The attack was quickly losing its momentum. A little to the south, the Berkshires seized the first line of defence, pushed on and passed over the German support line. Finally reaching the outskirts of the village the men saw Germans pouring down the perimeter trench from both flanks and a bitter contest ensued, the Berkshires holding their own until out of bombs. Thereafter, it was close-quarter work with rifle and bayonet.

The Suffolks had veered off course and joined the Berkshires. In the half light of morning the 9/Essex had strayed even further and found itself close to La Boisselle where about two hundred isolated Germans were taken prisoner. Two companies of the 6/Buffs (37th Brig.) followed the Kents in close support. Smoke from British and German artillery made it difficult for the artillery observers to ascertain what was happening. The men who had reached Ovillers were now virtually cut off, the Germans having laid a heavy barrage in no man's land and beyond as well as raking the whole path of approach with machine-gun fire. Realising their position was untenable, the units to the south of the village, were able to retire down the road to a small quarry on the road to La Boisselle. Here they were protected and remained until dark when they returned to the British line. The men of Cator's 37th Brigade in or at the outskirts of Ovillers and units of the 35th Brigade were in a perilous position. Hunted down, almost out of munitions, tired, hungry and thirsty, they were slowly overwhelmed.

In less than six hours the attack, which had started well, had finished with no permanent gain, the casualties for the 3rd July totalling 2,400 officers and men. The sad roll-calls of the 8th Division on the 1st July were repeated forty-eight hours later in the 12th Division. It was bitterly evident that a frontal attack on Ovillers could not succeed without the support and success of divisions on either flank. It has to be mentioned that the artillery barrage had not severely damaged the German defences in and around the village.

Having now made two costly and unsuccessful attacks, the next three days were spent preparing for the next assault. After the first two hot days in July the sky was clouding over, and some rain had fallen making the trenches uncomfortable. The men worked hard in the warm sultry conditions clearing, cleaning and extending the trenches past the former limits and were now within 300 meters of the enemy front line. The next attack was planned for the 7th July, to be preceded by a sixty-minute bombardment at the end of which a smoke screen would be released to cover the advancing infantry. Zero hour was set for 8.00am. Brig.-Gen. Armytage's 74th Brigade (25th Div.) had been attached to the 12th Division and the 13/Cheshire along with the 9/Loyal North Lancs led the attack from the south across Mash Valley. The weather conditions not being favourable, the discharged smoke screen served no useful purpose. Immediately exposed to a hail of machine-gun fire, the leading battalions could get no further than the first trench. The assembled 36th Brigade, waiting for its signal to advance at 8.30am, suffered heavy casualties from the hostile barrage Undaunted, the brigade moved forward close behind the British barrage. After some hard fighting the

A captured trench at Ovillers, hence the provisional firestep. The British soldier mounts guard while his friends sleep as best they can.
[A. Perret]

7/Royal Sussex and 8th & 9/Royal Fusiliers secured the first three lines in Ovillers. Their proximity to the creeping barrage had created an element of surprise and some prisoners were sent back. This was the best result so far at Ovillers. However, the three battalions of 36th Brigade had sustained over 1,400 casualties, their ranks too thinned to hold on to their gains. Receiving no support from the 74th Brigade on the right, the Fusiliers and Sussex men retired to the German second trench where they consolidated. The men were tired but the German prisoners also talked of their fatigue and heavy casualties from the British barrage. Rain in the afternoon did not enhance the situation, the trenches quickly becoming sticky and miserable, creating difficulties both in collecting the wounded and bringing up supplies. The gap between the 36th and 74th Brigades was filled by the 8/South Lancs (75th Brig.) during the night.

During the night of the 7/8th July the 36th Brigade in the Ovillers trenches was reinforced by 7/East Surrey (37th Brig.) and 9/Essex (35th Brig.). It was a hard struggle for these men to get forward through the sticky mud. There was stern physical and fighting work ahead. The men moved forward into the village and secured a further 200 meters. The 74th Brigade moved up from its position of the previous day and, reinforced by the 2/Royal Irish Rifles of the same Brigade, fought and bombed its way forward through the mud-clinging trenches, finally securing the trench which led towards the ruins of the church. In the early evening, 74th Brigade attacked again, the 11/Lancs Fusiliers securing a forward trench. The 2/Royal Irish Rifles and 13/Cheshire were sent up to consolidate. Having been in the Ovillers area since the 2nd July and with casualties of 4,721(i), the weary men were relieved by the Brig.-Gen. Compton's 14th Brigade of 32nd Division, control of the sector now being passed to Maj.-Gen. Rycroft, CO of that division. The 14th Brigade moved into the ruins of Ovillers, constructed strongpoints and took advantage of the deep German dugouts. The 97th Brigade moved up later.

The two Bazentin villages and ridge of the same name lay 5,000 meters to the east of Ovillers and, still further, lay the village of Longueval with Delville Wood nestling on the east side of the village. A major dawn attack was planned in this area on the 14th July. Prior to this attack the British artillery was fully engaged in softening up the German front line and beyond, leaving pratically no artillery to support the final capture of Ovillers. This meant that no major assault could be made to oust the remaining Germans from the village. Between the 9th and 14th July small, piecemeal attacks were made incurring useless heavy casualties. The 11/Borders were ordered to make a night attack on the 9/10th. Not all the men left the trenches, many reporting sick when pressed to go forward. They were received sympathetically by the Medical Officer (MO). These men were not cowards - they had thrown themselves against the Leipzig Redoubt on the 1st July and lost 516 men, they had given their all. Their losses had not been replaced. Rycroft was displeased when informed of the incident and transferred the MO away from the front. A commander's single duty is to win his battle, there is little room for sympathy. There are no court-martials recorded - Rycroft simply got rid of the cause. The 11/Border was to be engaged later in fierce fighting both south and north of the Ancre, its reputation justly intact.

Another night attack on the 13/14th July was undertaken by a host of battalions from 25th, 32nd and 48th Divisions. The 96th and 97th Brigades attacked from the north-west, the 14th from the west, 75th from the south, 7th from the east and south-east. The Worcesters, Warwicks, Dorsets, Borders, Cheshires and other units seized their objectives but could not hold the positions, their losses being too great.

A further assault was made at 2.00am (15th July) but was checked yet again by stubborn German resistance. During the evening, 32nd Division was relieved by Brig.-Gen. Nicholson's 144th Brigade of 48th Division. After midnight, another determined assault was made to encircle the German garrison. Again, a whole array of units was engaged. The 5/Royal Warwick (143rd Brig. attached to 25th Div.) advanced from the north, the 74th Brigade came in from the east and the 144th from the east and south. This time, there was no escape and in the evening, two proud officers and 126 men surrendered to the 11/Lancs Fusiliers and 2/Royal Irish Rifles of 74th Brigade. One of the German officers stated he was down to thirty men, had received no reinforcements and had many wounded. These brave German soldiers had put up a magnificent defence and it is reported their wounded were well looked after by British medical staff and soldiers alike.

(i) O.H. Vol. 2, page 42

OBJECTIVES 18 & 19
CONTALMAISON, QUADRANGLE, CATERPILLAR WOOD, MAMETZ WOOD, FLAT IRON COPSE, BERNAFAY WOOD, TRONES WOOD, MALTZHORN FARM

Objectives 18 and 19 have been grouped together since they constitute the preparatory actions necessary before the dawn attack scheduled for the 14th July. Before this attack could take place, Contalmaison and the above mentioned woods and trenches needed to be secured, including Maltzhorn Farm. (Calvaire de Maltzkorn)

Bernafay Wood was the first of these objectives to be taken. Patrols sent to reconnoitre the wood had established that it was virtually empty. Lt.-Gen. Congreve, GCO XIII Corps sought and obtained permission to occupy the wood. Accordingly, after a short bombardment, the 12/Royal Scots and 6/KOSB of 27th Brigade (9th Div.) entered and took possession of the whole wood with only six casualties. Seventeen German prisoners were captured along with three field and machine-guns. The prisoners stated that an earlier barrage had caused the evacuation of the wood. It was a good start. The occupation of other woods would prove to be an entirely different matter.

MAMETZ WOOD

The 38th (Welsh) Division, not having taken part in the Big Push of the 1st July, moved down from its training area during the night of 1/2nd July from Toutencourt to Acheux, just over five miles west of Beaumont Hamel. They had undergone extensive training, designed to overcome the many obstacles they might meet on the field of battle. However, this did not include any training or advice on how to capture a wood by a frontal attack in which the enemy was firmly entrenched with numerous machine-gun nests guarding the various approaches. After a night at Treux the division relieved the 7th Division in the front line from Caterpillar Wood to Bottom Wood. The former had been occupied in the morning of the 4th July by the 10/Essex (53rd Brig. 18th Div.) while the latter had been secured by units of the 7/Border (17th Div.) and 21/Manchester of the 7th Division.(i)

The following day the 17th Division took over Bottom Wood and the area a little to the north while the 38th occupied Marlborough Wood and the western half of Caterpillar Wood, 350 meters to the south. It is proposed to follow the action of the 38th Division to its conclusion and then turn our attention to the 17th Division's attack on the trenches to the west of Mametz Wood. Brig.-General Evans' 115th Brigade was to lead the assault. The 16/Welsh and 11/South Wales Borderers (SWB) moved forward, taking cover at first from the northern confines of Caterpillar Wood. As soon as they moved into the open they were met with enfilade fire from Flat Iron Copse and Sabot Copse, small areas of woodland a few hundred meters to the north-east of the Hammerhead, so named through its shape as it formed, and still does, the re-entrant on the eastern side of Mametz Wood. Although the advance had been well covered by trench-mortar and machine-gun fire the men had to go to ground. Following further bombardments at 10.15am and 3.15pm, the 10/SWB moved up in support but could not get within 250 meters of the wood. Rain in the afternoon caused a deterioration of ground conditions. The three battalions were ordered to withdraw and two companies of the 17/Royal Welsh Fusiliers (RWF) remained in position to hold the line Marlborough-Caterpillar Wood.

Orders came through to make a lodgement in the southern part of the wood and at 2.00am on the 9th, a platoon of the 14/RWF (113th Brig.) moved forward in the dark. The half-inch of rain during the previous day had turned the communication trenches into a sticky mess. Leaving the trenches, the men tried again to get forward over the open ground but muddy conditions, shell holes filling with water, loose wire and debris brought an end to their attempt.

Following the two failed attacks on Mametz Wood, Major-General Watts, GCO 7th Division, was put in temporary charge of the 38th Division. Watts intended to engage the whole division rather than continue the previous piecemeal attacks. The dawn attack of the 14th July was fast approaching - time was beginning to press. Everyone was aware that the Germans had had ample time to prepare the

Objectives 18 & 19
General map of the area

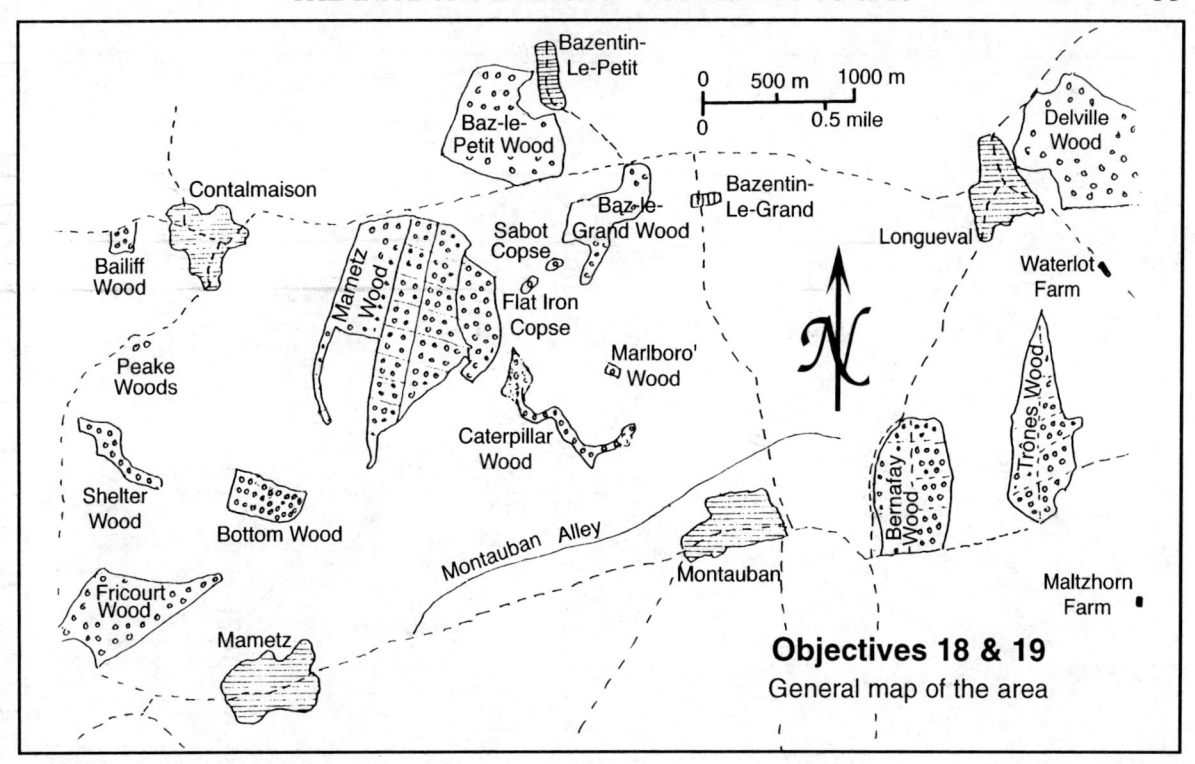

defences and the edges of the wood were now brimming with strongpoints. The attack was preceded by a combined creeping barrage by the 38th and 7th Divisions' artillery units beginning just after 4.00am, assisted by trench mortars. The attack was made from White Trench, Price-Davies' 113th Brigade on the left and Marden's 114th Brigade on the right - their objectives being the narrower southern part of the wood to the first transverse ride and the Hammerhead. Each brigade was allotted a field company of Royal Engineers and a company of the 38th Divisional pioneers, the 19/Welsh. The 13th and 14/Welsh led 114th Brigade's attack - they had further to go than their comrades to the left. Initially on a reverse slope giving good cover, they came into full view of the enemy as they descended Caterpillar Valley and with still 600 meters of open ground to cross. Close on the heels of the creeping barrage they nevertheless gained entry into the wood and pushed on towards the first ride. Here they dug in. These tough men from Swansea and the Rhondda valley were no strangers to pick and shovel work, many being ex miners. But here, under enemy fire, in the heat of battle, digging themselves cover amongst the fallen branches and roots when almost every stroke of the tool would hit some obstacle, it was a trying and dangerous experience. The support battalion, the 15/Welsh attacked the Hammerhead which resisted the most courageous attacks and could not be taken. The C.O. Major Anthony was amongst the many dead. Leading the 113th Brigade's attack, the 16/RWF, having started a little late, was caught by heavy fire from Strip Trench and other German trenches to the west of Mametz Wood. Its commander Lt.-Col. Carden was killed leading his men. Joined by the 14/RWF, they pushed on to their first objective. Some casualties were caused by friendly fire and, communications cut, runners were sent back to seek further support. The reserve battalion, the 15/RWF was sent forward while the 10/Welsh (114th Brig.) took over the area between the two brigades which allowed them to send reinforcements to both wings. About a hundred German prisoners were taken and a further group surrendered to the Royal Engineers engaged in wiring the first objective.

An hour had passed since zero and now, suitably reinforced and consolidated at the first objective and with seemingly little German opposition directly facing them, the two brigades sought permission to continue north through the wood. They could pass well to the west of the German-held Hammerhead. The complicated artillery programme could not be changed easily and the men were ordered to hold until they could move behind the creeping barrage. No one was aware that the Germans were seriously considering an evacuation of the wood. This unexpected respite gave them the opportunity to reorganise their defences.

Objective 18
Mametz Wood & Contalmaison

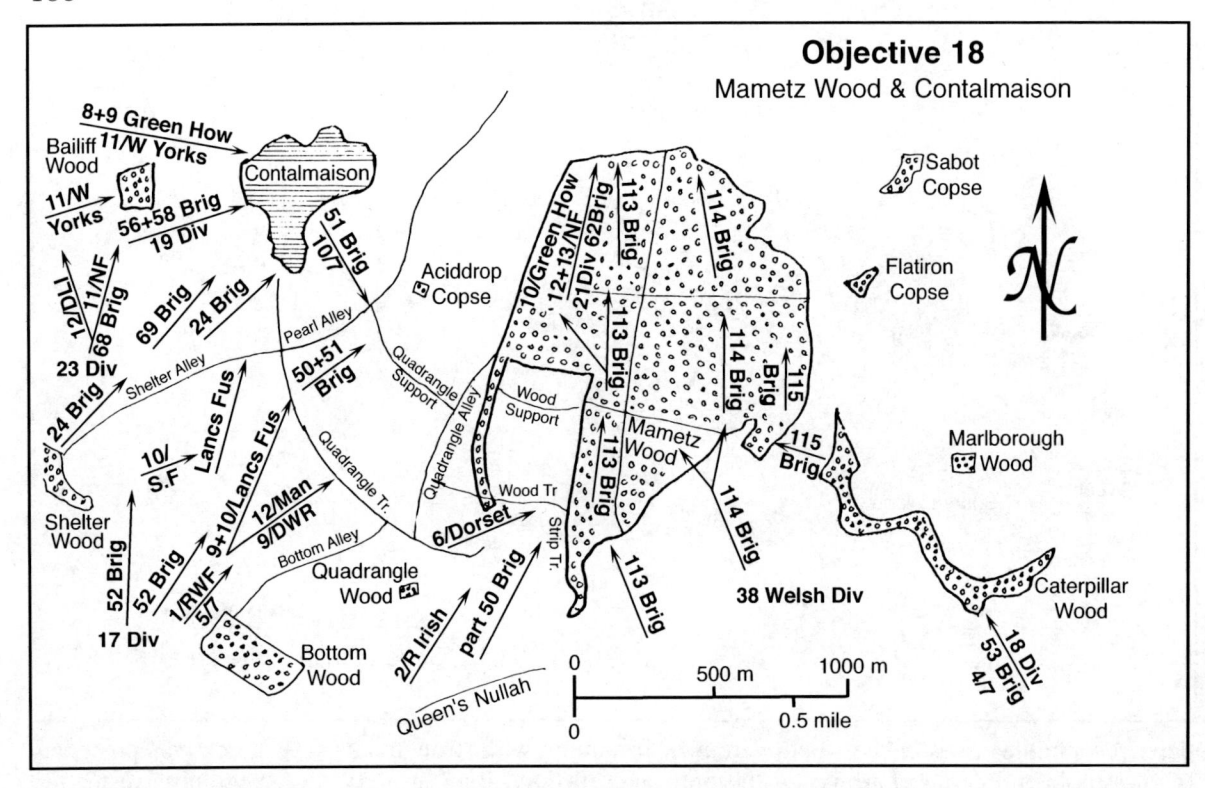

At 6.15am precisely, the units formed up to make their advance over a front twice the width of the earlier attack. This caused some delay in the formation of the units and the barrage, like some pre-set unstoppable monster, adhered to its strict timetable. The 15/Welsh threw themselves at the Hammerhead to protect the right wing of the main attack and, in spite of initial gains, were pushed back during a counter-attack. All attempts towards the western side of the wood attracted heavy German fire - it was evident they had taken full advantage of the unexpected pause. Fighting was confused in the now shattered remains of the wood. Any movement was difficult, even more so when sudden bursts of fire came from hidden machine-gun nests. The German artillery was accurately ranged causing horrendous casualties on friend and foe alike. Limbs were torn from bodies, not just by the explosion of shells, but through flying fragments of torn branches and tree trunks. Men were firing wildly at no specific target. The enemy could see but was not seen. When the two antagonists did meet, hard hand-to-hand fighting ensued and no quarter was given or received. Many Welshmen witnessed the gruesome sight of two adversaries, locked in mortal combat, each pierced by the other's bayonet, motionless, still standing as though death had united them. Shells had destroyed all communication cables and the only way of getting messages to the rear was by runner. The Official History records that almost all 96 runners became casualties in just one brigade. Brig.-Gen. Price-Davies went into the wood to assess the situation. At the same time, Lt.-Col. Hayes (14/Welsh) took over command of the right wing.

The 115th Brigade had been in reserve and Watts sent the 10/SWB to reinforce the 114th Brigade and the 17/RWF to the 113th. The Hammerhead was subjected to a heavy bombardment by Stokes mortars which, the area being small and well defined, produced some good results and the 114th Brigade wrenched the Hammerhead from the Germans. While the eastern side looked better, the west was still presenting many difficulties. With Germans in the wood and trenches to the immediate west, it was indeed a hard task which faced the attackers. Even so, the 13/RWF was ordered to take Wood Support Trench which ran from the left arm of the wood across the re-entrant to the main part of the wood. Aided by bombers from 50th Brigade (17th Div.), they advanced from the west, the Germans surrendering after a short struggle.

A general advance using all three brigades was timed for 4.30pm, the 10/SWB, supported by the 13th, 14th and 15/Welsh and the 13/RWF, to secure the area to the east following the fall of the Hammerhead. Spasmodic hard fighting began to push the Germans towards the north of the wood. At the same time the 17/RWF picking up small groups of men from mixed battalions, worked their way north to within 200 meters of the northern edge of the wood. Machine-gun fire from the German 2nd Line prevented further movement and so the men dug in. The pioneers and Royal Engineers were working hard to consolidate and wire the advanced position. The men were in desperate need of rest, food and water. During the night of the 10/11th July the 11/SWB and 16/Welsh, both of 115th Brigade, moved up to relieve the exhausted men.

Watts ordered the wood to be completely cleared and, a little after 3.00pm, the 11/SWB with the 10th, 15th & 16/Welsh advanced north, passing the line occupied the previous night and stopping 60 meters from the northern edge, keeping them out of site from the German 2nd Line. This position became untenable when the inevitable German barrage descended on the position and they withdrew some 140 meters. In the early hours of the 12th July, the exhausted and depleted battalions were relieved, Rawling's 62nd Brigade (21st Div.) taking over the northern part of the wood while the 7th Division took over the eastern side including Flat Iron Copse and linking up with the newly-arrived 1st Division. The 12th & 13/Northumberland Fusiliers (NF) and 10/Green Howards (62nd Brig.) finally cleared all German resistance in the wood. The Official History gives losses of the 38th Division as nearly 4,000 officers and men including seven battalion commanders.

Three days later, following the successful dawn attack along the Bazentin Ridge, Mametz Wood was again calm except for the occasional hostile shell but there remained the urgent task of burying the dead, many of the corpses already in the first stages of putrefaction in the summer heat. The 12th and 13/NF were engaged in this unenviable work, the 12th Battalion on the eastern half of the wood while its sister battalion worked on the opposite side. It was a harrowing experience. Harry Fellows was in the machine-gun section C Coy.,12/Northumberland Fusiliers and formed part of one of the burial parties. In an audio tape recording of his recollections and impressions of the essential but gruesome task facing him, he often speaks in a faltering, emotional voice of 'that wood'. Harry had taken part in the relief of the 115th Brigade and had already experienced the horrific sights in the wood. His sombre voice tells us "...all the branches which had fell from the trees had been pulverised into tinder and amongst all this lay dead bodies. It was impossible to walk more than three yeards in that wood without striding over a dead body. Some of these bodies had been laying there for days. It was mid-July, the weather was warm and the stench was unbearable. When once you'd got the stench of rotting bodies in your nostrils, it took months to clear it away."

And now Harry and his mates had to go back into the wood. He recalls, "on the morning of the 15th July we received the order to clear up the wood and this meant 'burying the dead'. We had only one young officer left in our company and he took over the operation. He divided the wood into sections, placing a party of eight men with an NCO in each section. I took over the right hand corner at the bottom of the wood. It was here that the fighting had been most severe and the dead were extremely numerous. We dug graves eight foot long, six foot wide and about two feet deep. We found the roots of the trees to be an encumbrance but one of our boys found a heavy pair of wire-cutters and this helped us extremely well. Then the task of moving the bodies to the graveside. We had no stretchers and the only thing we could do was to use our groundsheets with the pull-throughs from our rifles looped through the eyelets of the groundsheet, roll the body on to the groundsheet and then use it as a sledge."

A view of Mametz Wood taken from the high ground at Danzig Alley Cemetery. [Author]

British troops ready to move forward under the shelter of the sunken road to the west of Bailiff Wood. [A Perret]

Army regulations were fortunately very precise regarding burial procedure which no doubt facilitated later identification. Harry continues, "the drill for burying the dead - the first instruction for this was that the equipment was to be removed from the body with as little damage as possible. If the body was lying on its back, this was an easy task. The lapels from the tunic would be cut, the belt undone and equipment would come away. If the body was lying face down, each man carried in his haversack one of the old-style cut-throat razors and this was used to sever the belt at the back When the body was at the graveside, the drill continued. The first thing to do was to sever the string holding the identity disk round the neck. Then, the pay-book removed from the right-hand breast pocket. This pay-book contained all the man's particulars including his next of kin and every man had to make his last will and testament in this book. In the left-hand breast pocket most men carried a wallet and some of them had the habit of carrying what we called 'French postcards'. They are nothing like the pictures we see on page three these days but we destroyed these for fear they fell into the hands of the relatives. Some men also had a few francs in the wallet and I confess that we could not see the sense of burying good French money. This was put into a kitty and shared out afterwards. I know it seems like robbing the dead but I must confess that this was done. When the body was in the grave, the paybook and wallet were tied together by the string of the identity disk. A bayonet would be stuck in the ground near the head of the man in the grave and the little parcel was hung on the boss of the bayonet with a steel helmet over the top to protect it from the weather. We buried them head to tail, six to a grave. The German soldiers, we buried in separate graves. It must be realised that some of these bodies had been laying there for several days. The bombardment had been intense and shells made no distinction between the quick and the dead and their condition was such that they could not be removed. The only thing we could do was to dig a trench alongside the body and then three men with shovels would remove the remains into the trench. There was no ceremony or service and I think if anything had crossed our minds, it would have been, 'There, but for the Grace of God, go I'. That day, I helped to bury five hundred Welsh, English and German bodies in that wood. We were only in the right-hand sector. Our 13th Battalion were in the left-hand sector. They had the same task as ourselves so it would seem that upwards of a thousand bodies were buried in that wood on the 15th July" The days work made a deep impression on Harry and was to influence his own burial arrangements seventy-one years later.

THE QUADRANGLE
During the night of 4th/5th July the 1/RWF and 2/Royal Irish (22nd Brig. 7th Div.) with the 9/NF and 10/Lancs Fusiliers (52nd Brig. 17th Div.) after leaving the forward trenches before zero hour and creeping towards the German front line over the rain-soaked ground, dashed forward and seized their first objective, Quadrangle Trench (see map) and Shelter Alley which runs north from Shelter Wood. The Royal Irish, attacking Wood Trench were held up by uncut wire and driven back by machine-gun fire. At the same time the 69th Brigade (23rd Div.) was giving support both to the 7th Division and the 19th Division on the left wing.

A glance at the map will show how well the Germans had foreseen the defence of the ground between Mametz Wood and Contalmaison. Although Quadrangle Trench was secured, there were many obstacles before any assault could be made on Contalmaison. The next task was to secure Quadrangle Support Trench and Pearl Alley and this fell to the17th Division. Following a bombardment of the Quadrangle system of trenches and Contalmaison village, Clarke's 52nd Brigade with the 51st in support, moved forward to be met by intense fire across flare-lit ground. Harassed also by British shells falling short, the 9/NF and 10/Lancs Fus. (52nd Brig.) had to cope with rain-soaked ground and uncut wire. Their attack ground to a halt. They tried again with no improved result. The 10/Sherwood Foresters (51st Brig.) moved up to support but the combined forces could not enter Quadrangle Support. However the Lancashire Fusiliers on the left had forced an entry into Pearl Alley which acted as a springboard for an entry into Contalmaison. Too few in number, they were eventually forced to retire following a counter attack.

A heavy barrage was laid at 7.20am on the 7th July which attracted a retaliatory German shelling of the British front line. The two remaining battalions of 52nd Brigade, the 12/Manchester and 9/Duke of Wellington (DWR) moved forward. Having encountered a short delay in leaving the front line trench, the Germans were waiting and the men were cut down by fire from three sides. Leaving a few outposts to deal with German bombing counter attacks, the men withdrew as best they could. On the right the 7/East Yorks and 6/Dorset (50th Brig.) fared no better. They sustained heavy losses from fire from Strip Trench on the south-western edge of Mametz Wood. A further attack was made at 8.00pm by the remaining companies of the Yorks and Dorsets along with the 10/Sherwood Foresters from 51st Brigade. It was a futile attempt against impossible odds.

In the early hours of the morning of the 8th, bombers of 50th and 51st Brigades experienced much difficulty in forcing a way through the clinging mud as they made their way north towards Quadrangle Support. In equally bad conditions, the Germans defended their position and little progress was made. A further attack in the late afternoon by the 6/Dorsets, 7/Green Howards and 7/East Yorks of 50th Brigade was repulsed. The long day's work for the Dorsets was not yet finished. Detailed to take Wood Trench running from Quadrangle Alley to Strip Trench in order to protect the left flank of 38th Division's attack on Mametz Wood, the day ended with Wood Trench being secured for over half its length. Quadrangle Support remained inviolate in spite of repeated attacks. A night surprise attack with the bayonet was undertaken by the 7/Green Howards (50th Brig.) and 8/South Stafford with bombers from the 7/Lincoln (51st Brig.). On the left the Staffords leaped into the

Above: Contalmaison Château before its total destruction [J Bommeleÿn]

Right: The only parts of the château not completely destroyed were the steps to the main entrance and access to the cellars. [Michelin]

The provisional cantine at Montauban. There was an inherent risk of fire from the walls, made from tarred cardboard! [J Bommeleÿn]

trench killing or wounding the occupants but, unsupported on either side, had to retire after losing 219 officers and men. On the right, half a company of Green Howards moved forward a few minutes later than the Staffords on the left. Intense fire brought them to a standstill. One company from 7/E Yorks and 6/Dorset went to support but could not get forward. With the fall of Contalmaison on the 10th, the 51st Brigade bombed its way into Quadrangle Support via the eastern exit of the village. The 50th Brigade fought its way up Strip Trench and then into the east end of Quadrangle Support. The Germans were cut off, attacked from both ends but they fought desperately and only after the most bitter hand-to-hand contest, did the 17th Division finally take possession of this trench which had been the cause of so many casualties.

CONTALMAISON

To the left of the 17th Division, the 23rd and 19th were planning a joint attack on Contalmaison. Due to 17th Division's failure to secure Quadrangle Support and Pearl Alley, the 23rd Division had to delay its attack. Consequently, the 19th Division was unsupported for the time being. Nevertheless, the 19th advanced north from a line between La Boisselle and Contalmaison, keeping close to the creeping barrage, the 9/Welsh (58th Brig.) and 7/King's Own (56th Brig.) leading the attack. Casualties and confusion were encountered as the men found themselves under friendly fire. The 6/Wiltshire (58th Brig.) was hurriedly sent up in support and the three battalions took their objective, capturing over 400 prisoners. Consolidation began immediately and strong points were made on the arrival of the 56th Machine Gun Coy. The right wing was secured by the 9/RWF (58th Brig.) while waiting for the 23rd Division.

At 9.15am, the 11/NF (68th Brig. 23rd Div.) attacked Bailiff Wood, 600 meters west of Contalmaison. Arriving at the southern edge of the wood, they came under severe fire from Contalmaison and German positions to the north and were obliged to withdraw a few hundred meters. To protect the left flank it was important to link up with the 19th Division and the 12/DLI (68th Brig.) advanced in that direction under heavy fire and occupied a trench close to the 19th Division.

Meanwhile, the 24th Brigade, still attached to 23rd Division, made its assault on Contalmaison. The 1/Worcester and 2/East Lancs advanced at the same time as the 17th Division attempted to take Pearl Alley. Movement was difficult in the muddy conditions, wounded men were lying everywhere. The Worcesters moved north from the southern section of Pearl Alley to the edge of Contalmaison losing many men from machine-gun fire on the way. From here, they cleared the ruined houses and cellars, often engaging in hand-to-hand fighting and, some thirty minutes later, they were in the vicinity of the church. The Lancashire men on the left were more exposed, experiencing much difficulty in getting forward under fire from Contalmaison and Bailiff Wood. Not being able to enter the village, the Worcesters were unsupported. The German artillery indiscriminately shelled the village killing both their own men and some Worcesters, who, now out of bombs, had to abandon their hard-earned gains. The two battalions had lost almost 800 officers and men.

In the evening of the 8th a further attack by the Worcesters failed to make any progress. In the valley below the 2/Northampton came under heavy fire as soon as it left the cover of Peake Woods. The 10/DWR (attached to 24th Brig.) sent bombers forward to establish a strongpoint and mount machine-gun posts which could rake the whole of the area facing them. The 12/DLI (68th Brig.) attacked Bailiff Wood at 8.15pm and with bomb and rifle, securing all but the extreme eastern edge,

THE INFANTRY ATTACKS - OBJECTIVES 18 & 19

guarded by a trench. The Durhams barricaded the trench and were able to repulse the enemy counter attack. In the afternoon of the 10th, the 69th Brigade advanced close behind the creeping barrage, the 8th and 9/Green Howards moving forward from north of Horseshoe Trench. The leading companies came under fire and had to negotiate uncut wire, but succeeded in securing the trench in front of Contalmaison, driving the Germans into the village. The 11/West Yorks having previously moved forward to Bailiff Wood, had a good field of fire over the Germans retreating north towards Pozières. The Green Howards and Yorkshiremen linked about 5.30pm and gradually cleared the village. Some strong resistance was met but mostly, the enemy preferred to try to make its escape. The Wellingtons did good work from their newly constructed strongpoint. Contalmaison had fallen at last. Consolidation was urgent in order to protect against the German policy of quick retaliation. A box barrage was placed around the village while the defence of Contalmaison was put in order. By noon the following day, the 11th July, the 1st Brigade of Strickland's 1st Division relieved the battle-worn 23rd. The Official History quotes casualties of 855 officers and men in the 69th Brigade during the attack on Contalmaison and 3,485 casulaties for the 23rd Division to the 10th July.

Montauban windmill. [J. Bommeleÿn]

TRONES WOOD

The Commander in Chief, General Douglas Haig had stressed the importance of taking Contalmaison, Mametz Wood and Trônes Wood before the planned dawn attack on the 14th July. Trônes Wood (Bois des Troncs on your IGN map) was vulnerable to fire from the German occupied villages of Longueval to the north and Guillemont to the east. This presented particular difficulties in so much that if the Germans felt under threat, they could evacuate the wood to allow a saturation barrage of the new occupier. Of course, the British had the same advantage prior to the attack, the difference being that the British desperately needed the wood whereas the Germans, still holding Longueval, the Switch Line, Guillemont and Ginchy could all temporarily cede possession if this was absolutely necessary.

The first attack on the 8th July came from the Allies, the French to secure the southern section of Maltzhorn Trench while the British took the northern end. Afterwards, the French were to secure Hardecourt while the British took over the ruins of Maltzhorn Farm. The British would have to advance 1,500 meters over open ground from the Brickfield, 800 meters south-east of Montauban, due east to Maltzhorn Farm. In view of this, it was decided to secure the southern part of Trônes Wood, thereby

The main road at Montauban. [J Bommeleÿn]

Contalmaison Château Cemetery, situated a few meters from the original château, just behind the trees. [Author]

Inset: the same cemetery in 1916 [Michelin]

affording some protection for the proposed advance. Following a bombardment of the objectives including Longueval, the 2/Green Howards (21st Brig. 30th Div.) left the shelter of Bernafay Wood and advanced on Trônes Wood with covering fire from 26th Brigade (9th Div.) on the left flank. Cresting the gentle rise, the Green Howards were met by a murderous fire from Trônes Wood and Maltzhorn Trench and although a few men reached Trônes Wood by way of Trônes Alley, they were not seen again. The remainder were recalled. The 2/Wiltshire of the same brigade attacked at 1.00pm. In spite of heavy losses including Lt.-Col. Gillson who was severely wounded and his replacement Capt. Mumford who was killed, the Wiltshires penetrated the southern edge of Trônes Wood and

Montauban railway station, situated near the southern edge of Bernafay Wood. [J P Matte]

busied themselves digging a trench northwards to protect their flank. Units from the 18/King's and 19/Manchester (21st Brig.) and later the 18/Manchester (90th Brig.) moved up to help to consolidate the entry into Trônes. With at least the southern part of the wood secured, a successful assault on the lower half of Maltzhorn Trench was made early next morning by the 2/Royal Scots Fusiliers (90th Brig.) which then secured the ruins of Maltzhorn Farm. Working their way back north, they occupied the remainder of the trench taking over a hundred prisoners. The 17/Manchesters (90th Brig.), assembled in Bernafay Wood, had to put on gas masks and encountered much difficulty crossing the wood. Eventually debouching from its eastern edge they crossed the open ground, entered Trônes Wood and linked up with the Fusiliers. Patrols sent north reported only a few German strongpoints remained in the wood. The occupiers found themselves in the situation described in the introduction to Trônes Wood. The remaining Germans withdrew and the wood was subjected to a heavy barrage inflicting many casualties on the Manchesters. Only the extreme south-eastern corner of the wood escaped the barrage. It was evident the German barrage would be followed by a strong counter attack and the 17/Manchester was ordered to return to Bernafay Wood. One detachment did not receive the order and remained in the wood. The Royal Fusiliers and one company of the 18/Manchester remained in the south-eastern corner of the wood, the rest of the 18/Manchester retiring to the Brickworks. The Germans launched their counter attack at 3.30pm from the east covering a front from Maltzhorn Farm to the north of Trônes Wood. The 18/Manchester and Royal Fusiliers took a heavy toll on the advancing enemy. However, only the isolated group of 17/Manchester blocked an entry from the north. These men defended their position very courageously, but were overwhelmed. The German barrage now ranged on Bernafay Wood causing further casualties. The 16/Manchester attacked north from east of the Brickworks and was able to cover its comrades on the right. The Germans having re-established strongpoints and snipers in the trees, the Manchesters dug in for the night just south of the wood.

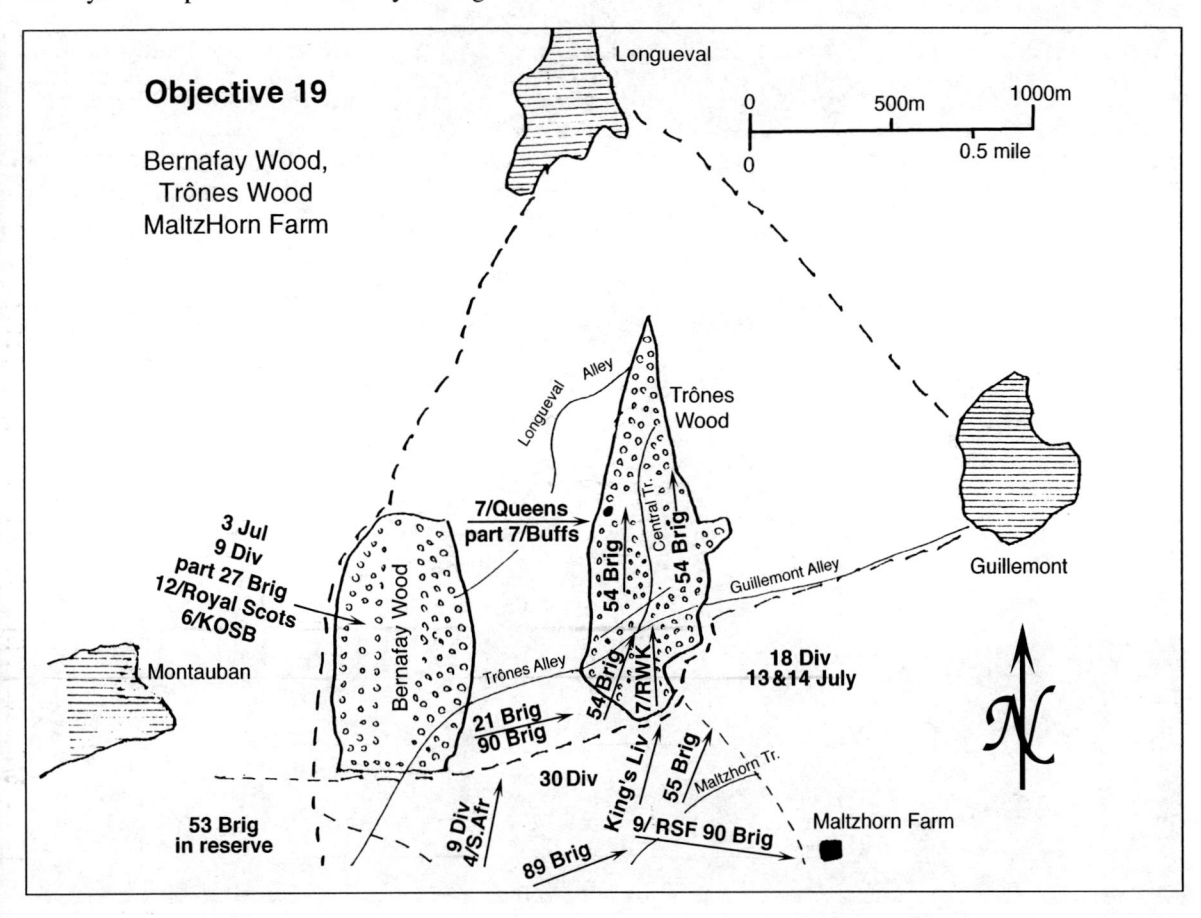

Patrols before dawn on the 10th by the 16/Manchester and the 4th South African Regiment (9th Div.) found the wood severely damaged making movement very difficult. Many were disorientated and lost their way. Others continued northward meeting little German resistance. It was thought the enemy had expected a retaliatory barrage and had evacuated the wood, but this was not the case and fighting took place in Central Trench which traversed the wood and in Longueval Alley linking Trônes Wood to Bernafay Wood. German reinforcements hurried down from the north-east, entered Trônes and captured a number of the 16/Manchester. The patrols were too few in number and the Germans took possession of all the wood except the south-eastern corner. The 90th Brigade, having lost almost 800 officers and men, was relieved by Stanley's 89th Brigade, the 20/King's took over Maltzhorn Trench while the 2/Bedford entered the Sunken Road east of the Brickworks.

At 2.40am all British troops were withdrawn from the wood which was then subjected to a saturation barrage. An hour later the King's successfully bombed a passage up Maltzhorn Trench taking two machine-guns and some prisoners but stopping short of the planned link at a strongpoint with the Bedfords. The latter advanced from the Sunken Road, German fire causing the right companies to veer to the right forcing them away from the link-up point with the result that the junction was missed. However, the Bedfords entered the wood but were unable to get past the strongpoint. The remaining companies kept to the plan of attack and entered Trônes from the west sending patrols north and east. Their progress was arrested by the arrival of German reinforcements from Guillemont which forced the Bedfords to retire.

The French had passed on a message found on a captured German officer which gave details of a counter-attack. As a result, the XIII Corps directed a barrage to be laid covering the area east of Trônes Wood as far as east of Guillemont. The 17/King's (89th Brig.) made a

Left: Trônes Wood, ravaged by war. Below: An early memorial at Trônes Wood to the 58th Brigade of the 18th Division. [J Bommeleÿn]

The 18th Division Memorial which replaced the earlier Cross. This was erected long before the wood had recovered from the battle. [Michelin]

further attack at 10.30pm and consolidated in the south-eastern corner. The following day, 12th July, the King's and Bedfords made contact via a newly-dug trench. Another German counter-attack on Maltzhorn Trench and Trônes was repulsed by combined French and British artillery. The dawn attack planned for the 14th July was less than twenty four hours away. Rawlinson ordered that Trônes Wood must be taken "at all costs" before midnight of the 13/14th July. The dreaded order had come. With casualites of over 2,300 the 30th Division could not be expected to succeed. Maxse's 18th Division was ordered to take the wood after a two hour bombardment by 30th Division's artillery. Brigadier-General Jackson's 55th Brigade led the attack, the 7/Buffs bombing its way up Maltzhorn Trench. Like the 20/King's before them, the Buffs could not take the stronghold but erected a barricade to give some protection..The 7/Royal West Kent lost its way amongst the innumerable obstructions in the wood, fire from Central Trench causing heavy losses. In spite of this, some 150 men reached the eastern edge of the wood but further efforts to advance to the north were unsuccessful. The 7/Queen's on the left, reinforced by a company of the 7/Buffs, advanced from Longueval Alley but were brought to a halt by an accurate enemy barrage. The Queen's losses of 478 all ranks at Mametz on the 1st July had not been replaced and with a further loss of 200 men at Trônes, the battalion existed virtually in name only. The survivors withdrew, leaving one small party in the northern part of the wood. Time was running out.. Maxse ordered Brig.-Gen. Shoubridge to engage his 54th Brigade. There was no time to plan a coordinated attack or request a preparatory barrage. In less than three hours, British troops were to pass within 300 meters of Trônes Wood. Shoubridge opted to advance quite simply from south to north paying particular attention to the defensive flank on the eastern edge. Due to difficulties in getting the leading battalion, 12/Middlesex forward, the 6/Northampton led the attack at 3.00am after crossing a thousand meters of open ground under a heavy German barrage. Seizing most of Central Trench, the Northamptons continued north, arriving at a copse which juts out from the eastern perimeter. Thinking they had reached the northern limits, they stopped. The Middlesex followed, meeting up with other units within the wood, most completely disorientated by the desolation. The whole attack, so urgent, was fast becoming disjointed. The situation was saved by Lt.-Col. Maxwell GCO 12/Middlesex who entered the wood, took stock of the situation and reorganised the men. He sent a company of his battalion to take the stronghold on the Guillemont road which was still causing problems. With the help of the 7/Buffs, the objective was taken. Maxwell instructed officers to move to the western perimeter of the wood using a compass to guide the men. Eventually this was accomplished. The men then advanced in line with fixed bayonets, firing from the hip at unseen targets, having to overcome a machine-gun post as well as natural obstacles. Men were falling from sniping fire but the sight of a huge line of men advancing shoulder to shoulder, bayonets fixed, was sufficient to induce a retreat of all but the most hardy defender. As the wood narrowed towards its apex the men swung round to complete the capture of the wood. During this final sweep, Sergeant W. E. Boulter of the 6/Northampton won the Victoria Cross for his courageous actions in silencing a machine-gun post. At last, six hours late, the whole of Trônes Wood was in British hands, the Germans fleeing towards Waterlot Farm and Guillemont. The Middlesex, Northamptons and 7/Buffs took a heavy toll of the retreating enemy.

(i) See Objective 15

OBJECTIVE 20
THE DAWN ATTACK ON THE 14TH JULY BAZENTIN-LE-PETIT, BAZENTIN-LE-GRAND VILLAGES AND WOODS, BAZENTIN RIDGE AND EAST TO LONGUEVAL

The front of the dawn attack extended from Bazentin-le-Petit village and wood, Bazentin-le-Grand village and wood and the area east towards Longueval and Delville Wood. Rawlinson's bold plan was not well received by Haig who had misgivings about many aspects of the plan, not the least being that of assembling secretly several thousand men in the dark. The 5,000 meter front was taken over by the XIII Corps on the right with the 3rd and 9th Divisions, the XV Corps in the centre with the 21st and 7th Divisions and the III Corps with the 1st Division on the left wing. The preparatory barrage commenced on the 11th July concentrating on the objectives as well as areas beyond to hinder German reinforcements. Zero hour was set for 3.25am on the 14th. As 22,000 men began to assemble in the dark and while the 18th Division was in the latter stages of the capture of Trônes Wood, the 1/Black Watch (1st Brig. 1st Div.) was busy occupying Lower Wood just to the north of Mametz Wood and at 3.45am on the 14th, they captured Contalmaison Villa to the north-east of the village.

In spite of the general state of anxiety and apprehension, the men managed with surprisingly few problems to take up their positions, following pre-laid tapes to guide them. They were within 500 meters of the enemy front line and before zero, men were creeping forward into no man's land, some to within 50 meters of their first objective. An hour before zero the VIII Corp's artillery in the Beaumont Hamel area opened fire on the German front line facing them with the view to diverting German attention away from the main attack.

As the barrage lifted at 3.25am the leading battalions comprising the 8/Black Watch, 10/A&SH

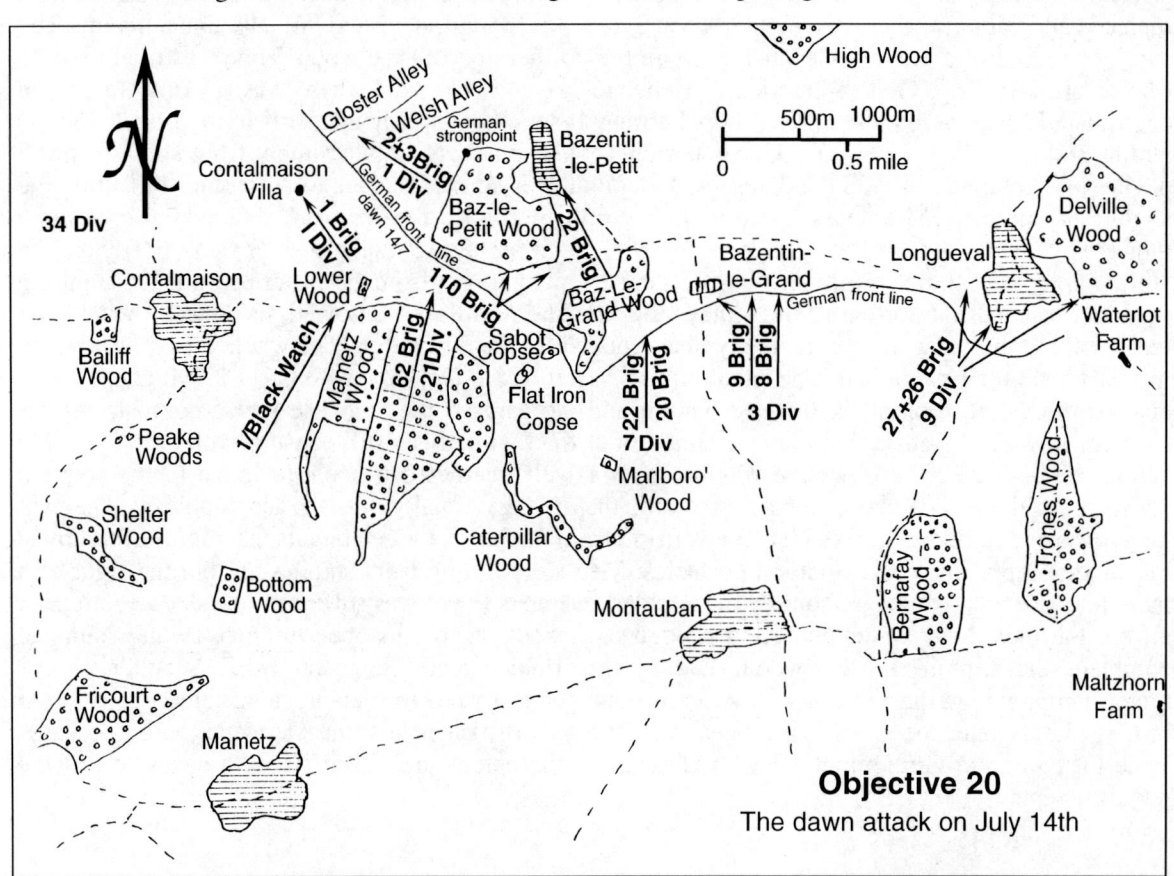

Objective 20
The dawn attack on July 14th

(26th Brig.), 11/Royal Scots, 9/Scottish Rifles (27th Brig.), 8/East Yorks, 7/KSLI (8th Brig.), 13/King's, 12/West Yorks (9th Brig.), 2/Border, 8/Devon (20th Brig.) and the 7th and 8/Leicester (110th Brig, exchanged with 63rd Brig. 7th July) moved towards their first objectives. In a few seconds the men were in the enemy front line surprising the occupants and prisoners were taken. On the right the 10/Argyll had to cut a way through the wire but eventually broke through and the four battalions from the 26th & 27th Brigades were soon over the next obstacles. Now realising they were under a major attack, the Germans quickly mounted their machine-guns but the Scottish infantry was already fighting in the burning village of Longueval and had secured a lodgement in Delville Wood. The 7/Seaforth and 5/Camerons (26th Brig.) hurried forward in support and a German strongpoint in Longueval was taken. The Germans were pouring machine-gun fire from the orchards in the northern part of Longueval making it impossible to secure the whole village. However, the 9th Division was able to consolidate and construct strongpoints before the Germans launched a counter-attack.(i) The Seaforths and Camerons were withdrawn to Longueval Alley and were thus in touch with 18th Division.

The 3rd Divison's 8th Brigade encountered severe problems with uncut wire, the first line of wire being hardly cut and the second line undamaged. The advance came to a halt until a company of the 2/Royal Scots equipped with a machine-gun went to the left and broke through the wire with the 9th Brigade and then bombed east down the trench quickly overwhelming the Germans. The 9th Brigade made excellent progress from the start, overcoming stubborn resistance. On they went over the crest of the hill under fire from Bazentin-le-Grand, and seized the trenches. Following a bombardment with trench mortars, the 1/NF (9th Brig.) took the lead and although faced with machine-gun fire, captured the village. A fine, successful attack executed with great elan but which incurred a number of senior officer casualties.

The 7th Division formed the right-hand division of Horne's XV Corps. The 2/Border and 8/Devon (20th Brig.) led the attack northward between the Bazentin villages. The bombardment in this sector had destroyed the wire and front line trenches allowing the leading battalions to continue their rapid advance. Taking the 2nd German line, the retreating enemy came under rifle and machine-gun fire as they desperately sought the shelter of High Wood, some 2,000 meters to the north-east. The Border and Devons, after waiting for the barrage to lift off Bazentin-le-Grand Wood, quickly occupied the wood without serious opposition. The men from opposite corners of England had good reason to be pleased as they dug in and made strongpoints. The 22nd Brigade now took up the chase led by the 2/Royal Irish with the 2/Royal Warwick supporting the attack on Bazentin-le-Petit. By 6.30am they were at the southern entry to the village. Aided by the 6/Leicester of 110th Brigade coming from the

Bazentin-le-Grand on the summit of Bazentin Ridge. (J. Bommeleÿn)

left, the three battalions captured the village taking over 200 prisoners. The northern limit of the village was lost following a counter-attack but the Royal Irish, reinforced by the 2/Gordon which had moved up from Bazentin-le-Grand, retook the German gain. Consolidation work was quickly undertaken to ensure the defence of the village. The cemetery and quarry 400 meters east of the village were secured.

The attack of the 110th Brigade, attached to 21st Division, went well almost along the whole of its frontage, apart from the left-hand battalion which was held up by accurate machine-gun fire. Bombers silenced the guns but Lt. Col. J. Mignon, C.O. of the 8/Leicester was killed leading the attack. The Leicesters could now move forward and, joining the other battalions of 110th Brigade, secured the 2nd line. Their next objective, Bazentin-le-Petit Wood was taken fairly easily, the only severe resistance coming from the north-western corner which was not cleared until 7.00pm. With the wood cleared, the 21st and 1st Divisions linked up. Royal Engineers were sent up to help with the consolidation in both the village and wood. The day had gone well, particularly in the centre and left-wing of the attack. 42 German officers and 1,400 men were being escorted to the prisoner's compound.

Reports had been received that Longueval had fallen. This was not the case, the 26th and 27th Brigades had been fighting all day and were exhausted. The South African Brigade(ii), hitherto in reserve, was brought up and entered the struggle but it was becoming clear that Longueval and the adjacent Delville Wood would need the attentions of more than one fresh brigade. As things stood, with success on the centre and left, it was decided not to take further action at Longueval and Delville Wood until the following day. There has been some criticism that the successes were not exploited. Sufficient troops were available to be able to counter any retaliatory action from the Germans. High Wood stood intact and unoccupied on the crest of the ridge. However, at 9.45pm General Rawlinson ordered the continuation of operations on the following day. This delay allowed the Germans sufficient time to bring up reinforcements and prepare for the next attack. The Germans re-established their strongpoint in the north-west corner of Bazentin-le-Petit Wood. A subsequent attack on the strongpoint by mixed units was brought to a halt by machine-gun fire.

The 33rd Division was to pass through the 21st Division on the next day, the 15th, but Brig.-Gen. Baird, acting on his own initiative, sent the 9/HLI and 1/Queen's (100th Brig.) forward in the evening of the 14th to plug the gap between Bazentin-le-Petit and High Wood. As this movement resulted in an engagement at High Wood, the action is described in Objective 23.

Maj.-Gen. Strickland's 1st Division attacked in a north-westerly direction from Bazentin-le-Petit Wood. The 1/Loyal North Lancs (2nd Brig.) moved up the battered German second line trenches and into the support lines where they came under heavy fire. With little cover, they came to a halt. A further attempt was made at 5.00pm with the help of the 2/Welsh (3rd Brig.). This too was halted by machine-gun fire. As daylight turned into night the 3rd Brigade established outposts and linked up with Ingouville-Williams' 34th Division. The 2/Welsh was in action again at 2.00am bombing its way up the German trenches. The clinging mud hampered both the movement of the Welsh and the men bringing up supplies. The men were recalled when Strickland decided to request a further barrage to be followed by an attack at midnight. This time the artillery cut the wire well although the ground conditions were worsening. Following right on the heels of the creeping barrage the 1/Gloucester and 2/Royal Munster Fusiliers (RMF), both of 3rd Brigade, set off in the rain in a north-easterly direction, the 2/Welsh supporting on the right. The Germans recoiled before the bayonet and fled, leaving their dead and wounded in the trenches. The captured German trenches were appropriately renamed Welsh Alley and Gloster Alley. The 1/SWB (3rd Brig.) on the left established strongpoints in Black Watch Alley. Rather surprisingly, there was no counter-attack on the 17th although during the morning, a hostile barrage was laid over much of the Fourth Army front.

The 19th Division just north of Bazentin-le-Petit was relieved by the 34th Division, now commanded by Major-General C. Nicholson following the death of the former GCO, Major-General E. C. Ingouville-Williams(iii). Almost all the objectives had been taken and General Rawlinson's bold plan did much to allay the heavy criticism placed against him following the 'Big Push' on the 1st July.

(i)   Longueval and Delville Wood are described in fuller detail under objective 21.
(ii)   Raised by the Union of South Africa, was composed of regiments from Natal and Orange Free State, Transvaal and Rhodesia and one South African Scottish regiment.
(iii) Major-General E. C. Ingouville-Williams, affectionately known at "Inky Bill" was killed just to the south of Mametz by a stray shell on the 22nd July.

OBJECTIVE 21
LONGUEVAL, DELVILLE WOOD AND TRENCHES NORTH TOWARDS FLERS AND NORTH-WEST TOWARDS HIGH WOOD

The 9th Division had already been engaged at Longueval in the dawn attack on the 14th July and had forced an entry into the village as well as the southern edge of Delville Wood. On the 15th, the 5/Camerons and 4th South African Brigade had taken possession of the ruins of Waterlot Farm (in reality, a sugar factory) thus neutralising a dangerous strongpoint on the approaches to Longueval from the south or west. The cavalry had attacked between High Wood and Delville Wood in the evening of the 14th when enemy fire forced them to dismount and they were forming positions in the high corn field between the two woods.

Major-General Furse's 9th Division was ordered to complete the capture of Longueval and Delville Wood, the latter having an area of over 150 acres. Behind the northern edge of the wood the ground descended gently towards Flers making a useful reverse slope by which German reinforcements could be brought up in safety. Following an artillery and mortar barrage on the village, the 12/Royal Scots (27th Brig.), reinforced by the 1st S.A. Brigade, bombed their way up North Street clearing the ruined houses and good progress was made until reaching the northern part of the village when fire from the orchards prevented any further movement. The Royal Scots, now in diminished numbers, made a further attack at 7.30pm but with no better result.

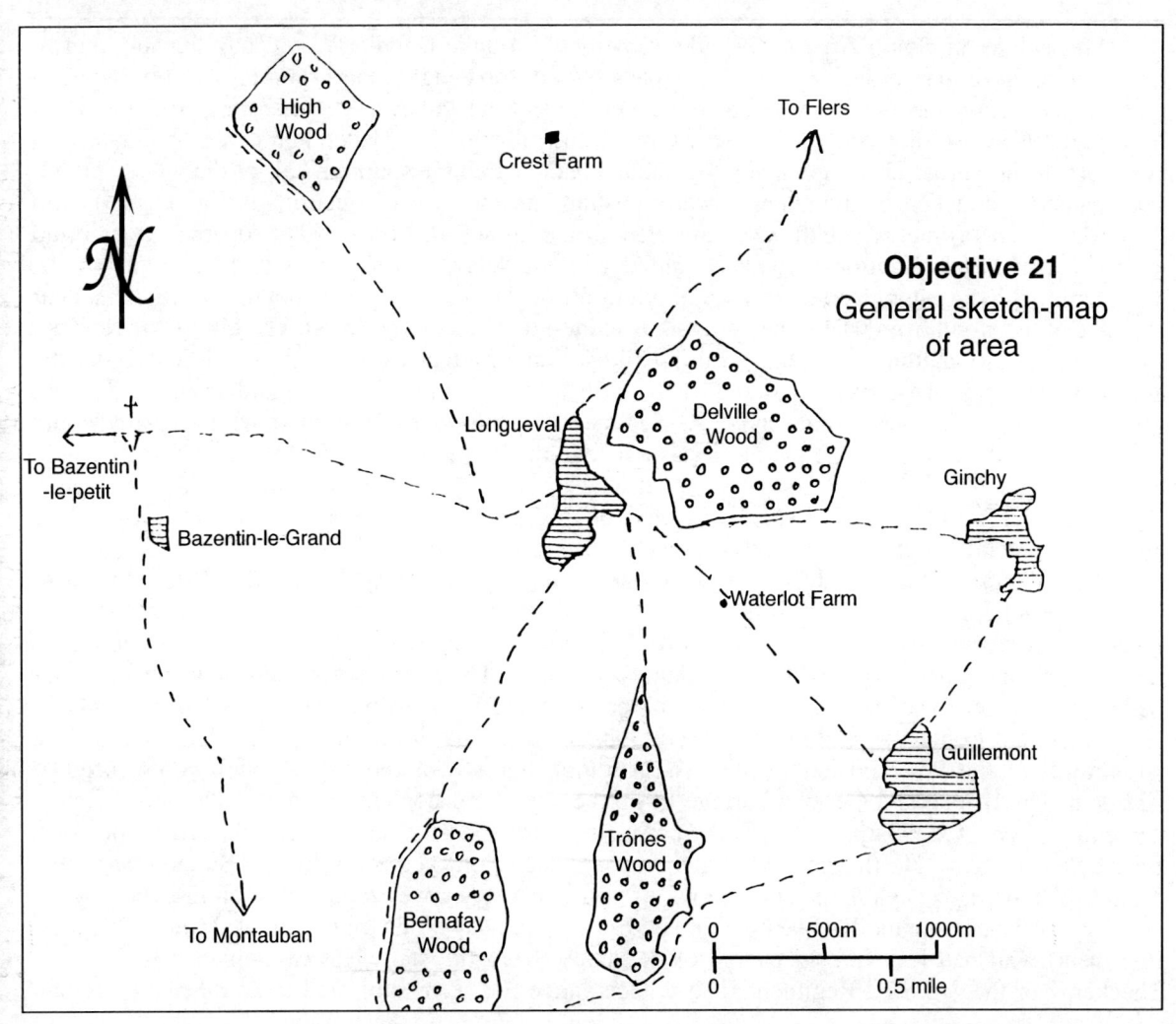

Objective 21
General sketch-map of area

Objective 21
Longueval and Delville Wood

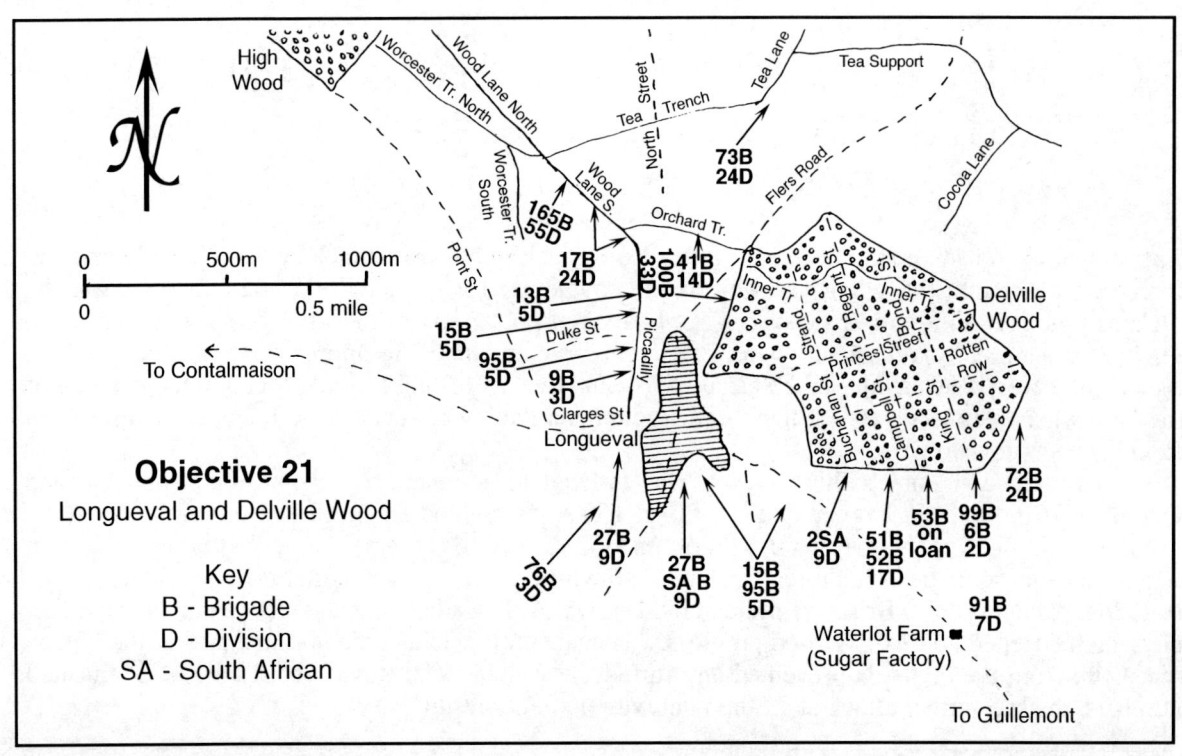

Key
B - Brigade
D - Division
SA - South African

Meanwhile the South African Brigade was ordered to take Delville Wood "at all costs", a grim reminder of a similar order received at Trônes Wood forty-eight hours previously. Following a preliminary barrage the attack was led by two companies of the 2nd S.A. Regiment. Entering the wood from the south-west, they fought their way through the entangled trunks and branches having also to negotiate the numerous shell craters and by 8.00am had cleared the southern half of the wood. After a short pause the attackers continued northwards pushing the enemy back, securing the wood apart from the north-western corner where the Germans were well entrenched. About 140 prisoners were escorted to the rear and consolidation work was started without delay. As in Trônes and Mametz Woods, digging through the entangled roots was a laborious affair. The Germans were not in full flight but near the edge of the wood from where they poured machine-gun fire into the wood. The South Africans not engaged in consolidation were engaged in repelling counter-attacks which continued throughout the afternoon, although these were hindered by the British barrage ranged beyond and around the wood. The remaining companies of the 2nd S.A. Regiment went forward to support while the divisional pioneers, the 9/Seaforths, wired the north-eastern edge of the wood. As night fell, the Germans laid a barrage on the wood prior to another counter-attack. The enemy too, was under orders to retake Delville Wood "at all costs" but each attack, although delivered with great resolve, was repulsed. A hostile barrage laid on the village and wood continued throughout the night.

At 10.00am on the 16th July, after a mortar barrage, the 11/Royal Scots (27th Brig.) advanced well to the left of North Street with the intention of attacking northern Longueval from the west. Meanwhile the 1st S.A. Regiment attacked from the western end of Prince's Street within the wood. Intense machine-gun fire brought both ventures to a halt. The mortars restarted their barrage after which the attackers made some progress. An order received from Lt.-Col. Congreve's XIII Corps H.Q. stated that both Longueval and Delville Wood must be taken by dawn on the 17th. This entailed the withdrawal of the 9th Division to allow a saturation barrage of the wood which commenced at 12.30am. The British and German barrage lit up the sky, the cacophonous roar of the guns reached deafening pitch. At 2.00am, with less than five hours in which to secure their objectives, the 27th Brigade advanced up North Street while the South Africans attacked from Prince's Street towards the Strand. (see map). Yet again, the courageous efforts of these men were doomed to failure, the barrage not having silenced the hidden machine-guns. Losses were heavy including Lt.-Col. Tanner (2nd S.A. Regiment) who had been put in charge of the attack from the start. He was replaced by Lt.-Col. Thackeray of the 3rd S.A. Regiment. The village and wood were subjected to an all-night German

The Main Square at Longueval before the war.
(J. Bommeleÿn)

bombardment with H.E. and gas, making conditions even worse for the exhausted infantry.

Congreve issued another "at all costs" order to take the objectives. This was to be undertaken by the 76th Brigade of Haldane's 3rd Division which had moved eastward and taken up positions to the immediate south-west of Longueval. After a preliminary barrage the 1/Gordon and two companies of the 8/King's Own and 2/Suffolk (76th Brig.) assaulted Longueval from the west, securing the village apart from the stubborn northern extremity where the Germans were still entrenched in the orchards. Meanwhile, the South African brigade had advanced to the northern edge of the wood. The enemy placed a saturation barrage on Longueval and Delville Wood, the whole area becoming a blazing inferno. The South Africans suffered horrendous casualties. The 76th Brigade was ordered to withdraw to their front line trench near the mill. A light drizzle had turned to more persistant rain, adding to the discomfort of the troops. A strong infantry counter-attack was brought to a halt by British artillery and fire from the South Africans in the south-eastern corner of the wood. The 153rd Regiment attacked from the north with orders to sweep right through the wood. Pushing the surviving South Africans back through the wood, the Germans swept all before them, with just a few South Africans remaining in the eastern part of the wood. However, on leaving the southern perimeter, the Germans suffered heavy casualties as they came under fire from British artillery and machine-guns in Longueval. Yet a further counter-attack came from the north-west. After bitter hand-to-hand fighting the 27th Brigade (9th Div.) was forced to the southern edge of the village. The situation was critical and the 5/Cameron Highlanders (26th Brig.) counter-attacked with the utmost

The Main Square at Longueval today [Author]

*Church Street Longueval, with the Château behind the trees.
(J. Bommeleÿn)*

vigour, resulting in the re-occupation of Longueval as far as Clarges Street. Major-General Furse's plans to continue the attack with the 19/Durham L.I. (106th Brig. 35th Div.) was postponed due to the weight of the enemy bombardment.

At 7.15am on the 19th July, the 8/Norfolk of Higginson's 53rd Brigade on loan to the 9th Division, moved forward into the wood and pushed the Germans out of the southern part. Reinforced later by the other three battalions of the same brigade, the 8/Suffolk, 10/Essex and 6/Royal Berkshire, good consolidation work was done although no further advance could be made. Both sides had suffered heavily from the barrages, the Germans losing many officers. In order to retain their hold in the central and northern parts of the wood, the exhausted Germans were in desperate need of support. The British troops were in no better shape. Accordingly, the 3rd Division relieved the 9th apart from the South Africans and 53rd Brigade still in Delville Wood. The 76th Brigade engaged its 2/Suffolk at 3.35am. They should have been supported by the 10/RWF of the same brigade, but the guide lost his way in the dark, leading the battalion into the British barrage during which most officers became casualties. The exposed Suffolk pushed home its attack with great courage but its ranks were decimated. Arriving ten minutes late, the Welsh Fusiliers attacked vigorously. The fact that the Suffolks and Welsh Fusiliers stood little chance of success did not diminish the ardour and it was during this action that Corporal J. Davies and Private Albert Hill, both of the 10/Royal Welsh Fusiliers, gained the Victoria Cross for their outstanding courage in Delville Wood. Two further awards of the Victoria Cross were made. One to Private W. F. Faulds of the 1st South African Regiment for his courage in bringing in wounded men under fire on two separate occasions and the other to Major W. La. T. Congreve, Brigade-Major of 76th Brigade, for his leadership and courage displayed on many occasions during July. He was killed by a sniper's bullet on the 20th July.(i)

The same evening, Lt.-Col. Thackeray brought out the terribly few survivors of the South African Brigade. They had resisted an almost continuous barrage, innumerable counter attacks with machine-guns, rifle,

Longueval Château on the edge of Delville Wood.(J. Bommeleÿn)

bayonet, bombs and gas shells, and never once, did they falter. They came from the great African continent to fight for the cause and covered themselves in glory. Today, a magnificent and worthy memorial stands in Delville Wood to the memory of these valiant warriers. The Official History statistics bring home the stark truth of the sacrifice. On the 15th July the brigade numbered 121 officers and 3,032 other ranks. At the roll call on the 21st July, only 29 officers and 751 other ranks answered the call. During the action between the 2nd and 20th July, the 9th Division lost 7,500 all ranks. Again, the lessons of Trônes Wood could be seen. The occupier will always be at risk if the enemy artillery still has a clear field of fire.

The north of Longueval and part of Delville Wood were still under German control and Potter's 9th Brigade (3rd Div.) took up the task of the relieved 9th Division. The 1/NF, closely supported by the 13/King's and 12/West Yorks, followed the barrage, advancing from Pont Street. The north countrymen made good progress at first but under heavy fire from front and left, they withdrew to Piccadilly, and, a little later, to Pont Street where they were heavily shelled. Meanwhile, the 95th Brigade (5th Div.) assaulted the orchards in the north of Longueval which had resisted all previous attempts. Units from the 1/East Surrey and 1/DCLI also made initial progress, even crossing the Longueval/Flers road, seizing a strongpoint with the help of a party of Durham Field Engineers. A very determined counter-attack forced both battalions back to join other units at Pont Street.

The Commander in Chief, General Haig, after conferring with his Corps commanders on how best to secure Longueval and Delville Wood, decided that the XIII and XV Corps would make a combined attack, zero hour being fixed for 7.10am on the 27th July. The 3rd Division was relieved by Major General Walker's 2nd Division which was to attack Delville Wood. The 5th Division took up positions in south Longueval in preparation for its attack to clear the village and the western part of Delville Wood. During the early hours of the 27th, a maximum artillery barrage concentrated on both objectives. Patrols sent forward to establish the result of the bombardment sustained some casualties from a German counter barrage. However, the British barrage demoralised many Germans who hurriedly surrendered to

Right: one of the two former windmills at the western exit of Longueval.

Below:South Street, which connected Delville Wood to Ginchy had deteriorated to such a degree that the whole length was overlaid with railway sleepers.(both J. Bommeleÿn)

Troops undertaking difficult trench digging in the shattered remains of Delville Wood. [A Perret]

Kellett's 99th Brigade. At zero the l/KRRC and 23/Royal Fusiliers advanced through the blazing furnace of Delville Wood, making their way round the innumerable craters and the general abbatis of the shattered wood. On arrival at Prince's Street, the main central east-west ride, the attackers found many destroyed machine-gun emplacements and dead Germans. The surviving enemy either fled north or surrendered to the Rifles or Fusiliers. The supporting companies of the two battalions moved forward to consolidate. The 1/Royal Berkshire also moved up to protect the flanks, but the eastern extremity of the wood between the end of Prince's Street and Rotten Row was not consolidated, it being particularly vulnerable to enemy fire. However, the Rifles and Fusiliers had, in under two hours, arrived within 50 meters of the northern edge of Delville Wood. It seemed as though only German dead and wounded remained.

Meanwhile the leading battalion of the 15th Brig. (5th Div.), the 1/Norfolk with the 1/Bedford in support moved parallel with the 99th Brig. on their right. While waiting for zero hour, the left-hand company of the Norfolk battalion suffered heavily from the German barrage and the Bedford's general supportive role changed to one of immediate support. The Norfolk advanced up the western edge of the wood, followed closely by the Bedford which quickly made contact with 99th Brigade. The attack on Longueval itself proved more difficult, the preliminary barrage not having destroyed the hidden machine-gun posts in the ruins, the 16/Royal Warwick along with other units having to fight hard to make progress towards the north where the orchards were still obstinately defended by the enemy.

The inevitable counter-attack was delivered about 9.30am, the courageous German bombers defying the British barrage. The enemy was held for some time and although he sustained heavy losses, the surviving bombers resolutely continued the attack putting many machine-guns out of action. The Germans gradually infiltrated the wood just north of Prince's Street causing the British right flank to withdraw some way into the wood. Sergt. A. Gill, a platoon leader of the 1/KRRC died while displaying outstanding bravery defending his position. He was posthumously awarded the Victoria Cross. Virtually all communications had been destroyed by the barrage resulting in false, uncertain and alarmist reports being made.

During the night, the 99th Brigade's front was taken over by the 17/Middlesex and 2/South Staffordshire of 6th Brigade (2nd Div.) The following day brought an intensive German barrage during which the Middlesex and Staffordshires not only held firm but repelled a counter-attack by German bombers in the late evening. The 15th Brigade was relieved during the night by Gordon-Lennox's 95th Brigade (5th Div.). Duke Street, which crosses Longueval about half way north through the village was occupied on the morning of the 28th July. The next day, after a short bombardment, the 12/Gloucester secured a further 500 meters north of Duke Street while the 1/East Surrey made progress to the west of the wood. It was established later that the Germans were in a sorry state, having had hardly any sleep for several nights, the British barrage also preventing evacuation of the wounded.

The 13th Brigade (Brig.-Gen. Jones) was now brought into the action in the evening of the 30th July. Advancing from the gains of the previous day, it attacked the German strongpoints in the north

Left: Devils Trench in Delville Wood.
(Michelin)

Below: The shattered trees and cratered landscape of Delville Wood.
(Michelin)

of Longueval including the entry from the north-west. The preliminary barrage had achieved little and the 2/KOSB suffered casualties from the German barrage on the north-west corner of Delville Wood. One company of the Borderers managed to dig in and hold its position until relieved two days later. The support of the depleted 1/Royal West Kent did not enhance the situation. The Borderers and Kents were shelled so severely that the 1/Bedford of 15th Brigade moved up to take over the right flank. The 14/Royal Warwick (13th Brig.), in spite of keeping close to the barrage, was halted in front of heavy fire and had to go to ground in the shell craters. When the 16/Royal Warwick was brought up from reserve, there was some confusion amongst the mixed units having to endure the never-ending German bombardment and the rest of the day was spent in reorganising the positions in Longueval. In the evening the 15th Brigade relieved the forward units of the 13th Brigade.

Apart from barrage and counter-barrage, there was little activity on the 31st July which permitted various troop movements to be effected. Major-General P. Robertson, the new GCO of 17th Division, relieved the weary 5th Division which had sustained casualties of over 5,600 officers and men. A barrage was laid on Orchard Trench which exits Delville Wood from the north-western edge and links up with Wood Lane South. The barrage was intensified during the last five minutes but the leading companies of the 12/Manchester and 9/NF (52nd Brig.) were met by machine-gun fire, high-explosive and gas shells. The losses were heavy and with communications cut, some four hours passed before divisional H.Q. learned of the failed attack. Following a further barrage, Brig.-Gen. Trotter's 51st Brigade attacked in the afternoon of the 7th August to establish posts beyond Delville Wood. Heavy enemy fire forced units of the 7/Border and 8/South Staffs to go to ground in the wood but under cover of night, the 10/Sherwood Foresters managed to establish some posts in front of Longueval.

The British artillery continued its barrage upon the enemy positions while the corps commanders were considering the next plan of attack. Supplies were brought forward, the wounded evacuated and general reorganisation of units in the area was undertaken. This took several days but by the 18th August, the 41st Brigade of Couper's 14th Division was in position for its attack on Orchard Trench. While the 7/KRRC went forward, the area from Prince's Street to the Flers road was kept under heavy trench-mortar fire. The Rifles, close to the barrage, arrived at Orchard Trench which was occupied with little loss. A new trench was dug north to the Flers road. To the left of the Rifles, the 7/Rifle Brigade (41st Brig.) advanced, its left flank coming under fire from Wood Lane. It was evident 33rd Division's attack on Wood Lane North had failed and the Rifle Brigade managed only to secure a small section of Wood Lane at its southern end. On 21st August the 8/KRRC of the same brigade, moved up

Above: Delville Wood Cemetery, Longueval, shortly after the battle. The tree-stumps of what is left of the wood can be seen on the horizon.
(J. Bommeleÿn)

Left: Delville Wood Cemetery today
[Author]

in an effort to establish strongpoints in the German line in Delville Wood but this was repelled by heavy enemy fire.

The 9/HLI and 2/Worcester (100th Brig. 33rd Div.) led an attack between Wood Lane and the Flers road on the 21st. The Worcesters were late in getting away on the right, the HLI advancing alone, was beaten back by machine-gun fire. Three days later the same brigade attacked again with the 2/Worcester, 16/KRRC and 1/Queen's. All three battalions came under fire from the enemy barrage while waiting for zero hour. On the left the Queen's, covered to some degree by a smoke screen, stormed the German trench running east of Wood Lane. Some gain was made but a hostile barrage hindered the consolidation work.

Rain on the 25th turned the whole area into a quagmire causing much difficulty in supplying the troops. The limestone soil quickly turned to a sticky, slimy substance which clung to everything. Ground conditions deteriorated even more following a violent thunderstorm on the 29th. The bad weather also hindered the artillery work causing a postponement of the original attack planned for the 25th to the 30th August, then to the 1st September and finally to the 3rd. The last day of August saw the start of a short, three-day dry period. The RFC had reported much enemy movement to the north, a sign that reinforcements were on the way. The enemy barrage had also increased in density - all the signs were that a counter-attack was imminent. Capper's 24th Division took over the front line, the troops far from being in a good mood. They were tired and wet through as a result of their efforts to come up to the front line and now, were the subject of the German artillery's attention which seemed to increase in density by the minute. A British counter-barrage had no effect. The 1/North Stafford on the right flank of 72nd Brigade, dug forward to escape the barrage. The 8/Royal West Kent of the same brigade in position near the eastern edge of the wood had to withdraw under the barrage to Inner Trench. The German storm troopers seized a strongpoint in Cocoa Lane which runs north from the north-east edge of the wood. The left flank of the Kents and the 9/Royal Sussex (73rd Brig.) were assaulted by the 88th Regiment advancing down Tea Lane. The Germans were held off by machine-

Delville Wood with Longueval at the right-hand edge of the picture. [Author].

gun, rifle and artillery fire. The left hand battalion of the 73rd Brigade, the 13/Middlesex had sustained almost 400 casualties from the German barrage before moving forward. Advancing from their strongpoint in Wood Lane, the enemy forced the Middlesex, in spite of gallant resistance, to retire into Tea Trench and eventually to a point close to North Street. Other German units from the same regiment entered Orchard Trench from the north-west but were contained by the 2/Leinster, also of 73rd Brigade. It had been one of the most severe counter-attacks encountered by the Fourth Army. Virtually all communication cables had been damaged and it was some time before the true situation was known. When eventually, the result of the attack was established, plans were made immediately to recover the lost ground.

The losses had to be retaken and the 2/Leinster (73rd Brig. 24th Div.) bombed its way up Orchard Trench at dawn on the 1st September but superior numbers forced a retirement. A further attack was made a few hours later from the east of Pear Street but this also, was unsuccessful. It was vital that Orchard Trench and at least the southern end of Wood Lane were secured in order to prevent the Germans reinforcing High Wood to the north-west. At 6.30pm the 3/Rifle Brigade (17th Brig.) advanced under fire on a wide front and secured Orchard Trench and about 250 meters of Wood Lane, almost to the junction with Tea Trench. The attack ended here, the Rifles having sustained too many casualties to be able to continue against the well-fortified Tea Trench.

The 7th Division was entrenched between Waterlot Farm and the southern edge of Delville Wood. Two platoons of the 2/Queen's (91st Brig.) made a dawn attack on the east side of the wood but was checked by heavy fire. Another attack was made in mid-afternoon by the 1/N Staffs (72nd Brig.), forming the right flank of 24th Division. Although some progress was made by bombers working their way along Edge Trench, the gain could not be consolidated. The Queen's made a second attack on the 2nd but snipers and enemy bombers took advantage of their concealed positions. At the end of the day, the Germans still held the eastern corner of Delville Wood and the "Alcohol"(ii) trenches to the immediate east. This was to affect the imminent attack on Ginchy as will be seen later in the descriptions of the action at Objectives 25 and 26.

Only a small part of the southern section of Wood Lane had been secured and at noon on the 3rd, the 8/Buffs (17th Brig. 24th Div.) made a frontal attack while the battalion bombers attacked Tea Trench from the southern end of Worcester Trench. Fire from Tea Trench/Wood Lane junction prevented any progress on this and a subsequent attack four hours later. Under cover of dark, the 7/Northampton, on loan to the 17th Brigade, secured the extreme western end of Tea Trench. In the early hours of the 5th, Duncan's 165th Brigade relieved the 17th Brigade on the left flank of the 24th Division.

In the evening of the following day the 6/King's of the 165th Brigade attacked Wood Lane, but the bombers made little progress. However, its sister battalion, the 7/King's, after entering Tea Trench more or less at its centre point, made good progress bombing its way west towards the junction with Wood Lane. From this gain, under cover of night, the 7/King's established forward posts in readiness for an attack on Tea Support Trench (see Objective 27) which crosses the Longueval/Flers road, then heads first south-east and finally south towards Ginchy. During the 7th, the forward battalions were relieved by the 5/King's which attacked Wood Lane in the evening, securing another section of the lane. An enemy counter-attack after midnight on the 7th was repulsed by Lewis-gun fire. The 5th and 7/King's, having linked in Tea Trench, set about improving conditions in the battered trench. Patrols were sent up North Street, continuing to the Flers road, but the enemy was not sighted. Consolidation was constantly hampered by the enemy barrage and snipers.

On the 9th September the 5th and 6/King's extended their hold on Wood Lane. The northern sector of Wood Lane, included in the description of the action at High Wood under Objective 23, was presenting particular difficulties and each attack engendered heavy casualties. With Delville Wood and Longueval secured, it was important to maintain the hold in the southern end of Wood Lane to assist in the protection of 1st Division's right flank.

(i) Major "Billy" Congreve was the son of Lt.-Gen. Sir W. Congreve, GCO XIII Corps, also a holder of the Victoria Cross. The Congreve family had a long military history. Sir Walter's great-grandfather was the inventor of the Congreve Rocket used during the battle of Waterloo on the 18th June 1815. It has been reported the trajectory of the rockets was not too reliable and was received with little enthusiasm by the Duke of Wellington. A hundred years later, Sir Walter still carried the nicknames "Squibby" and "Squibs".
(ii) The group of trenches to the east of Delville Wood bearing names which brought back memories to the troops of a visit to the local pub, has been named by the author for the sake of convenience as the "Alcohol Trenches"

OBJECTIVE 22
POZIERES VILLAGE AND RIDGE
MOUQUET FARM

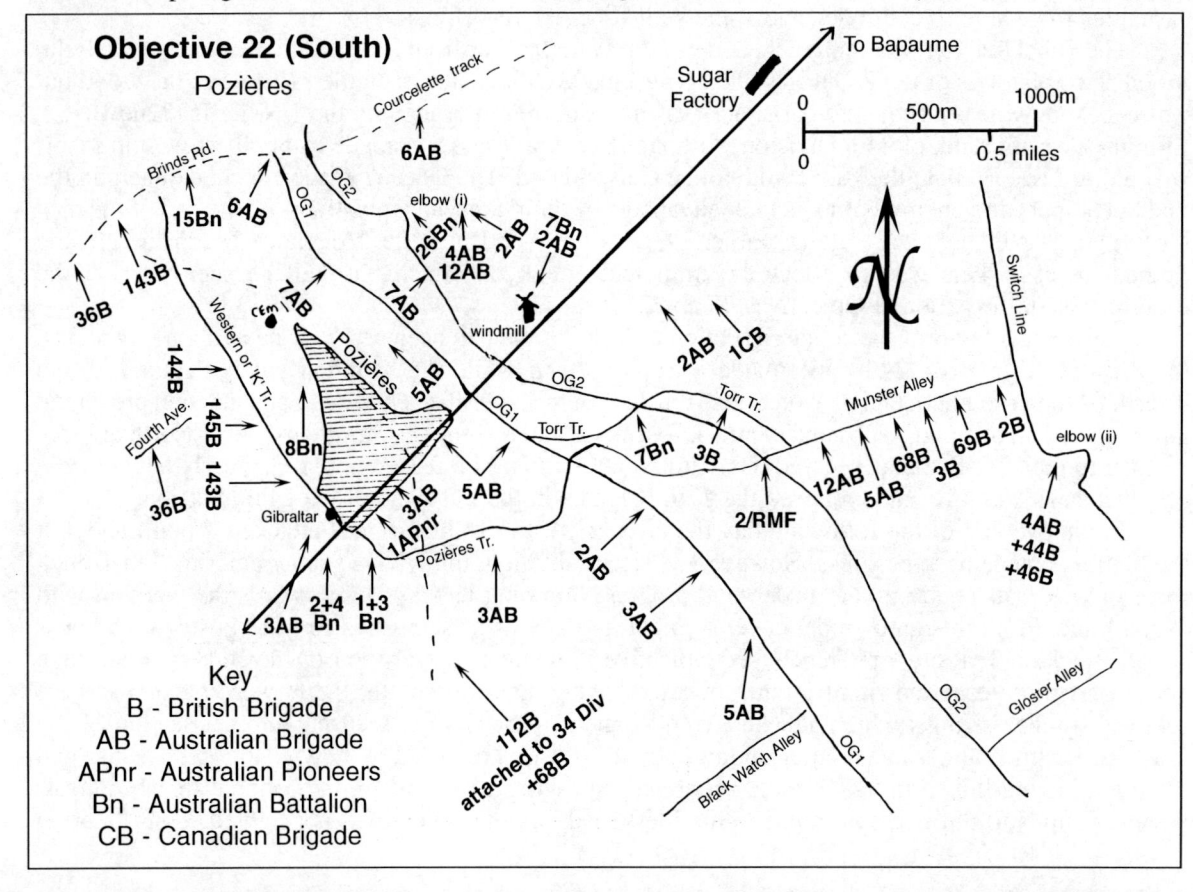

Lying on the main Albert to Bapaume road, the village of Pozières stood guard against any approach to Bapaume from the south-west. Following reports that Pozières had been evacuated, the 34th Division sent patrols forward in the early evening of the 14th July to verify the situation. On approaching the village, the men were met by fire and were forced to retire. The following day, after a sixty-minute bombardment a further attack from Bailiff Wood north towards Pozières was made by Brig.-Gen. Robinson's 112th Brigade which had been attached to 34th Division. Sixteen hundred meters of open ground faced the 8/East Lancs Regiment as it left the safety of the front line, the leading waves quickly coming under fire from hidden machine-guns. In ever diminishing numbers, they moved forward to within 350 meters of Pozières, just a few Lancashires managing to get a little further forward. Major-General Ingouville-Williams sought and obtained permission for a further barrage prior to a second attempt. This barrage did not silence the machine-gun nests. Even so, the East Lancs managed to get a little nearer to their objective. Brig.-Gen. Croft's 68th Brigade had replaced the 112th and, about 8.00pm on the 17th July, following an artillery and mortar barrage, the 12/Durham Light Infantry (DLI) attacked the German trenches protecting Pozières from the south. The Durhams were surprised by heavy machine-gun fire. It was obvious the barrage had not been successful. The men went to ground and when it was clear no further progress could be made, they were recalled. However, the 68th Brigade set to work improving the forward positions and making assembly trenches ready for the next attack.

With the fall of Ovillers, the way was clear to make an attack north towards Thiepval and east towards Pozières. The 48th Division which had been engaged at Ovillers, sent units from its 145th Brigade to reconnoitre the German defences west of Pozières but were met by machine-gun fire. The 144th Brigade made better progress, moving up the communication trenches on the higher ground north of Ovillers. In the early hours of the 21st, both brigades were engaged in a struggle lasting almost until dawn.

The 1st Australian Division, recently arrived on the Somme, had taken over the area of Black Watch Alley which formed the boundary between Gough's Reserve Army and Rawlinson's Fourth Army. Major-General Strickland's 1st Division was on the immediate right of the Australians. The 48th Division at Ovillers was to assist the Australian Division in its operations. The 2/RMF (3rd Brig.) took over the junction of Munster Alley and O.G.2 (2nd old German line) but were expelled later by machine-gun fire. A number of strongpoints were established between Bazentin-le-Petit Wood and Munster Alley. At dawn on the 20th July, an attempt to retake the previous gain of the Fusiliers was unsuccessful.

The preparation of assembly trenches from Black Watch Alley was undertaken at night and an attempt to move forward under cover of night to OG1 and OG2 by the 9th Queensland Battalion (3rd Brig.) was repulsed by machine-gun fire and bombing attacks.

After the preliminary bombardment which began at 7.00pm on the 22nd July, and while units of the 1st and 2nd Brigades of the 1st British Division were attacking the Switch Line West to the south of Martinpuich, the 1/Loyal North Lancs, (2nd Brig.) was engaged in an assault on Munster Alley. No progress could be made against the strong German defences.

The 1st Australian Division was now in position for the attack on Pozières. Zero hour was at 12.30am on the 23rd July. The last five minutes of the intensified barrage was concentrated upon the western perimeter of Pozières as far as the cemetery sited at the extreme northern edge of the village. A heavy barrage had been laid on Pozières Trench to the south of the village and the two OG lines to prepare for an attack from the south-east. Under enemy fire in their assembly trenches, the Australians crept forward into no man's land. The leading battalions, the 9th Queensland and 11th W. Australia (3rd Brig.) and the 1st and 2nd NSW (1st Brig.) leaped forward at zero hour across the flare-lit land, the Queenslanders securing Pozières Trench with relatively few losses. However, on the right, heavy fire was encountered between the two OG lines. Entering the trenches, the Australians continued their attack with bombs. The 10th S. Australia was sent forward to support but each attack on the right flank was met by stubborn resistance, the attackers being subjected to machine-gun, rifle and sniper fire as well as bombing counter attacks. When dawn came, OG1 had been taken to its junction with Pozières Trench.

The attack on the village had fared rather better. The 3rd and 4th Battalions (1st Brig.) and the 10th and 12th Battalions (3rd Brig.) had reached the western edge of Pozières close to the Bapaume Road and by 3.00am were digging themselves in amongst the ruined houses. Patrols were sent out towards the north-west of the village. Six guns and about a hundred prisoners were taken. The cellars were jubilantly cleared by the Australians and it seemed as if there was little opposition left in the southern end of Pozières. However, as they made their way north, a good number of Australians fell victim to German snipers, perhaps through being a little too elated and carefree. A German counter attack about 5.30am was repulsed with little loss.

It has been mentioned that Maj.-Gen. Fanshawe's 48th (South Midland) Division was north of the Bapaume Road and to the east of Ovillers. Brig.-Gen. Done's 145th Brigade being on the right and

Nicholson's 144th Brigade on the left. The men were to bomb up the trenches following the release of "flame mortars", a new weapon consisting of small oil-filled drums which exploded on impact and having a range of about 200 meters. The 6/Gloucester was shelled in the assembly trenches and murderous machine-gun fire cut the men down as soon as they went over the top. Just a few bombers from this gallant battalion managed to enter the German line. The 4/Gloucester, wanting desperately to help its sister battalion, was ordered to halt, it being obvious that little or no good could come of a supporting action at this time. The 5/Gloucester (145th Brig.) was also met by heavy fire, the battalion suffering many casualties. However, the 4/OBLI and 4/Royal Berkshire to the right of the 5/Gloucester obtained a better result capturing a long section of trench south of the railway. Dawn saw the Germans retreating towards Pozières. The 1/Buckinghamshire (145th Brig.) attacked at 6.30am. Following close on the heels of the creeping barrage, the men were upon the Germans before they could mount their guns. About 150 prisoners were taken.

A further Australian attack was planned for 4.00pm but reports by the 4th Squadron R.F.C. gave indications that Pozières had been evacuated. As a result, it was decided that patrols should be sent forward, to be followed by infantry brigades. Without waiting for authorisation, the 2nd NSW incorporated a German strongpoint, the Panzerturm in its line. It was renamed "Gibraltar" by the Australians.(i) The 8th Victoria, hitherto in reserve, was brought up to consolidate between the main road and the cemetery, about half the length being completed before midnight. Clearance of the ruins and cellars continued well into the night.

The 143rd Brigade relieved the 145th and the 5/Royal Warwick on the right tried unsuccessfully to link up with the Australians in the village. The 24th July was relatively calm, only the 48th Division was engaged in some bombing activities. Maj.-Gen. Walker, commanding the 1st Australian Division spent the day planning the attack of his troops on the 25th. His decision was that the 2nd and 3rd Brigades should secure the OG lines, zero hour being 2.00am. An hour later, the clearance of Pozières would begin. The bombardment began in the late evening of the 24th attracting a retaliatory barrage from the German artillery. The intensity of the British barrage increased on OG1 a few minutes before zero. The 5th Victoria rushed forward and seized OG1 and entered OG2. Here, bombed on both flanks, the men had to retire to OG1, repulsing German attacks on this position. After a fierce struggle, the Germans retained OG1 as far as the railway while the Australians blocked OG1 about 200 meters further down. Further to the right, the 9th, 10th and two companies of the 7th loaned to 3rd Brig. from 2nd Brig. were also engaged in a bombing action trying to clear the extreme right of Pozières Trench connecting the two OG lines. The action was bitterly contested. The final result was that both sides remained in close proximity, the Germans remaining near the junction of OG2 and Munster Alley which Strickland's 1st Division had not managed to secure. The 10th S. Australia captured a strongpoint in OG1.

At 2.00am on the 25th July the 1/SWB (3rd Brig. 1st Div.) attacked Munster Alley. Initially making progress with bombs, it was obliged to retire against machine-gun fire from Pozières Windmill on the crest of the ridge. Early next morning the 2/Welsh of the same brigade advanced without a preliminary barrage and joined the Australians near Munster Alley. Maj.-Gen. Babington's 23rd Division relieved the 1st Divison apart from two companies of the 2/Welsh which remained near Munster Alley. The casualties of the 1st Division were over 3,000 all ranks during their engagements between the 11th and 27th July. The 2/Welsh advanced towards the Switch Line but in spite of taking a heavy toll of the counter-attacking Germans, the Welshmen were forced back.

At 3.30am on the 25th July the 12th S. & W. Australia advanced north towards Pozières with the intention of entering behind the Germans facing the 48th Division. The II Corps artillery had been shelling the eastern side of Pozières and later ranged its batteries on the windmill sited on the crest of the ridge, then veering north-west on to Mouquet Farm. In spite of the artillery support, the 12th Battalion came under heavy fire from OG1 near the main road and was unable to link up with the 11th. This battalion had its own problems being caught by fire from the 8th Victoria which, in the half-light, had mistaken the 11th Battalion for the enemy. The unfortunate 11th Battalion also suffered so severely from a German barrage that a withdrawal to the starting line was ordered.

The 8th Victoria Battalion on loan to the 1st Brigade, advanced steadfastly through the village, eventually arriving at the cemetery where it consolidated and occupied a number of outposts on the outskirts of the northern edge of Pozières. During this action, Private T. Cooke of the 8th Battalion fired his Lewis gun when all his gun-team was injured or killed, until he himself was killed. He was awarded a posthumous V.C. for his outstanding courage. The 1st Australian pioneers were busy digging a redoubt and the 3rd and 6th Battalions were brought forward to help with consolidation work. Many men were lost while digging under the relentless barrage. Meanwhile, the 4th Battalion was working its way up Western Trench, taking a heavy toll of the enemy fleeing over the open country. The 4th Battalion eventually linked up with the 8th at the cemetery.

A German counter-attack about 8.30am was stopped by the combined artillery of the 23rd Division and 1st Australian Division. The 3rd Australian Brigade did sterling work accounting for many of the Germans who, for the most part, turned and ran for shelter on the far side of the ridge. The enemy did however retaliate with a long heavy barrage using every type of shell which the Australians bore with great courage. Some of the exhausted Australians were relieved by Maj.-Gen. Legge's 2nd Australian Division., the 5th NSW Brigade taking over from the 3rd on the right flank. After midnight on the 26th, the 7/Royal Warwicks (143rd Brig. 48th Div.) finally linked up with the Australians north-west of the village. Activity continued throughout the night, the 20th NSW forcing an entry into OG1 near the Bapaume road but the hold could not be retained. Enemy shells were raining on all captured positions, hindering all movement and supplies. In the afternoon the 17th NSW along with bombers of the 18th were fighting in Munster Alley which joins OG2 to the Switch Line south-west of Martinpuich. The relief of the remainder of the 1st Australian Division was completed during the night, the left flank being taken over by the 6th Victoria Brigade.

During the 27th July, indecisive actions continued on both flanks. The 8/Royal Warwick of 143rd Brigade bombed north towards the Courcelette Track while the Australians on the opposite flank continued their fight to occupy Munster Alley. The British artillery managed to lessen the intensity of the German barrage which helped the men digging forward trenches.

Maj.-Gen. Legge finalised his plans for the next assault timed for 12.15am on the 29th July. These included a frontal attack on OG1 south of the main road, OG2 from the main road to Munster Alley while to the north, the continuation of the OG lines as far as the Courcelette Track. The carefully planned barrage ranged past the above objectives to include Mouquet Farm and the village of Courcelette. It was particularly important to cut the wire. The men were assembled before midnight and already making their way forward before the barrage lifted.

The 7th Brigade was to push through the centre of Pozières, east of the cemetery to attack the OG lines north of the main road. At zero, the 28th Battalion, followed later by the 25th and 26th Battalions, broke through the advanced German posts, occupied OG1 but were cut down before uncut wire in front of OG2. In spite of three courageous attacks, OG2 could not be occupied. The few men who did enter were not seen again. The enemy fire was causing horrendous casualties and the 7th Brigade retired to OG1. Too few in number and under continuous fire, they had to retire to their starting line. They had done all they could against an impregnable defence.

The 6th Brigade, moving north on the left of the 7th, reached the vicinity of the Courcelette Track with little loss. The Track had been completely deformed by the barrage and the 23rd Victoria advanced further, still looking for its objective. Catching up with the barrage the Victorias sustained many casualties before being ordered to retire. The failed attack on the right resulted in the formation of a defensive flank from Courcelette Track towards the cemetery.

The 48th Division had been relieved by Maj.-Gen. Scott's 12th Division on the 28th July. The 48th had been in the line for twelve days and its casualties amounted to almost 2,800 all ranks. The 11/Middlesex (36th Brig.) moved over open ground to secure the extension of Western Trench in order to keep in touch with the 23rd Victoria Battalion. The attack was repelled by grenades and rifle fire.

The Australian 5th Brigade's attack on OG1 south of the main road had been preceded by a mortar barrage which had not silenced the German machine-guns. The 20th battalion, in spite of a resolute advance, was brought to a halt by very heavy fire from the start. This created a hopeless situation for the 17th Battalion's assault on OG2. Both battalions suffered heavy casualties and the men were withdrawn. The British 23rd Division continued the fight in the area around Munster Alley. The 7th Australian Brigade was withdrawn to the reserve position. During the 2nd Division's short but bloody encounter, it had lost almost 3,500 men.

The area of the re-entrant of Munster Alley and the Switch Line had been obliterated by shell fire and now took on the appearance of a lunar landscape. Across this barren land, the Australians made a further attack, gaining a little more ground. The 23rd Division's 68th Brigade continued the fight during the night and gains were won and lost. During the night of 28/29th July, the 10/DOW (69th Brig.) joined in the attack but progress in Munster Alley was minimal in spite of the very useful manual help of the Australians with supplies and fortifications. Better progress was made in Gloster Alley to the south east, and by 5.30am the following morning, the men were within 25 meters of the Switch Line.

A little before midnight on the 3rd August the 8/Royal Fusiliers (36th Brig. 12th Div.) attacked Fourth Avenue securing the south-western section. At the same time the 6/Buffs (37th Brig.) neutralised a strong point at the bottom end of the trench. Bombers from both battalions then forced an entry towards the centre of the long Ration Trench (Gierich Weg on German maps) a little way to the north-west. Ninety two prisoners were taken. Patrols were sent out towards Mouquet Farm but found the main road from Pozières to Thiepval blocked by the enemy.

While the combined artillery of Anzac and II Corps laid a bombardment on the German positions, the Australian infantry, pioneers and engineers were busy digging forward assembly trenches. Many men were lost during a hostile barrage while engaged in this work. Two inch mortars were being used to cut the German wire. Zero hour was fixed at 9.15pm on the 4th August. The 18th Battalion of the 5th NSW Brigade attacked on the left from south-east of Pozières securing OG1 with little loss. The assault on OG2 was a more desperate affair involving hand-to-hand fighting and the supporting companies were rushed forward to help. The shell-damaged German line was entered, consolidation work and setting up of Lewis-gun positions beginning without delay. The 20th Battalion on the right formed a defensive flank in Torr Trench.

Brig.-Gen. Paton's 7th Brigade with battalions from Queensland (25th), Queensland & Tasmania (26th), South Australia (27th) and West Australia (28th) took up their front line positions, the 27th on the right, 25th in the centre and the 26th on the left. The 22nd Battalion of the 6th Victoria Brigade took part in the attack on the extreme left flank. The leading waves easily entered OG1, but successive waves suffered heavily from the German retaliatory barrage. There was some confusion in the failing light and the general chaos of battle causing the leading waves to separate. The 27th Battalion on the right was well under strength as they assaulted the OG2 lines between the Elbow and the Bapaume road. The other battalions had veered to the left but this allowed for occupation of OG1 towards the main road. From here the attack continued on to OG2 which was extensively damaged by artillery fire making recognition of the objective somewhat difficult. The windmill, or, more accurately, the ruins of the mill, on the highest point of the ridge, lay before them and although no immediate attack could be mounted, most of OG2 in this area was secured. The British barrage had been particularly effective, the Australians finding hundreds of German dead in the trenches. Parties of the Australian 6th Brigade on the left flank endeavouring to close the gap between OG2 and Courcelette Track, came under heavy machine-gun fire from the north and the attack faltered. The enemy strongpoint was not silenced until late evening on the 5th and until then, the 6th Brigade was pinned down. The barrage was terminated at midnight to allow reconnaissance to be made. Consolidation of the gains began under a hostile barrage followed by the inevitable infantry counter-attack. The Germans, advancing from Courcelette, were cut down by the hurriedly erected machine-guns and some came forward, hands high in the air to surrender. Many took shelter in the cratered ground - these were bombed by Stokes mortars. The 27th and 28th Battalions were now close to the highest point of Pozières Ridge and dug in around the mill and in OG2 near the Bapaume road while the 26th Battalion, established in OG2 north-west of the windmill near the Elbow, made a defensive flank to protect the other two battalions. The 26th Battalion repulsed an attack by about a hundred Germans before dawn and the 24th Victoria Battalion (6th Brig.) was brought forward from reserve.

Meanwhile the 36th Brigade of the British 12th Division advanced to capture Ration Trench to its junction with Western Trench, also known as K Trench. The 7/Royal Sussex and 9/Royal Fusiliers made a frontal attack while the 8/Royal Fusiliers came in from the left. The frontal attack resulted in a long and bitter contest and eventually Ration Trench was secured a short way east of the Thiepval road. Parties of the Sussex advancing north astride K Trench were halted by enemy fire which prevented the other battalions from securing the junction of Ration Trench and K Trench. The Germans defended the trench with the utmost vigour and for a time the result was indecisive. The lower end of Ration Trench had been successfully assaulted by the bombers of the 6/Royal West Kent and 6/Queen's, both of 37th Brigade. As daylight came, over a hundred Germans isolated in the craters, surrendered. A little later the Germans counter-attacked with flame-throwers and grenades driving the 9/Royal Fusiliers before them. Moving out of Ration Trench and quickly mounting Lewis guns, the enemy was brought to a halt.

The weary Australian 5th Brigade was relieved by Brig.-Gen. Glasfurd's 12th Brigade from the 4th Australian Division. The 45th NSW Battalion effected the relief to the south-east of the main road while the 48th S. & W. Australia replaced the 7th Brigade near the Elbow.

To the south the 13/Durham (68th Brig.) gained a further 50 meters in Munster Alley during the night of 4th/5th August but its attack failed over open ground against Torr Trench which links Munster Alley to the OG lines. The 69th Brigade took up the fight once again in Munster Alley, the 8/Green Howards gaining a further 150 meters including the junction with Torr Trench(ii). The 11/West Yorks now joined the attack and over twenty prisoners were taken. Again, the Australians freely gave their help in Torr Trench.

Maj.-Gen. McCracken's 15th Division relieved the 23rd on the 8th August. 12th Division's 36th Brigade was also relieved by the 35th Brigade on the same day. The 35th Brigade had little time to set up its defences, coming under attack from both sides of Ration Trench. The left held but the 5/Royal Berkshire had to retire some distance.

During the night of the 6/7th August, a particularly intensive enemy barrage was ranged on the gains of the previous two days causing heavy casualties and destroying many forward posts. The newly arrived 48th Battalion had to check a strong German counter-attack from the direction of Courcelette. Similarly, the freshly arrived 14th and 15th Battalions of 4th Brigade were called upon to check a German attack between themselves and the 48th Battalion near the Elbow. There was fierce close-quarter fighting involved in meeting the counter-attack, the Germans eventually being routed.(iii) At the same time the 45th NSW Battalion with units of the 69th Brigade (23rd Division) made further progress in Munster Alley.

A combined attack by the Australian 4th Brigade and the 35th Brigade from the British 12th Division was planned for 9.20pm on the 8th August. The Australian 15th Battalion advanced north, crossing Brinds Road, which forms part of the Courcelette Track, into Park Lane which in turn links Ration Trench to OG1. Park Lane was secured and an entry made into OG1. On the left flank the 7/Suffolk of 35th Brigade encountered uncut wire while going up Ration Trench which, along with fire from K Trench, brought the advance to a halt. The junction not being made, the Australians withdrew west of the lane leading to Mouquet Farm. Following a bombardment, an attack by the 16th Battalion at midnight secured the junction along with three machine-guns, a flame-thrower and about thirty prisoners. The Australians now had complete possession of Park Lane.

A joint attack by the 15th Division and Australian 4th Brigade in the late evening of the 12th August was made against the Switch Line between The Elbow(iv) and Munster Alley. The short bombardment was ineffective and the 12/HLI (46th Brig.) on the right could make no progress against the enemy machine-gun fire. On the left, the 7/Royal Scots Fusiliers and 6/Camerons (45th Brig.) occupied the bomb-cratered objectives and consolidation work began immediately. The Camerons were in touch with the Australians near the junction of Munster Alley and the Switch Line and men set to work, using "pipe pushers", making a new communication trench towards the new front line. Further attemps to gain more ground in the Switch Line took place on the 12th when the 7/Camerons (44th Brig.) made a successful frontal attack. Their casualties mounted under heavy enemy fire as well as counter attacks by German bombers arriving from the Elbow, some 500 meters to the south-west of the junction of Munster Alley and the Switch Line. Help came from the 8/Seaforth, also of 44th Brig. which attacked and secured the Elbow with the bayonet. The 11/HLI (46th Brig.) then pushed east down the Switch Line to a point about 120 meters east of the Elbow.

The next attack was planned for 10.30pm on the 12th August and the bombardment was already well under way. Under cover of the barrage the 13th and 16th Battalions (4th Brig. 4th Aust. Div.) moved forward from Park Lane. The 13th Battalion established a forward post to the south-east of Mouquet Farm and, although sustaining some casualties from a German bombing counter-attack, held on. The 16th Battalion advanced to the crest of an incline from where they could see the quarry, only 200 meters south of Mouquet Farm. Both battalions came under heavy artillery fire in the morning and although a counter-attack was stopped, the artillery had reduced the Australian gains to a few

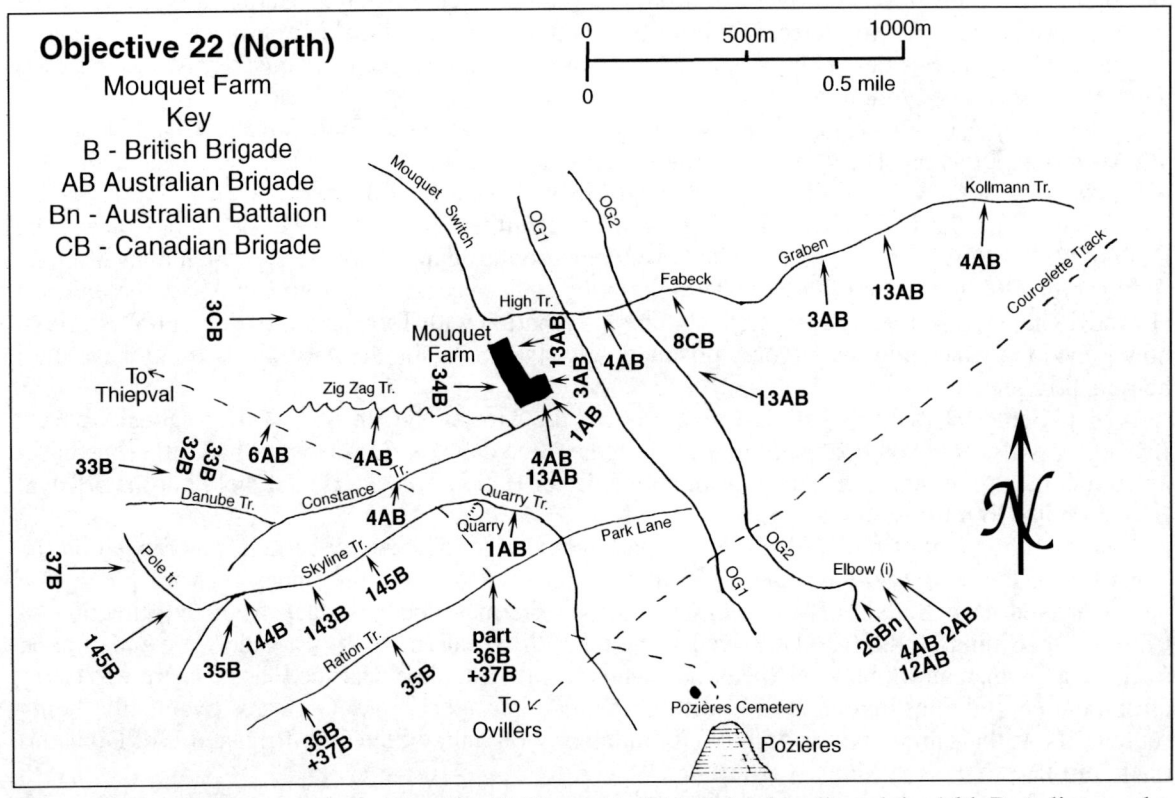

Objective 22 (North)
Mouquet Farm
Key
B - British Brigade
AB Australian Brigade
Bn - Australian Battalion
CB - Canadian Brigade

small strongpoints. The 50th S. Australia of Glasgow's 13th Brigade relieved the 16th Battalion on the left flank.

Zero hour had arrived and the 13th Battalion being already in place, consolidated its position while keeping in touch with the 50th Battalion on the left. This battalion set off from Park Lane and Ration Trench north towards Mouquet Farm. There was a little confusion, some platoons not having received their orders, but the left flank, in touch with 12th Division, reached the Thiepval road. One party reached the ruins of Mouquet Farm but, unsupported, was forced to retire.

The 7/Norfolk and 9/Essex (35th Brig. 12th Div.) advanced close behind the barrage from the south-west towards their objective, Skyline Trench. Arriving at the crest, or skyline from which the trench takes its name, the men from the East of England entered and secured the trench with little loss, taking some twenty prisoners. They had a commanding view of the valley below them, but the Germans had the exact range. Two strongpoints were made and manned in the trench, the remainder being ordered to withdraw down the reverse slope.

The 37th Brigade (12th Div.) had been engaged in attacks to the west to contain the enemy, the 7/E Surrey and 6/Royal West Kent suffering quite heavy losses in this "holding" operation. The 8/West Yorks (146th Brig. 49th), positioned near the Nab, assaulted a barricade in the German front line and gained a little ground.

While preparations were being made for the next night attack on the 14th, the 12th Division was relieved by Fanshawe's 48th Division and the 13th Brigade (4th Aust. Div.) replaced the 4th Brigade except for its 13th Battalion which

Mouquet Farm circa 1915
(J. Bommeleÿn)

remained in the line. The 51st W. Australia took over the positions in the OG Trenches.

The enemy placed a heavy barrage on Skyline Trench on the high ground. It has been mentioned the trench was only manned by sufficient men to meet the needs of the two strongpoints. Whereas this undoubtedly saved many casualties, it did leave the trench open to attack from the left flank which is what happened and two companies of the 4/OBLI (145th Brig.) were insufficient to withstand two counter-attacks, the battalion losing its hold on the trench apart from a small section near the Thiepval road. However, the OBLI, now in touch with the Australians, held on to the northern end. Further German counter-attacks on Ration Trench were repulsed but efforts to regain Skyline Trench failed against superior numbers. At dawn on the 14th, a further assault to regain Skyline Trench by the 4/Royal Berkshire of the same brigade was halted by heavy fire and with the Germans again in possession of the trench, any attack towards Mouquet Farm south of the Thiepval road was fraught with danger. As zero hour approached, the German barrage caused such damage to the communication system that it became impossible to co-ordinate the coming attack. Even so, the attack was made with dire results. The 51st W. Australia on the right was cut down by machine-gun fire and had some difficulty in rallying at its start line. However, the 49th Queensland Battalion improved the situation by securing ground to the east of the OG lines. In the centre, the 13th Battalion did well in establishing posts in a short section of the long Fabeck Graben between OG1 and OG2. These were good posts but untenable without support and the 13th Battalion had to make a fighting withdrawal before daylight. The 50th Battalion on the left suffered horrendous casualties from the barrage. This did not stop the Australians from trying to dig in near the quarry but it became evident they would have to retire.

At 10.00pm the 1/Buckingham (145th Brig.), moving north up the communication trenches leading off Ration Trench, set about re-occupying the lost Skyline Trench. With great determination and courage it had almost cleared the trench before dawn on the 15th. Meanwhile, the 6/Gloucester (144th Brig.) was engaged in similar hard fighting at the opposite end of Skyline Trench, but its efforts were not successful. The gains in the battered ground were subjected to yet another German barrage - it was becoming difficult to recognise trench lines, and the 1/Buckingham had to retire to the northern sections of the communication trenches through which it had passed earlier. The 5/Gloucester and 1/Buckingham plus the 4/Gloucester of 144th Brigade renewed the attack on the trenches at the south-western end of Skyline Trench and although some inititial progress was made, no permanent gain was achieved.

The tired troops of the 4th Australian Division were relieved by Maj.-Gen. Walker's 1st Australian Division, Walker taking over command of the Anzac front in the afternoon of the 16th August.

Preparations were then put in hand for the next attack at 5.00pm on the 18th August. During the intervening period the Australians checked a counter-attack during the night of the 16th. The 145th Brigade of 48th Division, holding Skyline Trench reported enemy movement to its front, called for and received a barrage which broke up the German units during assembly operations. On the left flank, aided by a smoke screen and machine-gun fire by the 49th Division from the Thiepval sector, the 5th, 6th and 7/Royal Warwicks, all of Dent's 143rd Brigade, advanced against stubborn resistance but the Warwicks advance lost none of its impetus. The enemy resolution faltered, many Germans surrendering to the elated Midlanders. On the opposite flank, the 4/Royal Berkshire (145th Brig.) supported the Warwicks from the high ground and further gains were made during the early hours of the 19th. German prisoners amounted to 425 all ranks and the new line now ran west from Skyline Trench to the original German front line north-east of Authuille Wood.

Meanwhile, the 1st Australian Division had been busy preparing advanced assembly trenches and saps beyond OG2 south-east of the Bapaume road in readiness for the coming attack. At zero the 8th Battalion (2nd Victoria Brig.) made three courageous assaults on the German line which was just over the crest of the rise. They were met by a curtain of impassable fire making it impossible to get to grips with the enemy. The 7th Battalion of the same brigade, advancing north of the light railway came under heavy fire from a German strongpoint near the Courcelette fork on the Bapaume road, the battalion suffering heavy casualties. All attemps to subdue the strongpoint ended in death for the assailants and the survivors of the 7th Battalion were back in their starting positions before dawn on the 19th. The 6th Battalion, also of the same brigade, had been busy digging north-west of the main road to link up with the Elbow, thereby creating a continuous line from the Elbow to the left flank of the 7th Battalion.

The Australian 1st Brigade's objectives were Fabeck Graben and the trenches immediately south, Quarry Trench and Mouquet Farm. The four battalions had been raised in New South Wales. Assembled to the east of the OG lines, the 3rd Battalion was held up waiting for the creeping barrage to move forward. In the centre, the 4th Battalion forced an entry into Quarry Trench and established posts near the farm track. Subjected to a number of counter attacks during the night, the battalion held this important position. To the left a line was consolidated between the quarry and Constance Trench.

A number of reliefs were effected during the 19th August, the Australian 3rd Brigade taking over from the 1st while the 144th Brigade of 48th Division replaced the 145th.

German reconnaissance aircraft had spotted the Allied preparations which were subsequently hampered by a hostile barrage. Notwithstanding, the work continued, the attack taking place at the scheduled time, 6.00pm on the 21st August. On the right, the 10th Battalion of the 3rd Australian Brigade advanced towards the eastern end of Fabeck Graben and made an entry in spite of heavy machine-gun fire from the east. Without support, the position could not be held and the battalion established a line just south of the track leading from Mouquet Farm to Courcelette. On the left, the 12th Battalion advanced north-north-west towards Mouquet Farm, some men reaching the ruins where clearance of the cellars began. The complex system of trenches and cellars with numerous exits and entries allowed the Germans to beat off these bombing attacks. Other units of the same battalion neutralised a strongpoint near the junction of OG1 and Fabeck Graben. Further assaults were made during the night and throughout the following day, all under a heavy barrage. These actions were largely indecisive. The Australian 6th Brigade took over the front in the evening of the 22nd, the 5th Brigade taking over the right flank during the following day.

Using heavy and medium howitzers, the British barrage shelled the enemy positions while final preparations were under way for the next attack scheduled for 4.45am on the 26th August. The 24th Battalion on the right could make no progress against the heavy fire and bombs coming from Fabeck Graben. Just left of centre, the 21st Battalion passed over Constance Trench and arrived at their objective, Zig Zag Trench which links Constance Trench near Mouquet Farm to the Thiepval road a little farther north. The men were surprised to find the trench unoccupied. The combined British and German barrages had changed the terrain to such an extent that almost all landmarks were unrecognisable. This, combined with the smoke, mud, flying debris and the deafening noise, caused the 21st Battalion to lose its way, now heading north with Mouquet Farm on its right front. The men linked up with a company of the 22nd Battalion engaged in establishing posts on the left flank and came under fire from Mouquet Switch to the north, as well as from Fabeck Graben and hidden machine-gun nests in the ruins at Mouquet Farm. The Australians fought bravely but were hopelessly outnumbered and the survivors were taken prisoner. The 4th Brigade of the 4th Australian Division took over from the 6th Brigade in the evening of the 26th. Twenty-four hours later, the bombers of the 14th Victoria Battalion assaulted two strongpoints, one at at junction of the lane leading to Mouquet Farm and the Thiepval road(v) and the second at the eastern end of Zig Zag Trench. The attacks on both points were met by fierce resistance and the bombers, after a bitter struggle, were obliged to retire. The Germans occupying the underground network of tunnels and cellars beneath Mouquet Farm enabled them to resist this attack just as they had done against the 3rd Brigade five days earlier.

The 143rd Brigade (48th Div.) was in action again in the evening of the 27th, the 8/Royal Warwicks advancing either side of Pole Trench in readiness for an attack on the western end of Constance Trench. The battalion was in front of its own barrage and veered to the right giving the enemy the opportunity to make an immediate counter-attack from both trenches. The Warwicks held out until diminishing numbers forced them to retire. On the left, bombers from the 5/ Berkshire and 5/Gloucester, both of 145th Brigade, successfully secured a series of small trenches to the south-west of Pole Trench.

Some of the brigades of the 2nd Australian Division were relieved by the remaining brigades of 4th Division on the 28th August. During the night of the 29th an attack was launched by the 13th and 16th Battalions (4th Brig.) against Fabeck Graben, Mouquet Farm and Zig Zag Trench. Heavy rain fell through most of the 29th during which time a heavy bombardment prepared the way for the 4th Brigade's attack. The trenches were filling with sticky, chalky sludge making any movement a test of physical endurance and at 11.00pm, the men struggled forward under the German counter-barrage. The Germans were also suffering from the same awful trench conditions but with the advantage that they were defending from fixed positions. The 13th Battalion, in spite of these obstacles, seized the partly constructed Kollmann Trench to the immediate east of Fabeck Graben. Under heavy fire from the waiting enemy in Fabeck Graben, other units pushed on, refusing to be beaten, and entered a short section of the Fabeck. It was difficult to use their rifles and Lewis-guns which were choked with white mud. The 16th Battalion's assault on Mouquet Farm and Zig Zag Trench was doomed to failure in the terrible conditions. The Australians did not lack courage, the odds were simply stacked too high against them. Just a few men on the right flank managed to enter Fabeck Graben, and these, along with the 13th Battalion a little further up the trench, were forced to withdraw against superior numbers. A proposed attack by the 12th Brigade which had relieved the exhausted 4th, was postponed due to the weather and it was at this time that the 1st Canadian Brigade (1st Canadian Division) began to take over the right flank of the Anzac front.

The Australian 13th Brigade made another assault on Fabeck Graben on the 3rd September, approaching this time from the west along the Courcelette Track. On the right, the 49th Queensland secured its section of Fabeck Graben after a bitterly contested struggle. The Germans did not surrender, they were killed or wounded. The centre was assaulted by the 52nd Battalion (S. & W. Australia & Tasmania). The bombers forced an entry whereafter the fight was with the bayonet and rifle butt. Here again, the enemy resistance was stubborn and the Australians could not develop their entry and did not, therefore, completely secure the area allotted to them. On the left, the 51st Battalion (W. Australia) dashed forward so quickly it arrived at Mouquet Farm before the defenders could man their guns and some 120 prisoners were taken. Unfortunately the three battalions were not linked and casualties had been heavy. Following a heavy barrage on the British positions, the German counter-attack was launched. The 49th Battalion repelled the enemy but the weaker hold of the 52nd in the centre could not withstand the assault, the Germans pushing the 52nd back and isolating two companies of the 51st in a trench beyond Mouquet Farm. The farm itself was lost, the Germans having entered it through undiscovered tunnels from the north-east. The remaining battalion of the 13th Brigade, the 50th (S.Australia) hurried forward to support the 49th in their hold on Fabeck Graben. During the afternoon units of the 13th Canadian Battalion reinforced the Australians. It must have been a strange meeting between men from opposite ends of the earth struggling to capture what looked like a heap of dust and rubble and whose name, Mouquet Farm, has gone down in history. Further Canadian replacements arrived the next day and the Germans were expelled from their gains by bombers. Fabeck Graben was then barricaded and a link-up made with the Australians astride the OG lines. All units of the Canadian 1st Division had now arrived and its CO, Maj.-Gen. Currie took over command from the 4th Australian Division on the 3rd September. The 13th Australian Brigade repulsed a number of counter-attacks before being relieved by the 3rd Canadian Brigade under enemy fire on the 5th September.

The 1st, 2nd and 4th Australian Divisions had attacked the bastions of Pozières and Mouquet Farm showing great courage, almost to the point of recklessness, but such is the nature of these lithe, strong and sun-tanned men from the great continent of Australia on the far side of the world. The Aussies or diggers as they were, and still are, affectionately called, cared less for authority and discipline than the will to conquer. If they could not always achieve their objective, it was never for lack of the will to do so but purely through the fact that human flesh cannot triumph without a disproportionate loss of life against established machine-gun and mortar positions. Without hesitation, they threw humble flesh and bones against the deadly fire and perished in their thousands. The cost of taking the small village of Pozières with a population today of about 250 and of capturing Mouquet Farm, known as "Moo-Cow" Farm to the Australians, just to the north was immense, 23,300 Australian officers and men were missing from the roll call.

After several attempts, the Germans retook possession of Fabeck Graben on the 8th September. Elsewhere, the continuous German counter-attacks were checked. On the 9th, the 2nd Canadian Battalion (1st Brig. 1st Can. Div.), south of the Bapaume road, was engaged in a bitter hand-to-hand fight before securing and consolidating about 500 meters of the German front line trench astride the railtrack in the direction of Martinpuich. About 60 prisoners were sent back and the battalion was in touch with the left wing of the 15th Division of Pulteney's III Corps. The Canadians were subjected to a heavy barrage followed by a number of strong counter-attacks. They stuck to their task and checked each German attack. During the defence of the trench, Cpl. L. Clarke of the 2nd Canadian

Australian War memorial in Freemantle, Western Australia (D. R. Wood)

Battalion, with just a few men attacked and routed a party of twenty Germans then, although wounded in the leg, took one of them prisoner. For this action Clarke was awarded the Victoria Cross.

Maj.-Gen. Turner's 2nd Canadian Division took over the right wing of II Corps and, a few hours later, Lipsett's 3rd Canadian Division took over Mouquet Farm from the 1st Canadian Division which went into corps reserve.

In the early hours of the morning on the 16th September the 1st CMR of the Canadian 8th Brigade had prepared strongpoints at the western end of Fabeck Graben and some twelve hours later, its sister battalion the 2nd CMR, bombed some dug-outs at Mouquet Farm.

On the same day Lt.-Gen. Woollcombe's 11th Division, in touch with the Canadians, was engaged to the west of Mouquet Farm, the 6/Lincoln (33rd Brig.) forcing an entry into Constance Trench which runs west-south-west of Mouquet Farm. By the end of the day the trench had been secured as far as the junction with the Pozières/Thiepval road. A German counter-attack to retake the trench was repulsed by the 9/Sherwood Foresters. The 6/Border assaulted and occupied part of Danube Trench and whereas the entry was made with little loss, the Borders were subjected to a heavy counter-barrage.

Plans were now completed for a major attack on Mouquet Farm as part of the general attack on Thiepval village and ridge. The whole area was subjected to a heavy bombardment from the 23rd September and zero hour fixed for 12.35pm on the 26th. Meanwhile, the heap of rubble which was once Mouquet Farm was attacked on the 24th by a company of the 6/York & Lancaster of 32nd Brigade but immediate action by the defenders usings bombs, machine-guns and artillery fire, forced the men to retire. While waiting for zero hour, fresh troops relieved those in the front line.

The situation at Mouquet Farm was unusual to say the least. While under attack by the British, the enemy laid an accurate barrage on the area, the German defenders quite safe in the deep cellars below the ruins. The defenders could then come out at any given moment by any of the many exits to mount their machine-guns. But Mouquet Farm had to be taken and the 11/Manchester of 34th Brigade (11th Div.) began the task of clearing the cellars. The two tanks which had been assigned to help in the attack had ditched on their way forward. However, the crew of one tank brought up their Hotchkiss guns to take part in the clearance. One company of the 5/Dorset and a party of the 6/East Yorks pioneers came forward to join the Manchesters and the tank crew. Bombs, rifle and machine-gun were used and finally, after throwing smoke bombs into the cellars, one officer and fifty-five other ranks of the 6 Coy. 165th Regiment, black-faced, haggard, tired but proud, surrendered and at last, ownership of this lunar landscape passed to the Allies. The Germans had fortified the farm well and defended it with the utmost gallantry. Until this day, the 26th September, Mouquet Farm had defied every assault by the Australians, Canadians and British. It had been a costly affair....

(i) The strongpoint, Gibraltar, can still be seen today a few meters from the 1st Australian Division Memorial at the southern end of Pozières.
(ii) Here, Private W. Short of the 8/Green Howards who, mortally wounded, continued to prepare bombs for his bomber comrades. For this action he was posthumously awarded the Victoria Cross.
(iii) Lt. A. Jacka of the 14th Victoria Battalion distinguished himself in this action when he and a few other men displayed outstanding bravery under fire. He had already been awarded the VC for his actions when a private in Gallipoli and many considered his actions at Pozières deserved a second award of the Victoria Cross.
(iv) Marked on the map as Elbow(ii) to distinguish between The Elbow, 400 meters north-north-west of Pozières Windmill which is marked on the sketch map as Elbow(i)
(v) On the site of the Australian Memorial unveiled on the 10th September 1997.

OBJECTIVE 23
HIGH WOOD, SWITCH LINE CENTRE INTERMEDIATE TRENCH, HOOK TRENCH, WOOD LANE NORTH

It has already been mentioned that Rawlinson hoped the attack planned for the 14th July would result in the capture of High Wood by the infantry but Haig opposed this and asked Rawlinson to modify his plans for the attack. On the 13th July Haig expressed the hope that cavalry, under the control of XIII Corps, could be used to exploit the expected breakthrough on the 14th and agreed that High Wood, along with Pozières Ridge and Martinpuich could be secondary infantry objectives.

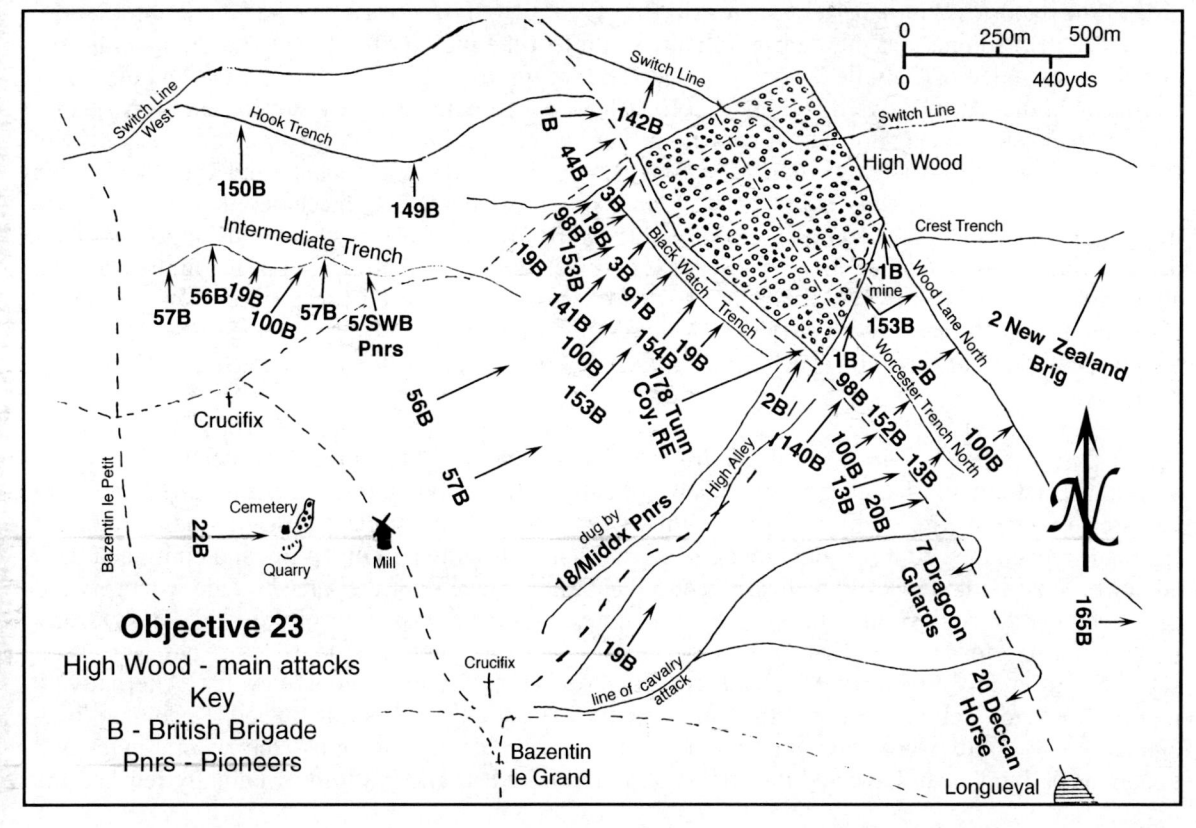

Flushed with success in the first hour and a half of the dawn attack, nearly all brigades had achieved their objectives, two brigades of 9th Division having reached Longueval. Only the 8th Brigade of Haldane's 3rd Division was held up to the east of Bazentin-le-Grand. This was to cause severe problems later concerning the attacks on High Wood. At this time, although unknown to headquarters, the wood was virtually empty and permission was sought by Major-General Watts, whose 7th Division had only committed a third of its available infantry, to take advantage of the situation and push forward to High Wood. Regrettably, his request was refused at XV Corps headquarters - High Wood was to be taken by the cavalry. The wood offered a safe haven for the retreating Germans. Also at this time, German units were arriving to relieve the 183rd Division on the Pozières to Bazentin Ridge. The relieving troops were now reinforcements and there was much action both in counter attacks and the completion of the partially constructed Switch Line, part of which ran through the northern corner of the wood. The dark mass of trees hid frantic fortification activities. It was later revealed that Brigadier-General Potter, 9th Brigade and an officer of the Royal Engineers had both, at different times, ventured close to the wood and were not challenged in any way. Both officers were convinced the wood could have been taken on the 14th July. Even with the advantage of hindsight it is difficult to say whether the wood could have been occupied successfully or not, the German infantry having been swelled by the newly-arrived reinforcements and others were available in the Flers sector. Longueval was in flames but only the southern part of the village was in British hands making any approach to High Wood from the east a dangerous operation. It would seem that the first troops to occupy the wood in large numbers would have the advantage and this probably lay with the Germans. However, Watts and others must have regretted the refusal which prohibited an attempt to enter High Wood. Its capture was to be a long and costly affair from now on ...

Not knowing if the Germans were in the process of occupying High Wood, Watts ordered the 2/Royal Warwick and 2/Royal Irish of 22nd Brigade, 7th Division to occupy the quarry and cemetery 350 meters east of Bazentin-le-Petit church. They achieved their objectives and thought themselves to be reasonably protected, the quarry and cemetery lying in a hollow. While reconnoitering the area under the false illusion of safety, they came under fire from a machine-gun sited at the windmill, only 300 meters to the east on the crest of the slope and a thousand meters from the south-west fringe of the wood. A brave attempt was made to silence the gun but with no success - Lt. Dean fell with most of his men. The dawn attack had generally gone well but the 2/Royal Irish had lost over 330 men and were pinned down a few hundred meters from High Wood. It was fast becoming obvious that the wood offered both excellent defensive and attacking features.

The cavalry had been ordered up to its assembly position but there now seemed little chance that it would arrive within the next few hours and, at 12.30pm, an order from Fourth Army headquarters was issued instructing XV Corps' artillery to fire on High Wood with its infantry following after the lift at 6.15pm. Horne, the XV Corps commander, concerned the right flank would be wide open to enfilade fire from Longueval, preferred to wait until confirmation was received that the village had fallen.

The cavalry was of course on its way and a brigade of the 2nd Indian Cavalry Division comprising the 20th Deccan Horse, 7th Dragoon Guards, 34th Poona Horse, a motorised machine-gun squadron and a veterinary section was originally designated for the attack on High Wood with other cavalry units on stand-to. The Rolls-Royce armoured cars moved forward but were bogged down in mud before they reached Montauban and were of no further use. Although the cavalry coped somewhat better with the terrain, shell-torn by bombardments since June, it did nevertheless, require the very best of horsemanship to find a way through the many obstacles to the assembly position.(i) At 6.30pm the Deccan Horse and Guards had arrived at their assembly position, the Poona Horse ordered to stand-to. With lances or sabres drawn the cavalry moved forward, the Guards on the left, the Deccan Horse on the right, their role being to protect the infantry to the left and later to raid the north-western edge of Delville Wood. They moved through the high corn, finding and lancing a few Germans on their way forward and about forty others, taken completely by surprise and terrified on seeing the cavalry bearing down on them with lances and sabres, immediately surrendered. Enemy fire from the newly constructed Switch Line took its toll of riders and mounts and the cavalrymen dismounted and took cover in the high corn. Without silencing the machine-guns, it would be impossible to remount. Several courageous efforts were made but the machine-guns continued their deadly work. Some wounded men were however collected under the protection of the high corn. The situation was saved by the arrival of a RFC pilot who not only fired tracer bullets at the hidden Germans thus informing the cavalrymen of the German positions, but also dropped a sketch of the enemy positions between High Wood and Delville Wood to an artillery unit to the south-west of Longueval.

Meanwhile, at 7.00pm, on the left of the cavalry, the 1/South Staffordshire (7th Div. 91st Brig.) set off from near the windmill, north of Crucifix Corner, on their 1100 meter advance towards High Wood, the first infantry battalion to be engaged directly at the wood. After only a few yards they were harassed by enemy fire both from the south-western edge of the wood and from small groups of Germans hidden in shell holes and other cover. The Staffords halted, returned fire and continued their advance. Their number was dramatically reduced and the 2/Queen's of the same brigade went forward in support on the right of the Staffords who, in spite of their losses were well forward. The Queen's made good progress for the first 500 meters, even capturing some abandoned guns but were soon subjected to enemy fire having lost the covering fire from the dismounted cavalry. Alternatively moving forward, then taking cover to return fire, they eventually reached the wood. Hesitantly, they entered the sombre wood, looking for signs of the enemy. It was quiet, the wood intact and undamaged. The Queen's awaited the arrival of their comrades. The Staffords eventually reached the wood and, with the evening light fading the two battalions entered the bowels of this awesome dark place. It was impossible to see more than a few paces in front and the bramble and undergrowth tugged at the mens' uniforms. At first, there was no sign of any Germans but spasmodic sniper fire caused some concern and fear of being suddenly affronted by the enemy in the wood in which the new occupants could easily lose their way. The two battalions had become separated, the Queen's eventually reaching the central ride running from the north-west to the south-east. Special flares were lit to signal their position to any RFC passing pilot. The Staffords were trying to make contact with the Queen's and gradually made their way north through the wood only to be swept by machine-gun fire from the Switch Line within the wood. After a series of bombing attacks, they eventually reached the north-eastern edge of the wood. It was now dark and the Staffords were still fighting for their lives near the Switch Line. The two cavalry units to the east had been beaten by the German machine-guns, and, like the Queen's, were digging in. The night came as a blessing to Major Witte, the German officer in charge of the defence of High Wood - he could now quickly bring in reinforcements, certainly more quickly than was possible for the British, and Witte took full advantage of the dark. New well-placed machine-gun posts were constructed, trenches completed and preparations made for a counter attack. At least, the 22/Manchester of 91st Brigade managed to get into the wood to help the weary Staffords.

The task of the 91st Brigade could have been easier if it had been supported by the 33rd Division which had been sent orders to attack the Switch Line to the west of High Wood. Unfortunately, Baird, the C.O. of 100th Brigade, 33rd Division had received no such order, but on his own initiative in the evening of the 14th, had sent the 9/HLI and 1/Queen's to fill the gap from the northern limit of Bazentin-le-Petit to the extreme western corner of High Wood to restrict German reinforcements entering the wood. The situation changed again quickly when Baird was informed the 91st Brigade had taken High Wood. Baird partially changed his plans and ordered his 9/HLI forward towards the south-western edge of the wood. Some 100 meters from the wood, the two leading companies were lit up by flares and attacked by machine-gun and rifle fire. Major Witte had been very busy and had now thrown his troops in an all-out counter attack from both inside and outside the wood. The Highlanders, particularly 'A' Company, fell thick and fast. The rest flung themselves to the floor digging frantically with head down to make some cover. Meanwhile, the 1/Queen's moved north towards the wood to form a defensive flank. Inside the wood the Staffords and Queen's, along with the freshly arrived Manchesters, were pushed back. The wood, which had been virtually empty a few hours previously, was now almost entirely in German hands. The Highlanders were still digging frenziedly just outside the wood. To the east, the cavalry endeavoured to protect the infantry on its left but, in reality, could do little and was withdrawn in the early hours of the 15th. The 100th Brigade was now in deep trouble and Baird ordered a short withdrawal in order to regroup. It had been a long night and at the first light of dawn 'C' Company made its way to the wood. The Highlanders entered cautiously, a few shots were exchanged with the enemy, but there was minimum contact. A mist began to form as the dawn temperature rose and scattered elements of the 2/Queen's, 1/Stafford, 22/Manchester and some R.E.'s were still in the wood as the Scots entered. There was an air of stillness, giving a false impression that the wood was almost empty. It was vital to establish to what extent the Germans had control of the wood. Patrols were sent out to the north towards the Switch Line, the cause of so much loss to the Staffords. They had to report that there were still large numbers of Germans, mainly north of the central ride. This was bad news for the 33rd Division which was to attack the Switch Line West in front of Martinpuich in a few hours, any attack towards this village would be exposed to flanking fire from the western side of High Wood. Baird made preparations for this attack instructing the 1/Queen's to attack from its overnight position near the Bazentin-HighWood lane and advance towards Switch Line West while the Glasgow Highlanders of the 9/HLI moved parallel with the Queen's on their left. The 16/KRRC (King's Royal Rifle Corps) was to advance from about 400 meters north-east of Crucifix Corner in support of the two leading battalions. The 2/Worcester was in brigade reserve but ready to move up if needed. The 98th Brigade was to attack on the left wing while the 19th Brigade would attack from Bazentin-le-Petit. Baird was convinced that the capture of High Wood was absolutely necessary if the attack on the Switch Line West was to succeed. His request to postpone the attack until the capture of High Wood was refused. The attack, timed for 9.00am was only a few minutes away. Requests by the Highlanders and Queen's for a barrage on High Wood had also been refused. The wood was yet again falsely reported to be in British hands. Zero hour, the men moved forward, the mist dissipated by the rising sun, the Switch Line before Martinpuich some 650 meters distant - they had a long way to go. The events of a few hours ago were monotonously repeated, the advancing infantry fell before intense fire from High Wood. Added to small-arms fire, the shrapnel shells accounted for many casualties and the Highlanders, closest to the wood, suffered terrible casualties. On their left, the Queen's were less exposed to enemy fire but were held up by uncut wire, always a serious problem when soldiers are obliged to group in the narrow openings making perfect targets for enemy fire. The 16/KRRC had now moved up in support but fell before the superior fire of the Germans in High Wood and installed in the newly-constructed Switch Line - Major Witte's fine organisation was paying handsome dividends. One company was instructed to enter the wood but were cut down immediately on entry. Similarly, the 2/Worcester sent a company towards the wood, the remaining companies taking up position on the lane from Crucifix Corner to High Wood. Baird made a desperate plea for a barrage on the Switch Line West to which the artillery responded unsuccessfully - the Switch Line being out of sight over the crest. The advance was brought to a halt, no-one could get forward under such fire and the men began the dangerous operation of withdrawing to their starting positions. This was the signal for the German counter attack during which many men, especially those near the south-western edge of the wood fell before the enemy fire. A few of the 2/Queen's managed to reach the eastern edge where they were somewhat more protected and, along with some Manchesters and units of the KRRC, prepared positions to counter any further attack.

The German barrage had cut wires and cables making communication possible only by runner, whose chance of delivering his message, let alone returning with an answer, was minimal, or by reconnaissance aircraft. One such aircraft had already done good work earlier and another aircraft from the same squadron was able to relay up-to-date information regarding infantry and German positions. The pilot had to report discouraging news of our infantry and also that the hidden Switch Line, only visible from the air, was packed with Germans. The attack had failed on all fronts and the 98th Brigade was ordered to withdraw to east of Bazentin-le-Petit. It was almost dark, another twenty four hours had passed, and the German flares lit up the battlefield making withdrawal difficult. The 91st and 100 Brigades did not receive orders to withdraw until the early hours of the morning and, having lost over 2,500 men between them, pulled back as best they could to clear the way for the expected British barrage. As they did so, the Germans took full possession of High Wood.

Whereas the dawn attack on the previous day had gone well, the 15th July action at High Wood had presented particular difficulties. The long Switch Line, part of which passed through the wood, would prove a nasty stumbling block. Brigadier-General Baird had spoken the full truth when he said the capture of High Wood was essential prior to any attack to the west or east of the wood.

Although rain postponed any immediate attack by the infantry, the artillery concentrated its shelling of High Wood and the Switch Line to the east and west. Having relieved the 100th Brigade, Mayne's 19th Brigade was ordered to attack the wood. The three leading battalions were augmented by some Royal Engineers and the divisional pioneers, the 18/Middlesex. Their orders were simply to take the whole wood and construct strongpoints on all faces of the wood starting from the southern tip. Elements of the 7th and 5th Divisions were detailed to attack the recently fortified Wood Lane (ii)

High Wood on the left, Delville Wood on the right. The cavalry charged between the two woods on the 14th July in a line just to the right of the large bush in the centre of the picture (author)

running from High Wood to Longueval. The 19th Brigade moved up from Crucifix Corner losing a number of men from hostile artillery fire but encouraged by the immense bombardment of High Wood with high explosive shells. As the British artillery lengthened its range the 6/Scottish Rifles (iii) and 1/Cameronians rushed into the blazing wood, the latter on the left towards the Switch Line at the northern end of the wood and the Rifles towards the north-east. The 20/Royal Fusiliers were in close support. Some dazed, shell-shocked Germans were taken prisoner and taken to the rear. Progress was good up to and past the central ride. The advance of the Cameronians took a drastic change for the worse as they approached the western edge when a German machine-gun crew, somehow having survived the shelling, took revenge. Fire was also coming from the adjacent Switch Line. Further to the right the Scottish Rifles were engaged in hand-to-hand fighting.

To the east of the wood the 8/Devon and 2/Gordon (20th Brig. 7th Div.), following the path of the cavalry advance on the 14th, were hampered by hidden machine-gun posts in the waist-high corn, from Wood Lane and from the Switch Line to the rear. In spite of heavy losses, particularly among the Gordons from a machine-gun at the eastern edge of the wood, they managed to reach their objective, Wood Lane. Here they were almost all killed while digging in. The intensive shelling of the wood had not silenced the enemy and some four hours later the British attack was flagging. However, those within the wood fared better than the Devons and Gordons whose every attempt to approach the wood was cut short by murderous machine-gun and rifle fire. These gallant men were eventually obliged to retire to the Longueval Road, helped by their comrades of the 9/Devon. The Royal Fusiliers had lost almost all their officers and the inevitable counter attack came in full force pushing the survivors out of the wood. The 18/Middlesex pioneers were detailed to dig a protective trench outside the wood, but it was costly work under constant shell-fire which not only harmed the pioneers but caused alarming casualties in the 2/RWF in reserve, some distance away. A third of the Fusiliers had become casualties but the remainder moved forward to High Wood. A further 200 Welsh Fusiliers were lost on the way forward but their arrival was received with great enthusiasm by their beleaguered comrades. The RWF entered the wood from the east and were joined by some Gordon Highlanders whose position, as already stated, was very precarious outside the wood. Moving forward through the abbatis, disturbed undergrowth, shell craters, trenches and the dead and wounded was terribly slow but, with great determination they made their way towards the northern corner to attack the Switch Line. Like their predecessors, they fell before the machine-guns but the survivors pushed on. The wood was at last taken but the Fusiliers were obliged to dig in to protect themselves against the German artillery and the guaranteed counter attack. Mayne's 19th Brigade had done all that had been asked of it but the ranks were so depleted that the C.O. had to make it very clear to headquarters that, unless relieved very quickly, it would be impossible to retain possession of the wood against any counter attack. In the intervening period many of the wounded were evacuated, and the dead, in the summer heat, quickly swelling and in the first stages of putrefaction, attracted thousands of flies and rodents. Consolidation of posts and trenches including barbed wire was under way but Mayne was worried; reporting his brigade was down to only 200 men plus about a hundred reinforcements who were sent up with a ration party.

The 18/Middlesex pioneers which has been mentioned earlier, had been forced by a severe hostile barrage to abandon the digging of a communication trench from Crucifix Corner to High Wood, were now able to continue their work. The situation of the remnants of the 19th Brigade in the wood became even more desperate when yet another heavy barrage fell upon them. The ranks were decimated, parts of bodies hanging from the torn trunks and branches. Nothing could survive such power and these heroes were forced back, ever thinning in number. The enemy re-occupied the western corner, the north of the wood including the Switch Line and the central ride became the dividing line, or, no man's land. Whereas before, Mayne had been asking for reinforcements, now he was desperate for his men to be relieved, they could do no more and were retiring. In the early hours of the following morning three relief companies of the 1/Queen's and 16/KRRC (100th Brig.) allowed the evacuation of the 19th Brigade whose four battalions had lost heavily, the Scottish Rifles having over 400 casualties. In turn 20th Brigade (7th Div.) was relieved by the 1/Royal West Kent and the 14/Royal Warwick of 13th Brigade (5th Div.). Patrols were sent out to establish the enemy's positions. The remaining two battalions of 13th Brigade, the 15/Royal Warwick and 2/KOSB, were sent forward. The high corn made it difficult to locate the enemy snipers. Added to this was the sickening stench of rotting horse and human flesh. The animals were buried at night to avoid sniper fire. It was necessary to cross Wood Lane to make any real evaluation of the German positions in the Switch Line east of the wood and this would be extremely difficult. Little could be done without aerial photographs.

Horne, the XV Corps commander was displeased with recent events at High Wood now almost fully occupied by the enemy when, only a short while previously, it had been reported in British hands. Landon would have the unenviable task of answering to the corps commander.(iv) Brigadier General Stewart's 154th Brigade (51st Div.) relieved Landon's depleted units, the casualties of his 33rd Division were 5,200 officers and men. The ground south and south-west of High Wood was but a field of corpses, an enormous graveyard. Major-General Harper's remaining two brigades of his 51st (Highland) Division were in support and reserve, the 153rd Brigade to the south of Bazentin-le-Petit and the 152nd in reserve south-west of Mametz Wood.

During the lull in the fighting, the Germans had yet again been busy fortifying the west with their new Intermediate Trench and particularly the eastern edge of the wood where a strong redoubt had been made. With such commanding positions, the Germans could repel any attack from the south-east, Wood Lane or from the north of Bazentin.

The 4/Gordon and 9/Royal Scots entered the wood from the south to find just a small part of this southern section still in British possession. To achieve their objective, the capture of the Switch Line within the wood and a few hundred yards of the Switch Line east and west, meant that they had virtually to capture the whole wood. This was impossible in daylight and a barrage was laid on the strong east side of the wood and on the fortified Wood Lane. Encouraged by the barrage, the Gordons and Royal Scots set off on their 'mopping up' operations. Instead, they were cut down, the tortured and twisted wood making excellent hides for machine-guns, infantry and, particularly, snipers. Some Royal Scots had been ordered to attack from along the southern edge. On leaving the perimeter of the wood, the barking of a machine-gun from the newly-built Intermediate Trench traversed their ranks.

Meanwhile the 13th Brigade (5th Div.) made its way at 9.50p.m. to the crest of the battered cornfield and was quickly over the top when the barrage lifted. German flares lit up the scene in the fading light and machine-guns, strategically placed in the redoubt on the eastern boundary of the wood, executed their deadly work. A platoon of the 1/R. W. Kents covering the left was virtually wiped out. The 2/KOSB moving forward to support the W. Kents suffered the same fate. The two Warwick battalions fared no better. The men of the 13th Brigade had done all that was humanly possible but man's flesh alone cannot triumph over a well-sited machine-gun. Again, the attack had failed, the artillery had not silenced the German fortifications or adequately destroyed the wire. None of the objectives had been achieved, the Germans still held the Switch Line, the north and east of the wood and had already put the new Intermediate Trench to good use. The weary survivors were withdrawn and Rawlinson, with his corps commanders had to consider what to do next.

Digging was now a top priority - the men needed better protection in order to approach the wood. The 51st Division had already constructed a long length of trench in front of the wood and were ordered to extend this westwards and eastwards and also to complete the trench from Crucifix Corner to the southern tip of the wood. The 19th Division (III Corps) on loan to Horne's XV Corps, dug furiously under fire throughout the long day in the flinty ground to make a new trench from the windmill towards the wood. Although slightly damaged by German artillery the new trench, which ran to within 50 meters of the wood, would provide welcome cover to the advancing troops.

Horne's stern order came down the line, both east and west corners of the wood must be taken. The Gordons and Royal Scots were in no shape to attack and under the cover of dark Stewart sent his 4/Seaforth Highlanders forward to the eastern corner. On leaving the trenches, they were cut down by fire from the machine-gun positions within the redoubt on the south-eastern edge. Their losses were so great, they went to ground and the attack ended. German artillery continued its barrage with shells and gas. Campbell's 153rd Brigade which had been in support and engaged in bringing up supplies, relieved the sorely-tried 154th Brigade. Brigadier-General Campbell had inherited the same problem as Stewart, how to take the east and west corners of the wood plus a last minute addition by Harper, the 51st Division's C.O., some 700 meters of Wood Lane.

19th Division, having finished its trenching work, prepared to attack Intermediate Trench. The 7/King's Own (attached from 56th Brig.) and 10/Royal Warwicks (57th Brig.) on the right rushed the eastern part of the trench as soon as the barrage had passed over taking the Germans by surprise, securing the trench before the defenders had time to mount their machine guns. Over thirty prisoners were taken. This was certainly one of the first successful attacks against a well fortified trench in the High Wood area. The divisional pioneers, the 5/South Wales Borderers went forward to help with the consolidation of the captured half of Intermediate Trench. However, the 10/Worcester and 8/Gloucester following moments later to the left of their sister battalion, suffered heavy losses from machine-gun fire and their attack was halted. This highlighted the possibilities and one could say, necessity, of advancing on the very heels of the creeping barrage, even at the risk of losing some men. These men who, for whatever reason, had not done so, had suffered terrible casualties. Aware of the difficulties encountered by the Worcester and Gloucester battalions, the pioneers barricaded the trench to the left of the Warwicks as a precaution against the expected counter attack.

Returning now to the 153rd Brigade of Harper's 51st (Highland) Division, the 7/Black Watch had the unenviable task of attacking the redoubt on the south-eastern edge which had so far repulsed all attack by both artillery and infantry. These kilted warriors, nick-named "Ladies from Hell" by the Germans, courageously attacked the redoubt and in spite of a preceding mortar barrage were mown down by the ever-present machine-guns and, although some small gain was made, eventually retired to their original trench. The 5/Gordon and 6/Black Watch had also a difficult task - an attack on Wood Lane protected from the north, west and east. Under cover of the remaining high corn, they went forward. What must have been in their minds as they advanced behind the creeping barrage in the relative safety of the slope? The time for such reflections ended as they rose above the crest and were met by a hail of bullets. Could nothing destroy the German defences? Although severely mauled by the British bombardment, enough Germans survived to do untold damage to the two battalions. Just one machine-gun, manned courageously, is sufficient to hold up an attack. The left flank of the attackers was punished by the same guns which had decimated the 7/Black Watch. The survivors took cover where they could only 200 meters from their starting point. It was a sad time at the roll-call of the 153rd Brigade. On being informed of the failed attack on Wood Lane and High Wood, Horne immediately ordered another attack - there is no room for a soft-hearted commander in the army, but the order was received too late to be put into effect. Harper insisted that the whole of High Wood would have to be shelled with high-explosive prior to any future assault. Horne deigned not to answer. The German artillery was now concentrating its attention on the support positions, harassing the supply of munitions and particularly of food and water. The ration parties were impatiently awaited by the troops desperate for drinking water and food on this day when the temperature rose to 82 degrees Fahrenheit. Horne had Rawlinson's full support to continue the attack on High Wood but Haig, after considering the results of more than two weeks fighting at High Wood, preferred to turn his attention to Guillemont.

Pelham Burn's 152nd Brigade, not engaged at High Wood during July, was sent forward to relieve the 153rd Brigade on the 1st August. No major attack was to be made but rather a series of small operations. It was vital to make a communication trench to effect a safer approach to Wood Lane. Accordingly, posts were excavated at night in front of the front line and these gradually linked to form a continuous trench.

Saps were to be made in High Wood, a difficult task digging through the debris, tree roots, rotting corpses and other obstacles. Brigadier-General Burn was well aware of the damage caused by the redoubt on the east side of the wood and sought permission to place a mine under the stronghold. Permission was granted and R.E. tunnelling units began work the next day. A new sapping device with the handsome title of the Barrat Hydraulic Forcing Jack, promptly nick-named 'pipe pusher', pushed armed iron pipes through the ground at a depth of about 1.20 meters which, when exploded, would provide at least some cover for the advancing troops in the wood. A number were used within the wood and one outside towards Wood Lane. Other devices to be used in future attacks were a large flame-thrower and burning oil drums. It was a risky business getting the long pipes and large parts of the dismantled flame thrower to the front. Under the cover of night the 152nd Brigade dug like fury, the digging drowned by the sound of artillery and in a few days were within 50 meters of the crest leading to Wood Lane. A return trench was quickly dug and the first one manned. Slow sapping progress was made in High Wood.

The 51st Division was relieved on 7th August by Landon's 33rd Division, now back in familiar, although changed territory where Baird's 100th Brigade replaced the 152nd. On his return, Baird noticed the immense change in the area, trenches everywhere, High Wood but a diamond-shaped tangle of broken trunks and branches. The previous losses had been replaced and Stormont-Darling's 9/HLI, the Glasgow Highlanders, assisted by some Royal Engineers, were quickly engaged in joining up the saps. The tunnelling of the mine gallery was presenting problems getting the iron pipes between the tree roots. At least, the communication trench, now named High Alley, from Crucifix Corner to High Wood was completed and men and supplies could be brought up. Mayne's 19th Brigade was also digging in the wood towards the Switch Line. Meanwhile, the 19th Division had wrenched yet a few more meters of Intermediate Trench from the Germans on the 11th August.

The next attack was fixed for the 18th August, zero hour set for 2.45pm, followed by a creeping barrage. The 2/A&SH of 98th Brigade moved forward to take up position in the forward trenches at High Wood under cover of night, arriving before midnight of the 17th. Some hours later, units of the 4/King's Liverpool with the 4/Suffolk on its left, were in the front line trenches near Wood Lane. After the long wait, zero hour arrived and the King's surged forward. Most unfortunately, the brigade's Vickers machine-gunners had not been able to get to their allotted saps some 50 meters forward of the front line to mount their guns and protect the Liverpool's advance. Without this protection, the King's were decimated by German artillery. None reached Wood Lane and the same sad story was being played out once again. The few who reached the barbed wire before the German front line were cut down. The remainder took cover where they could in the increasing number of shell holes. Situated between the King's and High Wood, the Suffolks fared rather better, passing through the enemy defences into the wood. Their intrusion into German territory was to be short-lived. Attacked on all sides by bombers and machine-gun fire, they sustained many casualties. The survivors, like those in the King's, sought cover. The Argyll & Sutherland Highlanders, having waited over fifteen hours in their front-line positions, were doomed even before zero hour. The creeping barrage had passed right over their position killing many men and damaging the flame throwers. The hardy Highlanders moved forward into the wood to be met, like all previous attempts, by a hail of fire which defied all human effort to penetrate. Caught in a vicious cross-fire, these gallant men fell, not even having the satisfaction of being able to hold on to a small gain they had made on the left of the wood. With numbers thinning every minute, they had no choice but to retire and began the dangerous manoeuvre of returning to their front line. The 'new inventions' had not been of much use. When the ammonal in one of the pipe-pushers was detonated, it exploded in the Highlander's line. The flame throwers were damaged beyond use. The new 'burning oil drums' proved to be ineffectual. The machine-gun section was ordered to pull back to the support line which was done under fire and the guns mounted to cover the edge of the wood to give some protection to the retreating wounded.

Prior to the main attack at 14.45, the 1/Black Watch (1st Brig. 1st Div) had attacked at 4.15am through thick fog towards the troublesome Intermediate Trench, some men losing their way in the fog and others falling victim of heavy shell fire. They overran the trench held mainly by bombers and continued their advance towards the western end of the Switch Line but were forced to withdraw under heavy fire. Some ten hours later the 8/Royal Berkshire, also of the 1st Brigade, attacked the

The infantry suffered heavy casualties from direct machine-gun fire from High Wood while crossing the vast expanse of open terrain in the foreground. Later, communication and assembly trenches were dug which afforded some protection.(author)

same trench, this time under cover of a smoke screen but the battalion was forced back. Rather better progress was made by two companies of the 1/Loyal North Lancs (2nd Brig.). Although the right company was virtually annihilated by friendly fire, the remaining company on the left entered Intermediate Trench which had been extended to High Wood and, aided by the 1/Northampton, bombed their way towards the wood. Due to the Highlander's failure, in spite of Herculean efforts, to take the north-west part of the wood, the 2nd Brigade's progress up Intermediate Trench was brought to a halt. Feelings in the Loyal North Lancs were running high on account of the heavy casualties caused by the British barrage. This is not surprising but the artillery had been in action since the 24th June, the guns were worn through overheating and continuous use, many guns damaged beyond repair plus irreplaceable losses of experienced gunners at the fixed-point positions of the artillery in Flat Iron Valley and Caterpillar Valley. There was now a drastic shortage of trained artillerymen.

Thirty-three days had passed since the first encounter at High Wood and the gains were virtually non-existant in the wood itself, just part of the Intermediate Trench had been taken. Small wonder there was disappointment, from Haig and Rawlinson at the top of the command structure down to the humble infantryman who, inspite of outstanding courage and perserverance, could not gain a permanent position in the wood. Major-General Landon ordered the withdrawal of the 98th Brigade, Mayne's 19th Brigade taking over the line. The 1/Cameronians relieved the King's and Suffolk while the 2/RWF replaced the 2/A&SH. The Fusiliers were almost at exhaustion point on their arrival having spent six days bringing up supplies, digging trenches and having a minimal amount of sleep. The losses of four weeks earlier had been replaced by men who had yet to face their baptism of fire and now, in the full knowledge of the difficulties which High Wood presented to any attacking force, and utterly wearied by their manual labours, moral was at an all time low. On receiving orders to attack the trench at the south-western edge of the wood, a mission of almost certain death, some Fusiliers from 'D' Company failed to leave the trench. Somewhat surprisingly, no action was taken and the engagement is described in the Official History as an 'abortive attempt to clear the trench on the western edge' (v). The 9/HLI (100th Brig.) which had been brought up facing Wood Lane earlier in the day, made a night attack on the right-hand section of Wood Lane and took possession of a partly constructed trench.

On the 22nd August the 2/RWF were withdrawn to a support position at Crucifix Corner. As the Fusiliers made their way down High Alley, the 1/Queen's, 16/KRRC and 2/Worcester, the remaining three battalions of 100th Brigade were moving into a position for an attack on Wood Lane, this time conceived in a different manner. The 100th Machine Gun Company comprising ten Vickers machine-guns installed in a trench giving a clear view of the German line at a distance of about 1,800 meters, was to concentrate continuous rapid fire during the whole of the advance and continue during the period of consolidation of the expected gains. The battery commander, Capt. G. Hutchison, faced a mammoth task of organisation, the most obvious problem being that of keeping the barrels cooled until they were changed. A variety of liquids was used for this purpose including soiled tea, water, any non-combustible liquid, even urine. Covered by a smoke-screen the three battalions moved forward while the machine-gunners continuously traversed the whole of the visible perimeters of High Wood and the adjacent trenches. The gunners stuck to their task for a period of twelve long hours. Their efforts paid dividends, driving the enemy back into the wood and the infantry took their objective along with some frightened prisoners.

Delville Wood, 1,600 meters south-east of High Wood, was at last taken on the 27th August in rainy conditions. In fact, it had rained every day between the 25th and 30th August and a little over one and threequarter inches of rain had made ground conditions very difficult. Water does not drain easily in the vicinity of High Wood.(vi) The 31st was dry and the Germans launched their biggest counter attack to date. The gains of the 100th Brigade were lost and the Germans made a foothold in Delville Wood but only after repeated and costly attacks. The German attack was met with relentless fire from rifle, machine-gun, trench mortars and artillery and eventually petered out.

The 1st Division was still on hand at High Wood, ejecting the remaining Germans in Intermediate Trench while the 1/Black Watch anxiously awaited the detonation of the mine under the infamous German redoubt on the south-eastern face of the wood. The 178th Tunnelling Coy. RE had worked tirelessly in terrible conditions and was at last placing the ammonal. A few seconds before zero hour the result of their hard work was seen when the ground lifted, the German machine-gun nests, their

The start of Wood Lane North which heads southwest towards Delville Wood. (author)

gunners and equipment blown to pieces. Hardly waiting for the debris to settle, the Black Watch bounded to the lip of the new crater and immediately dug in. Others advanced westwards along the trench which fed the former redoubt. The remainder were to sweep through the wood but the new devices, instead of helping matters, failed as before. The pipe-pushers backfired causing casualties and confusion, allowing the Germans to recover their senses after the mine explosion and the Scots were again under fire. Having mixed forces in the wood, it could not be shelled and so the artillery concentrated a creeping barrage on Wood Lane followed closely by the 1/Camerons and some 8/Royal Berkshire (1st Brig.). The redoubt silenced, they advanced free of enfilade fire from their left. Here, they lived up to their reputation of being in the 1st Brigade of the 1st Division by driving out the enemy from the trench with the bayonet, rifle butt, grenade or anything which was to hand and thus secured 200 meters of Wood Lane where it leaves the corner of High Wood. The carrying parties had not arrived - consolidation was urgent. The Switch Line north of Wood Lane was full of Germans and a counter attack seemed imminent. When it came, it was directed not at the occupiers of Wood Lane but at the Black Watch on the lip of the mine crater. The Camerons and Berkshires cut down the advancing Germans but with great determination the enemy pushed on to the crater, bombing out the Black Watch. Quickly mounting a machine-gun, they laid enfilade fire into the captured section of Wood Lane. The trench was untenable and the Camerons were obliged to abandon their gain. The end result of the day's work on this 3rd September was the same as all the previous attacks - no permanent gains. It had augured well at the start but even the seasoned regulars of the 1st Division could not take control of the wood.

The Corps would have to think again. Rawlinson was planning a general offensive for the 15th September which was to be known as the Battle of Flers-Courcelette. Some preparatory work would be necessary. At least, Guillemont had fallen and almost all of Delville Wood was in British hands While the Germans were in control of High Wood, the British artillery could not be moved forward for the 15th September offensive. High Wood must be taken. The 3rd Brigade of the 1st Division was ordered to secure the western half of the wood and at 6.00pm the 2/Welsh and 1/Gloucester, aided by two companies of the 9/Black Watch (44th Brig.15th Div.) moved into the wood. One of the two Welsh companies gained its objective near the central ride whereas the other could make no progress against the enemy fire. The Gloucesters fared badly, being short of bombs and having to clear Germans from wired defences at the point of the bayonet. The Black Watch from 15th Division successfully secured a trench running from the western corner of the wood, taking some prisoners. It seemed as if each attack provoked the Germans into making an immediate counter attack - it was crystal clear they wished to remain the keeper of this stategically placed wood. Under the counter attack, all three battalions had to retire as had all those which had preceded them.

Major-General Strickland, commander of the 1st Division had to organise another attack. The 10/Gloucester (1st Brig.) was to try for the western corner, with the 2/RMF (3rd Brig.) in the centre with the 1/Northampton (2nd Brig.) on the right to attack the eastern side including the German lodgement on the crater lip. The tunnellers had again been busy re-using the old galleries and placed a further 3,000 lbs of ammonal to be detonated as before, a few seconds before zero hour at 4.45am. With understandable apprehension the men went over the top, supporting fire from mortars giving a false sense of security. The Northamptons were met by a hail of bullets from the enemy trenches. Those who got as far as the wire gasped when they saw it was untouched. The mine had been exploded on time and destroyed the machine-guns and a few of the attacking Northamptons secured the crater but their numbers were so diminished they could not hold it against the German counter attack. The Royal Munster Fusiliers and Gloucesters fared no better - after entering the wood they were beaten back. A further assault on Wood Lane was led by the 2/Royal Sussex on the left and the 2/KRRC, both of 2nd Brigade, on the right flank.The British barrage, the retaliatory enemy bombardment, the ratatat-tat of the machine guns, the mortars, the whole resulted in an almost unbearable cacophonous din and through this hell the two battalions advanced, men falling left and right, the casualties mounting by the second but with grim determination the survivors pushed on and seized Wood Lane North. Further to the right, the 5th and 6th King's (165th Brig. 55th Div.) secured Wood Lane South.

Rawlinson was well pleased on hearing the news - the new line being essential to his plans for the general assault on the 15th September. Following tank demonstrations behind the front line, the possible use of this new weapon had been discussed for some while. There was some hesitation as to whether 'lanes' in the artillery barrage should be left to allow for tank movement or whether they should move forward like the infantry, i.e. just behind the creeping barrage. The general opinion was that the former suggestion would be better as the tanks moved very slowly and could perhaps hold up the infantry. The demonstration reports had evidently impressed a number of people, Lt.-Gen. Pulteney, III Corps Commander expressing the view that tanks would quickly pass through High Wood. Rawlinson questioned this view but Pulteney was adamant the tanks could do the job. Looking today at photographs taken of High Wood in 1916, it is unlikely that even today's modern tank could cope. Regarding the effectiveness of the new weapon in 1916 within High Wood, there is absolutely no doubt that it would be a death-trap for any tank trying to cross it, the interior being a mass of deep-rooted jagged stumps, dismembered thick branches very capable of tearing off the track running gear, plus hundreds of shell craters, many trenches and other obstacles. Ideally, the new tank was suitable for traversing over reasonably dry ground, crushing barbed wire and crossing over trenches which had not been subjected to intensive shelling. The commanders and crews were disgruntled at the large number of demonstrations they had to perform, leaving little time to service the machines and also causing unnecessary wear.

The site of Crest Trench which headed parallel to the potato field on the right towards High Wood. (author)

Major-General Barter arrived on the Somme with his 47th (London) Division which, as the name implies, was made up of lads from the big city or 'the Smoke' as it was often called. The battalions were much better known under their familiar or pals name and the City of London, Post Office, Civil Service, Poplar & Stepney, London Irish, First Surrey Rifle battalions along with the St. Pancras, City of London, Blackheath & Woolwich, Queen's & Clapham battalions were about to earn the unique honour of having their unofficial titles engraved on the vast piers and faces of the Thiepval Memorial.

The division arrived on the Somme at full strength after having replaced the losses during 1915 at Festubert and Loos and previous engagements and all three brigades had passed through intensive training to prepare them for the attack on High Wood - only the tanks were missings - these were on their way to the Somme out of sight under wraps. Moral was high. On arrival Barter describes how he was impressed by the devastation which had become almost routine to the 1st, 19th, 33rd, 51st and other divisions and adds that he had never quite seen destruction such as that at High Wood, the surrounding terrain being a complete wilderness and now it was his task to organise the taking of this seemingly impregnable bastion. Facing his men were the obstacles of High Wood, the Switch Line, which not only ran through the north of the wood but to the east and west as well, and the Starfish Line beyond. These were the objectives - others had tried and failed, not through want of courage - what could these Londoners achieve?

The 47th Division had the 2nd New Zealand Brigade on its right towards Delville Wood and on the left were the 149th and 150th Brigades of Maj.-Gen. Wilkinson's 50th Division. The III Corps had eight tanks at its disposal and four were allocated to High Wood. The 'male' tanks were equipped with 6 pounder guns, the 'female' ones with four Vickers machine-guns having the capacity of 2,400 rounds total on both sides. For identification purposes the tank companies carried a letter 'A' to 'F' followed by a number between 1 and 25. Barter reconnoitred High Wood to assess the value of the proposed tank assistance. He was not impressed and asked Pulteney for permission to withdraw his men slightly to allow a saturation barrage of the wood. This was refused, Pulteney having complete faith in the tanks. The 140th and 141st Brigades took up their positions and the 142nd Brigade withdrew to the support line, ready and on hand if needed. Meanwhile, 36 tanks laboriously made their way to the front, about 6 dropping out through mechanical failure. The four tanks allocated to High Wood arrived safely at their assembly position. The Londoners were crouched in their trenches while the artillery shelled the wood. The wait must have seemed interminable, everyone aware of the fate of the preceding attackers. What thoughts must have passed through their minds as some chattered to calm their nerves, others smoked quietly, others looked as though they were in another world, some cracked jokes, the Londoner has always been good at spinning a yarn. The rum ration was more than welcome just before zero.

A short time before zero hour, tanks D21 and D22 set off, negotiating with difficulty the terrain before reaching the southern corner of the wood where they were to enter and push north-north-east towards the crater. The outside temperature was not high before 7.00am but the heat and noise in the tanks was very uncomfortable and vision was limited. The drivers were not sure of their direction and stopped to enquire. Time was passing, daylight increasing with every minute as machine-gun fire began to sweep the Poplar and Stepney Rifles, the London Irish and Civil Service Rifles trenches, the tanks were already late. Twenty past six, no time for further thought. The infantry moved forward ahead of the tanks which were still behind schedule. Trying desperately to get into the wood, the men from Poplar and Stepney and the London Irish were cut down by a solid wall of fire. The Civil Service Rifles, moving up the east edge of the wood were caught in a murderous fire. 'B', 'C' and 'D' company commanders were down, Captains Davies and Roberts were dead and Captain Gaze wounded. Many NCO's were killed or wounded. The horror of these first few minutes was to haunt the survivors for the rest of their lives. The first attack had failed. The 19 and 20/London having already moved up ready to follow the leading waves were ordered to retire to the support line to allow room for the survivors in the assembly trench.

The 7/London (City of London) also moved forward at zero hour and, although mauled by fire, the British creeping barrage giving some protection, the men, advancing in four waves, seized the Switch Line immediately east of the wood. Meanwhile the tanks had entered the wood. D22 was forced off-course by the tree stumps, eventually coming out of the wood on the east side and ditching in Worcester Trench (see map). D21 survived for some ten minutes, a broken axle bringing it to a halt west of the crater. However, the tank remained active and its guns were used to good effect. D13 fared better, progressing as far as the Switch Line before a shell set it on fire. The crew climbed out of the burning machine and leaped into an occupied German trench where they took a few prisoners. Active service of the last tank, C13, was short-lived. Unable to negotiate the tree stumps, it had to be abandoned in the south-west of the wood. Maj.-Gen. Barter's fears were well-founded. Pulteney had been proved wrong at enormous cost.

It has been mentioned that the 50th Division's 149th and 150th Brigades were positioned to the west of High Wood. Brig.-General Ovens, who had been nominated to replace Brig.-Gen. Clifford killed by a shell on 11th September (O.H. page 333) sent forward the 4th and 7/Northumberland Fusiliers (NF) of his 149th Brigade to secure Hook Trench (see map) which they did with little difficulty. The advance to the Starfish line in conjunction with the 4/East Yorks (150th Brig.) was impeded by enfilade fire from the north-west corner of the wood. An hour later the 4/NF went forward to assist the units of the 47th Division in the wood. The Fusiliers were met by a German counter attack. The 5 & 6/NF rushed up in support enabling the 7/NF to reach the Starfish. Further details of the action on the Starfish Line, The Bow and Martinpuich are given under the title of Objective 28.

High Wood was still not secured, making any attack on Flag Lane, the next objective to the north-east of the wood, almost impossible. Assembly trenches were clogged with dead, wounded and those waiting to attack. In such a congested state, they would be soon untenable. The 19/London could not get to its forward position. The battalion commander, Lt.-Col. Hamilton left the rear trench with some of his men - they were mown down almost immediately. The Post Office Rifles' route took them on the eastern edge of the wood, following the ill-fated path of the Civil Service Rifles. They had received conflicting reports of the progress of the Civil Service Rifles and were not sure what to expect. Courageously they plodded on when, after only a few more yards, like the others, they too were cut down at close range, the machine gun bullets ripping bodies apart like bits of paper. With gritted teeth, the survivors somehow continued their advance. Three of the four company commanders were casualties, leadership being passed on to the platoon commanders.

The 6/London (City of London Rifles) was scheduled to move forward at 8.20am, its objectives being well beyond High Wood. The wood should have been cleared by now but as they waited in the assembly trenches they saw the wounded crawling back and heard reports of the massacre from returning survivors. It was not encouraging. The word came to leave the trench. At intervals of 100 paces, the waves advanced soon to be felled by a swathe of machine-gun fire. It was now obvious that neither the eastern edge of High Wood nor the Switch Line beyond had been taken. The New Zealand 2nd Brigade was also held up to the right. Waves of infantry fell to the ground. In spite of the very best efforts of Barter's 47th Division, the attack had failed. The men were disappointed and it is perhaps as well they were not aware of Pulteney's anger when the latter was informed that High Wood had not been taken.

Brig.-Gen. Lewis' 142nd Brigade, hitherto in reserve, was ordered north to attack the Starfish, a group of trenches to the north of High Wood and as its secondary objective, the Cough Drop, a further 400 meters to the north. This action is described under Objective 36.

If infantry could not dislodge the Germans, then a concentrated artillery barrage accompanied by trench mortars and a

Unveiling ceremony, 47th (London) Division memorial at High Wood on 13th September 1925. (J. Bommeleÿn)

Above: The original cross erected at High Wood on 21st December 1916 which commemorates the Black Watch Regiment on one side and the Cameron Highlanders on the other. (J. Bommeleÿn)

Left: The new memorial replaced the above cross in 1979 (author)

curtain of machine-gun fire would have to soften them up first. The trench mortar battery of Hampden's 140th Brigade had been standing by for some time, their mortars not being used in view of the expected surprise element of the tanks. The battery commander now ordered his teams to open fire on the eastern and western edges of the wood. This they did with great rapidity and effect. On the eastern side, the survivors of two Civil Service Rifles companies, encouraged by the effect of the mortar fire, leaped forward to attack. What was happening? Where was the German fire? Suddenly, grey-uniformed figures appeared. The tired but up to now invincible Germans began to emerge from the wood, hands high in the air. They had had enough. Similarly, from the west of the wood, Germans were streaming out, some being escorted by triumphant Londoners, others without escort. The prisoners were a pitiful sight - tired, gaunt, haggard, wounded and dirty, but some still displaying Teutonic pride. Their defence of High Wood had been magnificent, but now, at last, the steady stream of enemy troops continued to flow from the wood. After two long months of bitter struggle, the end was in sight, German resistance was crumbling. An hour later, it was all over.

The 47th Division, like all the divisions at High Wood, mourned its great losses, but was proud to have finished the work of all the courageous men who had passed before. On arrival at High Wood, these fine men from the capital of England were at the peak of fitness and there existed the same comradeship which was evident in all the pals battalions. The Londoners had great faith in their commanders and would have followed them into the jaws of hell - this they did and died in great numbers. Major-General Barter, after having successfully executed his order to capture High Wood, was unjustly sacked by Pulteney for 'wanton waste of men' and returned to England in disgrace. The

The water-filled crater in High Wood, scene of the most bitter fighting on the 3rd September by the 1/Cameron Highlanders and the 1/Black Watch. (author)

47th Division lost over 4,500 officers and men and yet was criticised for 'lack of push'. Although Barter was unofficially cleared of being responsible for the high casualties, his requests for an official inquiry were refused, and he died an embittered man in 1931.

The collection of the dead in High Wood presented special difficulties. Live ammunition was everywhere and, by the end of the war, the thick undergrowth made it both difficult and dangerous to search for bodies. Some were recovered but it was then decided to leave the rest where they had fallen. The wood follows the same contours of eighty-three years ago. The battered stumps have disappeared and today fine trees stand close and proud in High Wood. It still has the same impenetrable black, sombre appearance viewed either from a distance or close to. High Wood is private property and should not be entered but if you come to see this place of death, peer into the dark interior, pause and reflect a moment, what you see is the last resting place of over 8,000 British and German dead.

(i) This contrasts with Conan Doyle's account of the Deccan Horse and Dragoon Guards 'galloping' three miles from the rear. (British Campaign in France and Flanders 1916, Hodder & Stoughton, page 158) However, it is true the terrain was more suitable when the cavalry actually made their charge across the cornfield where, apart from some shell craters, the ground had not yet been subjected to a severe barrage.
(ii) Wood Lane has been divided into north and south, the northern part falling within objective 23 whereas Wood Lane South is dealt with in Objective 21, Delville Wood and Longueval. All references to Wood Lane in this section relate to Wood Lane North.
(iii) The 5th/6th Scottish Rifles are referred to as 6/Scottish Rifles in the statistical listings and text. The two battalions were amalgamated in June, 1916 (O.H., Vol. II, page 110)
(iv) No doubt Landon would have dearly loved to remind Horne that the same headquarters which was now seeking a scapegoat, was the one which refused permission to occupy the wood in the evening of the 14th July
(v) O.H., Vol. 2, page 198
(vi) It is interesting to compare the difference in the ground conditions at La Boisselle and High Wood. At the former, the large crater caused by 60,000 lbs of ammonal has remained dry. The smaller crater in High Wood, the result of two separate detonations of 3,000 lbs of ammonal, quickly filled with water and today, ducks can be seen on this site of death.

OBJECTIVE 24
GUILLEMONT VILLAGE, STATION & QUARRY, ARROW HEAD COPSE, TRENCHES NORTH-WEST TOWARDS DELVILLE WOOD

The site of Arrow Head Copse was just to the south of the road leading from Trônes Wood to Guillemont. The 35th Division, forming the right flank of the XIII Corps was ordered to secure the trenches between Maltzhorn Farm-Arrow Head Copse-Trônes Wood in order to clear the approach to Guillemont from the west. The half-hour preparatory barrage was not effective and a company of the 15/Sherwood Foresters (105th Brig.) suffered heavy casualties from shell, machine-gun and rifle fire on its approach to Arrow Head Copse. Another company of the same battalion near Waterlot Farm suffered a similar fate. The few men reaching the enemy trenches were quickly ejected with bombs. A second bombardment did not prepare the way for a better result and the 23/Manchester (104th Brig.) could make little progress against the enemy defences. About 450 casualties were incurred for no positive result.

A further attack made on the 22nd July by a battalion of the 35th Division between Maltzhorn Farm and Arrow Head Copse failed to secure the trenches. Similarly, units of the 3rd Division attacking Guillemont Station from Waterlot Farm also had to withdraw against the enfilade fire. The first assaults on Guillemont were disappointing to say the least, the approaches to the village being far from secure. It was hoped the carefully planned preliminary barrage for the attack on the 23rd July would result in a more protected line of attack. Brigadier-General Sackville-West's 21st Brigade (30th Div.) was to lead the attack. At 3.40am, three companies of the 19/Manchester progressed well from the eastern edge of Trônes Wood until they came upon uncut wire. They cut and hacked their way through the wire under shell, machine-gun and rifle fire, and resolutely continued their advance. They engaged and forced the enemy back, establishing machine-gun posts to block German reinforcements coming from the east. The Manchesters, having lost too many men at the wire, were forced to retire.

Meanwhile, the 2/Green Howards had started from Longueval Alley, heading south-east. Well to the north of the Manchesters, they lost their way and headed south due to some confusion caused by a smoke screen which had not fulfilled its purpose. The attack became disjointed, some men seized a trench south of the railway and retired later to the ruins of Waterlot Farm whereas most found themselves up against the same uncut wire as the Manchesters. These men withdrew to the trenches at Trônes Wood.

To the immediate north of the 30th Division, Haldane's 3rd Division had the task of clearing the area around Guillemont Station. Bombers from the 7/King's Shropshire Light Infantry (KSLI) and 8/East Yorks (8th Brig.) failed to secure their objective and retired to join units of the Green Howards from 30th Division. The 2/Royal Scots, also of 8th Brigade, tried to bomb its way south-east from Waterlot Farm with the aim of seizing Guillemont Station. Under enfilade fire from Ginchy, 1000 meters to the north-east, the Scots could not hold the few gains they made and they too retired to Waterlot Farm. The mixed battalions were now able to repulse a counter-attack later in the morning.

Responsibility for the capture of Guillemont still lay with the XIII Corps and the 89th and 90th Brigades of Shea's 30th Division took up their assembly positions, the 89th just to the south of Trônes

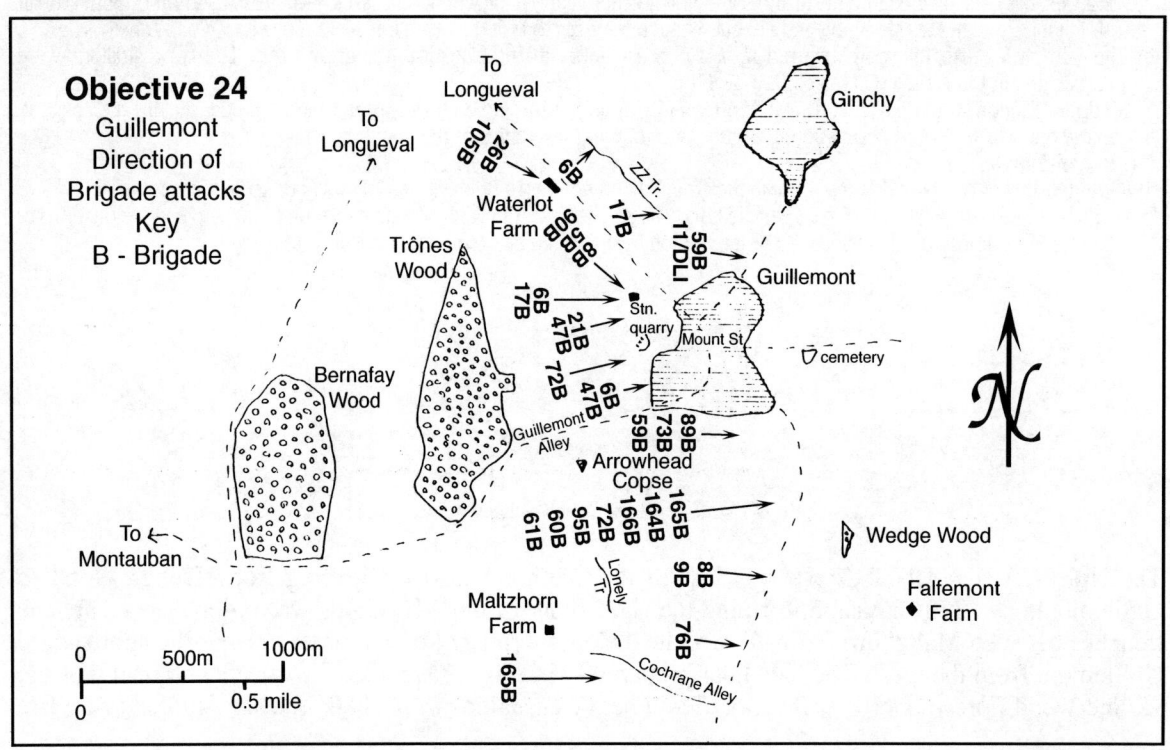

Wood and the 90th Brigade just to the north. They were to pass through the positions of the 35th Division. While waiting for zero hour, the 19/King's Liverpool (89th Brig.) were subjected to H.E. and gas as were some of the units of the 90th Brigade. A company of the 2/Bedford attacked Maltzhorn Farm, aided by the 153rd French Regiment attacking from the south. On the right flank, the 20/King's seized the front line trenches and continued on to the Hardecourt/ Guillemont road. Losing heavily from enemy machine-gun fire, they nevertheless dug in on the road. Parties of the 17/King's arrived to help with the consolidation. The 19/King's in the centre was also badly hit by enemy fire, only a few men reaching the road. A little further north, a company of the 19/King's succeeded in getting forward towards the south-eastern entry to Guillemont.

The 2/Royal Scots Fusiliers forming the right flank of 90th Brigade advanced either side of the track leading to Guillemont and entered the village with little opposition. A party of the 18/Manchester in close support helped to consolidate the entry. The Germans mounted a counter-attack from the eastern end of the village but this was repulsed. The Manchesters now advanced north, seizing the front line and taking some prisoners but machine-gun fire from the quarry and station prevented any further movement, the men eventually having to withdraw. A second attack by half the 16th and 17/Manchester was brought to a halt in front of the uncut wire.

The 2/Ox & Bucks Light Infantry (O&BLI) of the 5th Brigade (2nd Div.) advanced south-east from Waterlot Farm towards Guillemont Station. The battalion came under heavy fire but a few men overran the German front line positions and were close to the station. Here, the same strongposts which had caused so many casualties in the 90th Brigade, also checked the advance of the O&BLI, eventually causing it to retire, still under heavy fire.

The morning mist prevented communication by visual signals and almost all underground cables had been damaged. The only way of relaying messages to divisional headquarters was by runner. It would be a dangerous task once the fog had lifted. It was not practicable for the runner to use the crowded trenches and he had to cross the open ground between Guillemont and Trônes Wood over which German machine-guns at the quarry and station enjoyed an excellent field of fire. One such runner, CSM. W.J.G. Evans was awarded the Victoria Cross for his courage in performing this task. Five previous runners had been killed and Evans volunteered to carry the message. Wounded under the heavy fire he delivered the message and inspite of his wounds, rejoined his company.

Under the impression they were cut off, the 19/King's withdrew from the edge of Guillemont. Three companies of the Royal Scots Fusiliers were still holding on just inside the village. On the right wing, the Bedfords were frantically digging in to prepare for the expected counter attack. By mid-afternoon the Germans were observed advancing from Ginchy and Leuze Wood. The British artillery could not be registered on the village while the Scots Fusiliers still held out but a German barrage cut off the possibility of reinforcing or supplying the Fusiliers. The Scots fought tenaciously to hold their position but were finally overwhelmed by superior numbers.

In the afternoon consolidation was well under way in the only permanent gain of the attack - the line from Arrow Head Copse to Maltzhorn Farm where the 89th Brigade was still in touch with the French. The small gain had been made at a heavy cost, the 89th and 90th Brigades having casualties of 1,314 and 1,463 respectively while the 2/Royal Scots Fusiliers had staggering losses of 650 men of all ranks. During the night of 30/31st July the front line infantry of the 30th and 35th Divisions was relieved by Major-General Jeudwine's 55th Division.

And so the first month of the Battle of the Somme had ended. The month of August was to provide the most severe test of courage, hardship, moral fibre, as well as the physical and mental strength of the troops. The Commander-in-Chief and General Rawlinson were disappointed about the outcome of the attack on the important position of Guillemont. It was going to be a tough nut to crack and long discussions were held to decide the best course of action. The incessant German barrage eliminated any possibility of immediate action but men of the 2nd and 55th Divisions had been busy digging forward trenches in order to enhance the next assault on the village. It was trying work involving many casualties. The objectives for the coming attack were clearly defined - the 55th Division to secure the high ground south of Guillemont, keeping in touch with the French, and then to capture Guillemont village. The 2nd Division was to silence the strongpoints between Waterlot Farm and the village, particularly in the quarry and at Guillemont Station. The preliminary barrage was to include areas outside the objectives in order to confuse the enemy as to the direction of the attack. The preparations took a number of days and zero was fixed at 4.20am on the 8th August.

Promptly at zero the 5/King's Liverpool (165th Brig.) attacked along the spur adjacent to the French and made good progress at the start until enemy fire brought the men to a halt. The battalion bombers went down Cochrane Alley which runs south-east from the spur. The Alley was blocked by the 1/West Lancs Field Coy. using guncotton at a point about 400 meters east of Maltzhorn Farm.

The 4/King's Own Royal Lancaster Regiment (164th Brig.) attacking from Arrow Head Copse was stopped before uncut wire defending the entry to Guillemont from the south-west. The men tried to dig in to avoid the German barrage - wounded men lay all around. The position was untenable and a withdrawal was made to the assembly trenches.(i) The 8/King's (164th Brig.) attacked the German front line north of the Guillemont road, passing north and south of the quarry tucked in behind the front line and continued onwards fighting their way into west Guillemont. The 4/Loyal North Lancs of the same Brigade was sent forward to occupy and consolidate the German front line where the 8/King's had crossed. German bombers coming up from the south gave the Lancashires no chance to defend the trench and they were forced to retire leaving the 8/King's in a precarious situation.

The 6th Brigade of the 2nd Division attacked due east from north of Trônes Wood. The leading companies of the 1/King's swept across the German front line in a direct attack on Guillemont Station, hoping to link up with the 8/King's (164th Brig.) just to the south but that did not take place. The fourth company of the 1/King's, rushing forward to occupy the front line, found the Germans had reoccupied it and they could not effect an entry. The German defences, particularly on the west side of Guillemont were well conceived. Starting just below Waterlot Farm, their trench system passed just to the west of the station and quarry providing protection to these two strongpoints. The front line then continued south along the western edge of the village and finally south-south-west to a point just east of Arrow Head Copse. Many of the dug-outs had more than one entry/exit and it is thought this is how the Germans managed to reoccupy their front line north of Guillemont. The 17/Middlesex on the left flank of 6th Brigade, attacked ZZ Trench, 400 meters east of Waterlot Farm. After forcing an entry, the immediate task was to link up with the 1/King's just over 400 meters to the south. One company sustained the most awful losses while bombing their way down ZZ Trench.

The attack had now lost its impetus and, due to congestion in the trenches, orders for a further attack had to be postponed until the next day, 9th August. Zero hour was at 4.20am giving little time to organise an effective barrage. The wounded needed to be brought in. The replacements had much difficulty in getting forward. Brig.-Gen. Green Wilkinson's 166th Brigade was to relieve 164th. This was no easy task as the reoccupied German front line separated the two brigades. The 10/King's of the relieving brigade kept right on the heels of the barrage and approached the wire where they advanced against a steel curtain of fire. Twice they tried to close with the enemy - twice they were forced back. A most gallant attack during which most of the officers were casualties including Lt.-Col. Davidson. The 5/LNL of the same brigade, not being able to get started in coordination with the barrage, stood little chance and the Lancashires paid the price. The 7/King's attacked from the previous day's gain by the 165th Brigade but once again, no real progress could be made

The 13/Essex and 17/Middlesex, both of 6th Brigade (2nd Div.) could likewise make no progress when the former attacked the German line south of the railway and the latter made another attempt to gain ground south from Waterlot Farm. It was obvious that the hard-earned gains within Guillemont by the 1/Kings (2nd Div.) and the 8/Kings (55th Div.) on the previous day would be difficult to hold. Both battalions, the Territorials of the 8/King's and the Regulars of the 1/King's were completely isolated and in desperate need of supplies and support. While efforts were being made to achieve this, the Germans counter-attacked. With grim determination these courageous men met the foe. Soon out of bombs and now fighting hand-to-hand, they were overwhelmed. The Germans forced the surrender of some men at the quarry, while at the station, the fight continued until the men were out of bombs, ammunition, food and water. Utterly exhausted, they were killed or taken prisoner.

The 9th August had been a bad day - the attacks had been resolutely executed but none had succeeded, the situation in Guillemont remained the same. The proposal for a combined attack on the 11th August had to be postponed to allow sufficient time to complete the preparations. On the 10th August Lt.-General the Earl of Cavan took over command of the XIII Corps(ii). The attack was further delayed through mist and rain, zero being eventually fixed for 5.15pm on the 12th. The barrage had begun at 3.30pm concentrating on the area south-east of the village and the spur to the south of

Guillemont, objective of the 55th Division. Half the 9/King's Liverpool (165th Brig.) advanced up the spur while the battalion bombers attacked down Cochrane Alley. Against heavy machine-gun and rifle fire the King's fought their way to the Hardecourt/Guillemont road. The French attack to the south had gone well but their advance into the Maurepas ravine had failed making the King's position untenable. The King's were withdrawn in the evening. The bombers, however, had pushed a little further down Cochrane Alley, blocking and retaining their hold.

The spur was proving very difficult and Maj.-Gen. Deverell's 3rd Division(iii) was brought in to replace the 55th Division whose losses amounted to 4,126 all ranks. Brig.-Gen. Kentish's 76th Brigade was on the right flank and the 2/Suffolk secured Cochrane Alley as far as the Hardecourt/Guillemont road. Under enfilade fire from Lonely Trench less than 300 meters to the north, the Suffolks secured in part the trench on the Hardecourt road. Lonely Trench was indeed an obstacle - too near the British front line to be shelled, it had been subjected to a trench mortar attack which was only partially successful. When the 8/KORL advanced on the left flank of 76th Brigade, it was met by severe machine-gun and rifle fire. The same problem faced the 13/King's on the right flank of the 9th Brigade. The men went forward in short leaps, went to ground, tried again, and again but could not expel the Germans. The 4/Royal Fusiliers on the left flank of 9th Brigade met the same fire while attacking north-east from south of Arrow Head Copse. Further to the north, the 9/East Surrey (72nd Brig. 24th Div.) made an abortive attack on a German strongpoint just south of the Guillemont Track. Apart from the small gain in Cochrane Alley, no further permanent gains had been made. The French had had a bad day too - after initial progress towards Angle Wood, they were driven back by a strong counter-attack. In view of this, the 2/Suffolk was withdrawn from the trench on the Hardecourt/Guillemont road into Cochrane Alley.

On the 17th August, the troops were withdrawn sufficiently to allow a heavy bombardment of Lonely Trench. After the barrage, the 10/RWF (76th Brig.) and 12/West Yorks (9th Brig.) attempted a surprise assault only to find the Germans waiting. A further assault six hours later by mixed units from 9th and 76th Brigades including the 12/West Yorks, 2/Suffolk, 8/KORL and 10/RWF came to nought and Lonely Trench remained a thorn in the side of any attack from the north, south or east of Maltzhorn Farm.

Rain the following day made movement difficult in the cratered ground. The untaken objectives were subjected to a long bombardment, this time with no increase in tempo in the final minutes to avoid arousing the Germans to a possible attack. Zero was set for 2.45pm on the 18th August. Keeping very close to the carefully orchestrated barrage, the 1/Gordon and 10/RWF (76th Brig.) made their attack. The Gordons arrived at their objective near the Guillemont/Hardecourt road while the Welshmen assaulted the southern part of Lonely Trench, some arriving at the road. The 1/NF of 9th Brigade attacked the German front line south-east of Arrow Head Copse but were beaten back by heavy fire. With no cover on the left flank, the Welsh Fusiliers were forced to retire against continuous enfilade fire. The Northumberland Fusiliers made two further assaults on the northern section of Lonely Trench, one accompanied by the 8/East Yorks from 9th Brigade but were driven back by fire from the untaken German strongpoints. The few men who forced an entry were quickly overwhelmed.

The 24th Division, still in position west and north-west of Guillemont, attacked with the 73rd Brigade, the 13/Middlesex on the south of the Guillemont Track, the 7/Northampton to the north. The Middlesex, close to the creeping barrage, arrived at the German front line where the battalion was halted by cross-fire of machine-guns sited in the quarry and at strongpoints in the south-west corner of the village. In spite of the determined assault, it was impossible to force an entry. Just to the north, heavy fire also prevented the Northampton from entering the front line trench directly facing the main attack, but the left company forced a lodgement in the front line near the quarry. Here, the fighting was desperate and the Northampton was reinforced by the 9/Royal Sussex of the same brigade. Engineers also moved up to help with consolidation of the bitterly contested gain.

The 17th Brigade, assembled to the west of Guillemont Station, moved forward to assault the station. On the right flank the 3/Rifle Brigade, advancing dangerously close to the barrage, seized the front line, encountering several Germans coming forward to surrender. With hardly a pause they dashed forward, securing the station and capturing some more prisoners. But the Rifles had not yet finished their work - on they went to take their second objective, a section of the Waterlot Farm road. On the left flank, the 8/Buffs assaulted ZZ Trench to the east of Waterlot Farm. Making a frontal assault, the Buffs secured the German front line and, linking up with the Rifles, bombed their way into ZZ Trench, taking about a hundred prisoners. It appears the barrage had been particularly effective here.

The enemy, in typical style, was quickly organising his counter-attack. He may have lost ground but had every intention of retaking it. The German bombers, following a heavy barrage in the evening, drove the French back on the Maurepas ravine incorporating some of the Gordons from their position on the Hardecourt/Guillemont road. Patrols sent out the following morning established that the Germans had withdrawn to Falfemont Farm and the situation was less serious than at first thought. The Gordons, RWF and units from 8th Brigade went up and continued consolidation on the main road, linking up with the French who had also moved forward. The whole of Lonely Trench was now occupied. Major-General Pinney's 35th Division relieved the 3rd during the night of 19/20th August also taking over the right flank of the 24th Division.

The German strongpoint to the east of Arrow Head Copse held out against a dawn attack on the 21st August by units from the 35th Division. Meanwhile, the French had occupied Angle Wood to the south-east and the Allies had now linked up. Still under a heavy barrage, parties of the 3/Rifle Brigade and 8/Buffs (17th Brig. 24th Div.) occupied the southern end of ZZ Trench leading to Guillemont. Towards the end of the afternoon the quarry was assaulted by the 8/Queen's (72nd Brig.) where a bombing contest resulted in the Queen's withdrawal. Just to the north, one company each from the 1/Royal Fusiliers and 3/Rifle Brigade (17th Brig.) attacking in the area of Guillemont Station were met by accurate fire from the defending Wurttembergers and could not hold their position. The 24th Division had sustained casualties of just over 3,500 officers and men and was replaced on the 22nd August by Major-General Douglas Smith's 20th Division.whose objectives were defined as the northern perimeter of Guillemont and the enemy trenches north-east towards Ginchy. The 20th Division had hardly taken up its positions when, in the late evening, it came under a heavy barrage which, although countered by the British artillery, did not prevent the enemy from attacking the 11/KRRC (59th Brig.) south of the railway. The Rifles countered with machine-gun and rifle fire. The 11/DLI, 20th Division's pioneers, along with some Royal Engineers engaged in general consolidation work, made for a considerable degree of congestion in the trenches and just after midnight, a German barrage curtailed any further work.

Rain had fallen every day between the 14th and 20th August and ground conditions had deteriorated to a considerable extent, delaying the next attack on the spur south of Guillemont until the 3rd September. Brigadier-General Gordon-Lennox's 95th Brigade (5th Div.) was now formed up just south of Guillemont with the 12/Gloucester on the right and the 1/DCLI on the left. Leaving the assembly trenches at noon the leading companies of both battalions secured the battered German front line, taking control of some dug-outs. Continuing their advance, the Cornwalls and Gloucester seized the German 2nd position on the south-eastern edge of Guillemont, the latter coming under fire from Falfemont Farm. Pausing a moment, the men then advanced to the Wedge Wood/Ginchy road from where they consolidated and worked their way north to join up with the 20th Division at Guillemont, taking some 150 prisoners on the way.

Meanwhile, Smith's diminished 20th Division was still faced with the final capture of Guillemont. The British bombardment had shelled spasmodically during the lull in the fighting and forward trenches had been extended in preparation for the next attack. The village itself was a pile of rubble with craters everywhere but here, as within so many other villages in ruins, the deep cellars were still intact. The strength of Shute's 59th Brigade about to attack from the south was down to 1,650 men and was reinforced by the 6/OBLI (60th Brig.) and 7/SLI (61st Brig.). The remaining strength of 60th Brigade was so reduced it was deemed necessary to replace it with Brigadier-General Pereira's 47th Brigade from the 16th (Irish) Division. A complex barrage began at 8.15am to assist the infantry units to take their objectives in several stages.

The attack from the north was led by the 10/KRRC of 59th Brigade, which, keeping close to the barrage, took the Germans by surprise. To the north of Mount Street the 6/Connaught Rangers (47th Brig.), dashing towards the quarry in typical Irish style, swept all before them. Private Hughes, one of the Rangers, although wounded, single-handed captured a machine gun, and despite being wounded again, brought in some prisoners(iv). The support battalions followed at noon, the 10th and 11/Rifle Brigade and 10/KRRC (59th Brig.) quickly reaching the Hardecourt road. Now turning north towards Mount Street, detachments had to be left to clear pockets of German resistance left by the dashing

Guillemont Station, a severe obstacle for the attacking infantry.
(J. Bommeleÿn)

Irishmen whose sole concern was to push forever forward. From the left flank, the 7/Leinster (47th Brig.) attacked in a south-easterly direction from the east side of the Guillemont/Longueval road. One of the bombers, Lt. Holland, too impatient to wait for the creeping barrage, dashed through it, bombing the surprised enemy in their dug-outs.(v). Apart from the clearance of a few pockets of resistance, the first objectives had been secured. The second phase of the attack at 12.50pm met with stiff resistance but the leading battalions of 59th Brigade, now reinforced by the 6/OBLI and 7/SLI plus the 8/RMF of 47th Brigade, swept through the Rangers gradually taking control of the village against a weakening German resistance. The final phase began at 2.50pm when the 59th Brigade pushed forward with virtually no opposition to the Ginchy/Maurepas road. To the immediate north the Munster Fusiliers and 6/Royal Irish of the same brigade advanced with their pipers inspiring them to do great things. It is recorded that the ardour of some of the Munsters could not be restrained. Consolidation work was undertaken without delay, the 7/DCLI from 61st Brigade, the divisional pioneers, the 11/DLI and the 83rd Field Coy. Royal Engineers all engaged in the immediate urgent work which always follows a successful attack. The junction with 5th Division to the south was maintained while the 7th Division was fighting 1,100 meters to the north-north-east at Ginchy. Guillemont was now secure but the cost on the 3rd September was almost 2,500 all ranks including four Lieut.-Colonels (wounded - C.A. Blacklock, W.V.L. Prescott-Westcar, C.J. Troyte-Bullock and kllled, J.S.M. Lenox-Conyngham). The casualties in the senior ranks were becoming increasingly difficult to replace. Over 700 German prisoners were taken, and their dead lay as a thick carpet on the battleground..

In the late afternoon it was learned that the 7th Division had been pushed out of Ginchy causing some concern over the position of the left flank at Guillemont. Major-General Smith requested artillery cover for the open flank. Meanwhile, the 12/King's Liverpool of 61st Brigade was already taking up position on the Guillemont/Ginchy road to support the 47th Brigade and strengthen the flank. Two German attacks were successfully repulsed by Lewis-gun and rifle fire. Sgt. D. Jones of the 12/King's, in charge of a forward position, successfully held this post until relieved two days later, repelling a number of counter-attacks(vi)

The capture of Guillemont facilitated an assault on Ginchy from the south and also cleared the way for a general advance towards Leuze Wood, Bouleaux Wood and Combles, described under Objective 33.

(i) During the withdrawal 2nd Lt. G.G. Goury was in charge of half a pioneer company (4/South Lancs) digging a forward communication trench. Going forward in an attempt to rally the retreating infantry, he brought in at great personal risk the CO of the 4/King's Own, Major J.L. Swainson who unfortunately dicd later from his wounds. Lt. Goury was awarded the VC which he collected from the King at Buckingham Palace on 18th November 1916.

(ii) Lt.-General Sir Walter Congreve had been suffering for some while from a severe attack of gastro-enteritis. The new commander, the Earl of Cavan subsequently fell ill on the 17th August and was succeeded by Lt.-Gen. Sir Thomas Morland. Congreve must have been most disappointed in having to relinquish his command - not only had he lost his son "Billy" near Delville Wood, but now he was out of the action too.
(iii) Major-General Deverell had replaced Major-General Haldane, the latter having taken over VI Corps.
(iv), (v) and (vi) All three men were subsequently awarded the Victoria Cross.

OBJECTIVES 25 & 26
WATERLOT FARM AND "ALCOHOL" TRENCHES BETWEEN DELVILLE WOOD & GINCHY, GINCHY VILLAGE, QUADRILATERAL, TRIANGLE AND STRAIGHT TRENCH

Waterlot Farm, in reality, a sugar refinery, lies about half-way between Guillemont and Longueval. The farm was just a heap of ruins, nevertheless giving sufficient cover to install machine-gun emplacements having the capacity of enfilade fire on any approach to Longueval from any direction except the north. Consequently, an assault was made on the 15th July on Waterlot Farm by the 5/Camerons (26th Brig. 9th Div.) supported by half of the 4th South African Regiment, also of 9th Division. After a protracted struggle involving several attacks, the enemy was driven out but a German barrage delayed a permanent hold on the position until the 17th. Occupation of the site was eventually effected by the 16/Cheshire (105th Brig.) which inflicted heavy casualties on three separate German counter-attacks.

Waterlot Farm served as a base for attacks east to ZZ Trench, Ginchy and north-east to a group of so called "Alcohol" trenches and finally as an approach to Delville Wood from the south. On the 18th August while the 3rd and 24th Divisions were engaged in a bitter contest at Guillemont, Couper's 14th Division was in both Longueval and assembly positions facing the Alcohol trenches to the east. At 2.45pm, the 6/SLI (43rd Brig.), advancing close to the creeping barrage, caught the Germans by surprise and took a number of prisoners. The barrage had badly disfigured Beer Trench at its southern end. The attack of the 6/DCLI, on the left, lost its momentum under a heavy barrage, machine-gun and rifle fire. Taking cover for a short while, the men then continued in small groups entering and securing Edge Trench. German bombers counter-attacked from the north and reoccupied the trench, a useful asset on the edge of Delville Wood. Meanwhile, the Somerset was fighting its way west up Hop Alley using captured bombs. It was a desperate struggle, sometimes hand-to-hand, the men under fire from the enemy in Pint Trench, but Hop Alley was held and a barricade erected in Beer Trench. Consolidation began, enabling a successful defence against two counter-attacks from Pint Trench.

On the 24th August the 14th Division sent its 8/KRRC (41st Brig.) forward on the right flank. The battalion was halted by severe fire from Ale Alley. The 9/KRRC, 5/KOSLI and 5/OBLI, all of 42nd Brigade, advancing close to the barrage got well forward until the right flank came under heavy fire which obstructed the securing of Beer Trench. To the left, the Oxford & Bucks., in touch with the 33rd Division, secured and consolidated its objective, taking over 200 prisoners and some machine-guns. Early next day, the 9/Rifle Brigade secured the southern part of Edge Trench almost to Ale Alley.

Rain on the 25th made all movement of men and supplies difficult. In these bad conditions Maj.-Gen. Watts' 7th Division relieved the Rifle Brigade of 14th Division. In the afternoon of the 27th, the Germans were taken by surprise and driven out of Edge Trench by the 10/DLI (43rd Brig. 14th Div.), the Durhams then turning to their peace-time occupation of mining, erecting a barricade in Ale Alley. Twenty-four hours later the Durhams, aided by the 1/RWF (22nd Brig. 7th Div.) endeavoured to secure Ale Alley from its barricade to the junction with Beer Trench but were unable to force an entry into Beer Trench. During the night of the 30/31st August the 9/Rifle Brigade (42nd Brig.) had done good work in establishing posts south-east of Cocoa Lane into the northern section of Beer Trench and had also constructed a sap from Prince's Street. The casualties of the 14th Division were 3,615 officers and men.

Objectives 25 & 26
General Map

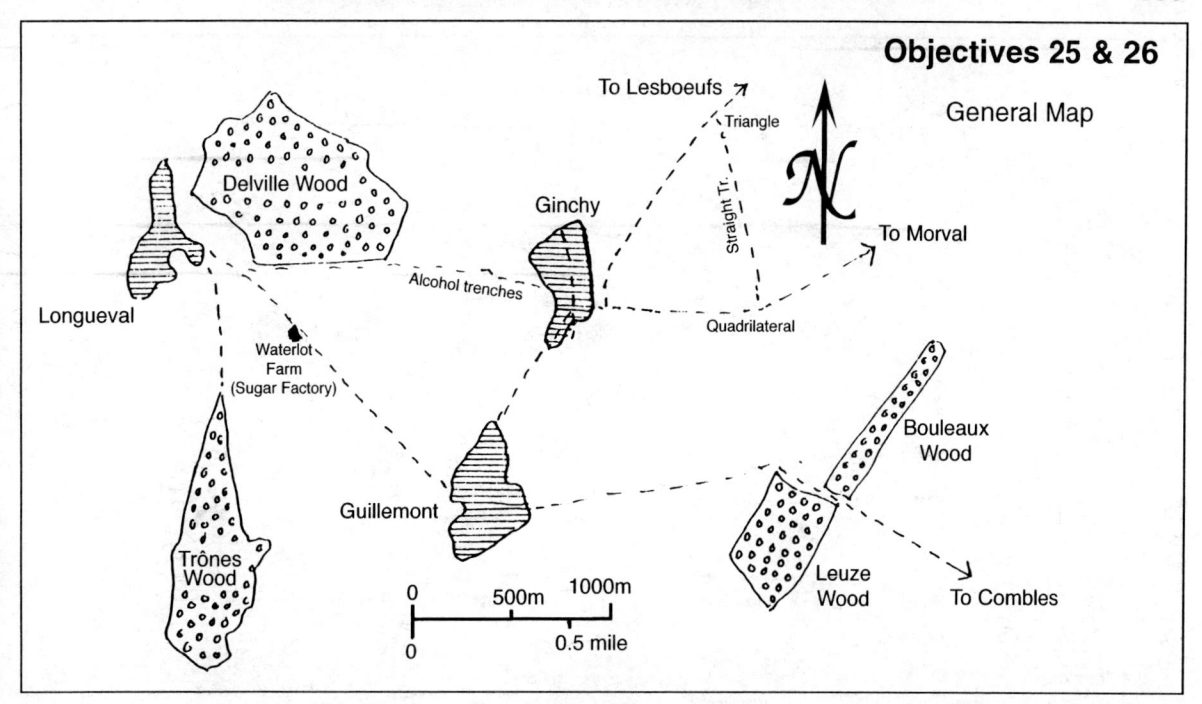

(Map showing Delville Wood, Longueval, Ginchy, Waterlot Farm (Sugar Factory), Alcohol trenches, Triangle, Straight Tr., Quadrilateral, Guillemont, Trônes Wood, Bouleaux Wood, Leuze Wood, with directions To Lesboeufs, To Morval, To Combles. Scale: 0-500m-1000m / 0-0.5 mile)

A long and intense German barrage announced a counter-attack which began at 1.00pm on the 31st August. The 1/South Staffs (91st Brig. 7th Div.) was assailed by storm troopers of the 35th Fusiliers bombing their way down Ale Alley towards Hop Alley. The Staffords stood firm, throwing the enemy back with accurate fire. A second attack proved more difficult but the Stafford's blood was up and after a bitter contest, the Germans were repulsed again. The Germans sought artillery help and the Staffords were bombarded for two and half hours following which, the stubborn Germans attacked again. The Stafford, equally stubborn, but with diminishing numbers, withdrew into Delville Wood and took up position in Edge Trench in touch with the 1/North Staffs (72nd Brig. 24th Div.)

The village of Ginchy lies 2,100 meters east of Longueval and the German defences formed a rough triangle starting from Ginchy, then 1,300 meters north-east towards Lesboeufs to a point known as the Triangle, then south along Straight Trench for 1,000 meters to the Quadrilateral on the Ginchy/Morval road, and finally, west for a further 1,000 meters back to Ginchy. Approximately half-way between the Quadrilateral and Ginchy, 150 meters to the north is the site of the Ginchy Telegraph, still shown on today's IGN maps at a height of 157 meters but which in fact no longer existed in 1916. It was the site of a semaphore station during the Revolutionary wars.

The 7th Division in XV Corps was faced with the task of capturing Ginchy and the triangular area to the east of the village. Brig.-Gen. Steele's 22nd Brigade relieved the 91st on the 3rd September in the trenches to the south and on the Longueval/Ginchy road. The Germans were still in control of most of the "Alcohol" trenches blocking the way to Ginchy and an attack was made a short time before zero by bombers of the 2/Queen's linking up with bombers of the 9/E Surrey (72nd Brig.) on the extreme right flank of the 24th Division. The Queen's bombed towards Hop Alley, the grenades attracting machine-gun and sniper fire. The Surrey could not get forward north of Ale Alley, and to make matters worse, received conflicting orders with the result that no progress was made.

Five minutes before zero the 20/Manchester and 1/RWF (22nd Brig.) moved forward from the assembly trenches. At 12.05pm the Manchesters on the right dashed across no man's land, disappearing into the southern part of Ginchy. The Welsh Fusiliers on the left came under enfilade fire from the untaken "Alcohol" trenches. Heavy losses were sustained in the attack on Hop Alley and Beer Trench. The Fusiliers advanced using shell holes for cover and eventually occupied the southern section of Beer Trench. The right company, overcoming the obstacles, entered Ginchy from the north but was not seen again. The support company had difficulty getting forward against the ever-increasing fire from front and flank but a few men reached the orchards to the north of Ginchy. Units of the 2/R Warwick, unable to take up their scheduled position in the unsecured Ale Alley, followed the path of the Fusiliers and joined them near the orchards. The Manchesters had indeed remained in the village

Objectives 25 & 26 (west)

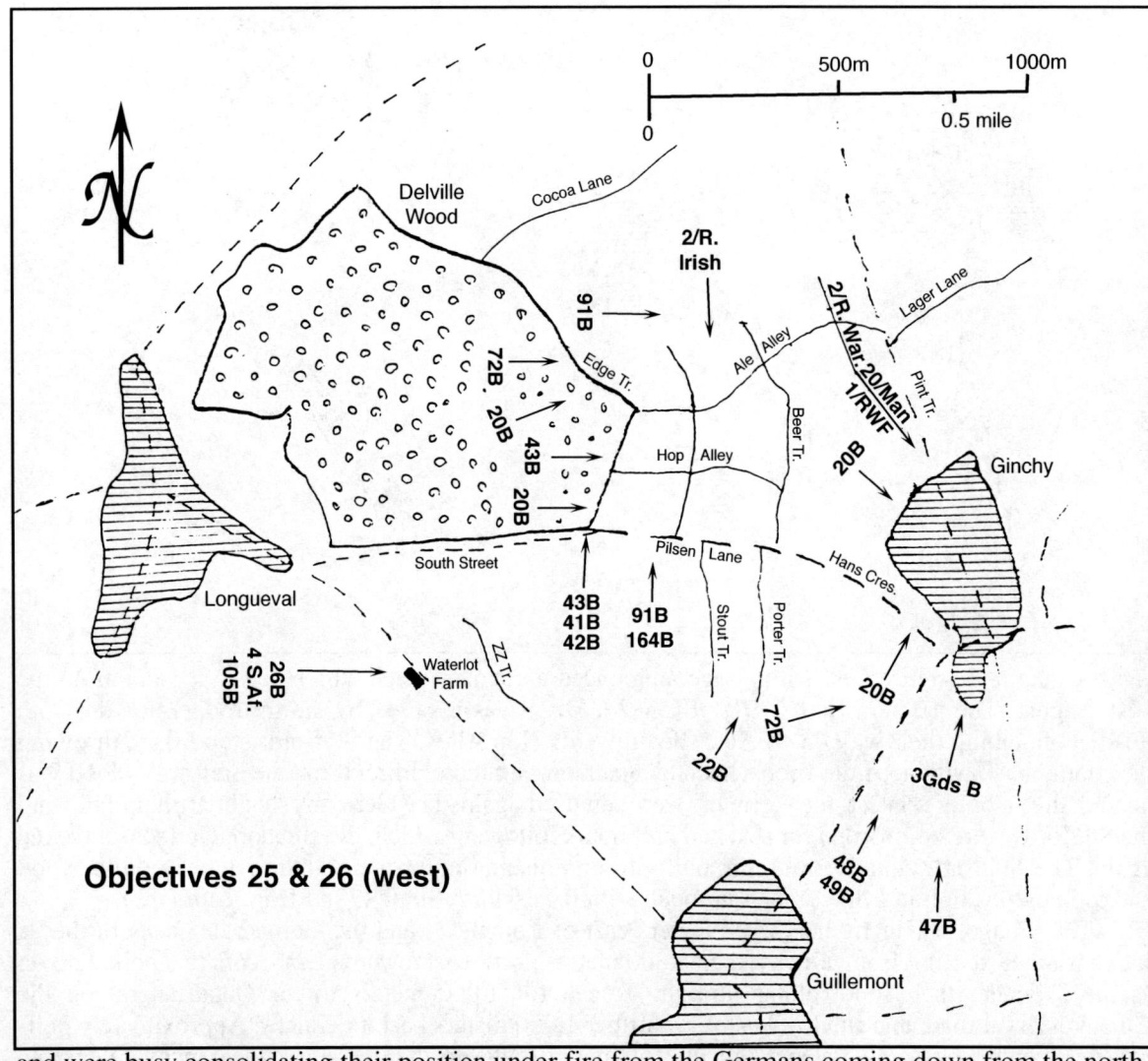

and were busy consolidating their position under fire from the Germans coming down from the north. A full counter-attack developed driving the survivors, including most of Warwicks, gradually west as far as Porter Trench where they prepared its defence. One party of Warwicks clung desperately to its hold on the sunken road just south-west of Ginchy, repelling a strong counter-attack in the afternoon.

Another effort was made to clear the troublesome left flank by the 2/Royal Irish, the last uncommitted battalion of Steele's 22nd Brigade with the view of continuing the attack on Ginchy afterwards. Little was achieved, the Royal Irish making two attempts across the open to enter Hop Alley while bombers worked their way up the eastern edge of Delville Wood. Contact was made with the Welsh Fusiliers pinned down in front of Hop Alley. A German counter-barrage hampered the support companies coming up to the front and machine-gun fire took a heavy toll on their approach to Stout and Porter Trenches.

No signals or messages were received from Ginchy leading to the assumption that the survivors of the 22nd Brigade must have pulled out. A heavy bombardment was laid on the German defences while plans were made for the next assault. The weakened 22nd Brigade was ordered to make another attack about 7.00pm while the 20th was instructed to hold the front line. The attack was subsequently cancelled when it was established that the enemy had re-occupied Ginchy and the 22nd Brigade was relieved. The 21/Manchester of 91st Brigade was posted to hold the left flank on the eastern edge of Delville Wood while the 20th Brigade made its attack on Ginchy.

At 8.00am on the 4th September the leading companies of the 9/Devon succeeded in reaching the outskirts of the village but further progess was impeded by accurate fire from the defenders, forcing the Devons back to Porter, Stout and ZZ Trenches. The 21/Manchester was immediately held up by the strongly defended triangle of Ale, Beer and Hop Trenches. One company of the Manchesters assaulted Hop Alley, attacking north from Pilsen Lane which runs east from Delville Wood to Ginchy, but these men also were forced to take cover from superior fire. The 2/Gordon relieved the far left flank at night and the next day was spent preparing for the next attack on the 6th. Meanwhile, the 2/Queen's of 91st Brigade took up the challenge of securing the trenches on the eastern side of Delville Wood. Following an ineffective two-hour mortar barrage, the Queen's moved forward against machine-gun fire and secured the edge of the wood to a point between Hop Alley and Ale Alley. Consolidation was undertaken without delay assisted by the 8/Devon. All units had to face an accurate enemy bombardment for many hours and losses were heavy, particularly in the Gordon Highlanders Regiment. Nevertheless, the leading companies of the Highlanders set off at dawn, losing their way to some extent in the dark and confusion of battle. A fresh attack was made at 5.30am, now supported by the 9/Devon. Delayed by bad ground conditions, the advance to the western edge of Ginchy was slow and dangerous. Against well-sited machine-guns, there was no chance of forcing an entry. At the same time the 8/Devon was pushing east from Pilsen Lane and strengthening its hold on the main road to Ginchy. Following a British barrage on Ginchy Farm and the area to the north, the Gordons attacked for the third time, the 9/Devon moving up from the south-west. Close behind the barrage, units from both battalions entered Ginchy, taking some prisoners. The enemy immediately laid a heavy barrage on the approaches from the west and south-west making it impossible to reinforce the Gordons and Devons in the village. The strength of the German counter-attack forced the men to abandon their hard-earned gain and withdraw to the British front line, which in itself, was no haven, coming under accurate enemy shell-fire. Maj.-Gen. Watts was disappointed - his 20th Brigade had been destined to take part in the forthcoming attack on the 15th September, but the Brigade had been badly mauled, the Division's casualties since the 23rd August amounting to 3,800 all ranks.

There was little action on the following day, the 7th, apart from an attack on the eastern corner of Delville Wood by the 2/Queen's. Preceded by rifle grenades, the Queen's were still unable to surmount the defences of Hop Alley and Ale Alley. The relief of the 7th Division was begun in the evening of the 7th by units of Hickie's 16th (Irish) and Jeudwine's 55th Divisions.

The 56th (London) Division was fighting in the Leuze Wood/Bouleaux Wood and Combles area in touch with the French, the 168th Brigade on the left flank now moving away from that area towards the Quadrilateral. The London Rangers (12/London) had made good progress from Bouleaux Wood towards their objective but were now, along with troops from 16th Division, held up by fire from the defences on the south-eastern perimeter of Ginchy. As darkness fell and under a heavy barrage, the Kensingtons (13/London) and London Scottish (14/London) moved up to

The sugar factory on the Guillemont road out of Longueval.[J Bommeleÿn]

Objective 26 (East)

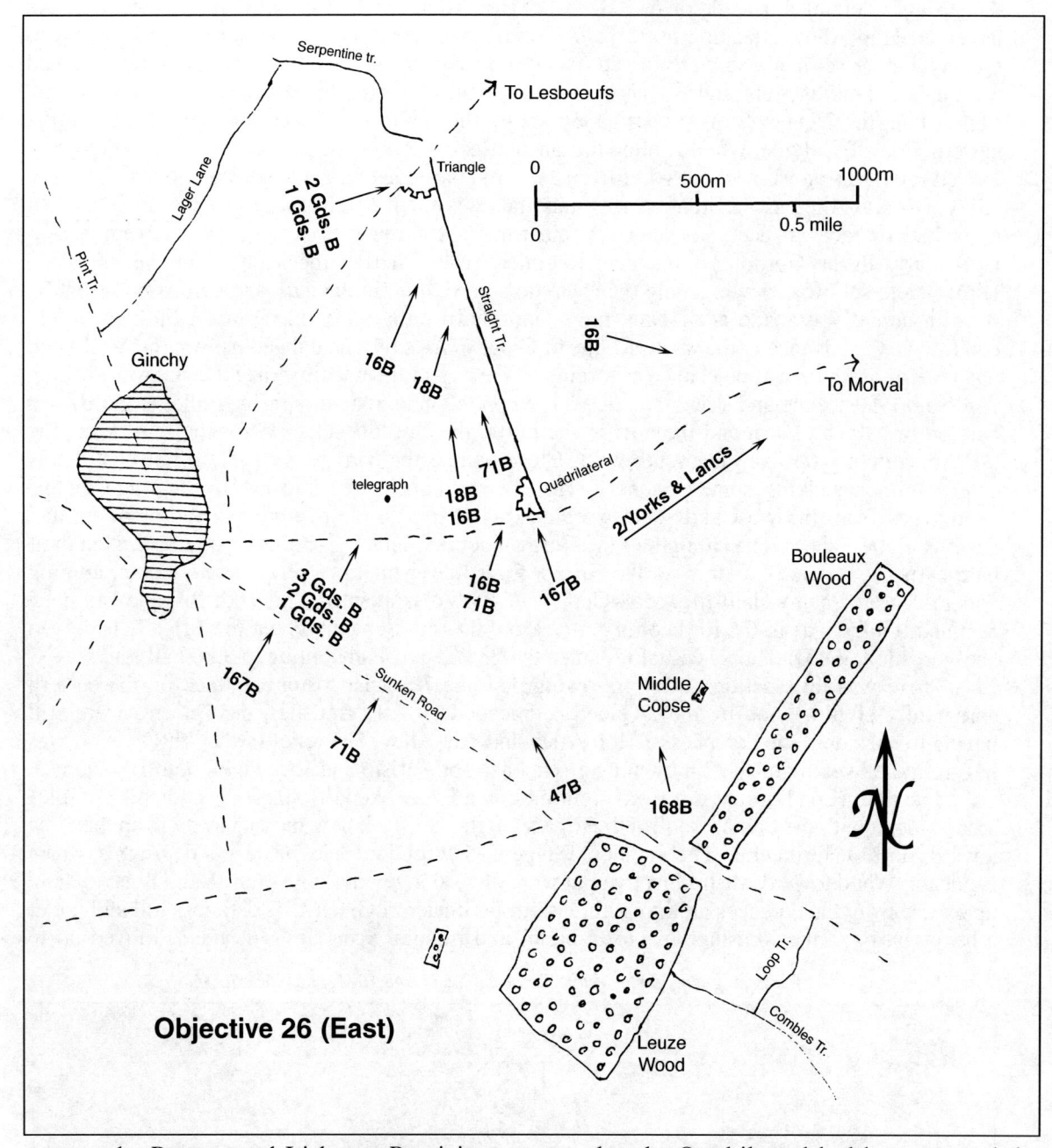

support the Rangers and Irishmen. Receiving a report that the Quadrilateral had been secured the London Scottish advanced after midnight, losing its way in the mist and darkness. At one time, they were attacked from the rear, but the crisis was quickly restored with the bayonet. As dawn broke on the 10th September, mixed units of the London Scottish, Rangers and Fusiliers found themselves in trenches west of Bouleaux Wood.

On the left of the 56th Division, the 47th & 48th Brigades of the 16th Division were forming to attack Ginchy. The 6/Royal Irish, 8/RMF and 6/Connaught Rangers in support (all 47th Brig.) were met by close-range machine-gun fire. Reinforced by the 7/Royal Inniskilling Fusiliers from 49th Brigade, another attempt ended in failure. The 48th Brigade on the left, now had its right flank unprotected. The 1/RMF wheeled right, routing the nearest strongpoint which had checked the 47th Brigade and then pressed on towards its objective. The Irishmen were now entering the village from the south and south-west. The 7/Royal Irish Rifles with units from the 7/Royal Irish Fusiliers (attached from 49th Brig.) were already across the Guillemont road and dashing with little opposition towards Hans Crescent and further on into the village between the Longueval and Flers road. Mixed units were now in and clearing the village. The 8/Royal Dublin Fusiliers coming up to support the 1/RMF, and joined by the 9/Royal Dublin Fusiliers, continued the attack through the village. A number of Germans surrendered, others fled towards Flers and Lesboeufs, chased energetically by the elated Irishmen. It was not without some difficulty that they were recalled to help the 156th Field Coy. R.E. with the consoldiation. The work was urgent as the enemy counter-attack was expected at any time. The Germans made a number of attempts to re-enter Ginchy during the night but these were repulsed with the help of the Royal Engineers. The 48th Brigade with the help of two battalions of the 49th, had captured the important village of Ginchy. The cost was heavy, Lt.-Col. H.P. Dalzell-Walton (8/Royal Inniskilling Fus.), Capt. W.J. Murphy, (commanding the 9/Royal Dublin Fusiliers) were killed and Lt.-Col. R.E.P. Curzon was killed by shell-fire the same evening.

Although it was very gratifying to have secured Ginchy, the triangle of German defences to the east of the village were still intact as were the stubbornly defended Alcohol trenches to the immediate east of Delville Wood. The 164th Brigade of Jeudwine's 55th Division made an unsuccessful assault on Hop Alley and Ale Alley. A later assault from Pilsen Lane brought no satisfactory result although the enemy had ceded possession of the eastern edge of Delville Wood.

During the night of the 9/10th September Brigadier-General Corkran's 3rd Guards Brigade started a difficult relief of the 16th (Irish) Division, sometimes under fire. The 1/Welsh Guards took over Ginchy from the 48th Brigade, the 4/Grenadier Guards relieving the 47th Brigade to the east of Guillemont cemetery a little later. There was a gap between the Welsh and Grenadier Guards causing some delay in the relief operation. The Germans mounted a counter-attack from the north in the early morning, their movements hidden for a while by the mist. There followed a bitterly contested struggle at the end of which the 1/Welsh Guards had ejected the enemy from Ginchy. But that was not the end of the matter, the enemy attacked twice more and the 1/Grenadier Guards which had been sent forward to fill the gap, joined in the defence of Ginchy. Both sides recorded heavy losses.

It was now necessary not only to retain the hold of Ginchy, but neutralise the Quadrilateral, Straight Trench and finally, the Triangle on the Lesboeufs road. The 1/Grenadiers and 2/Scots Guards took over the positions of their comrade battalions in the 3rd Brigade. A re-entrant was pinched out, some prisoners taken and the front line advanced. In the morning of the 12th September the 1/Grenadiers made some progress towards the site of Ginchy Telegraph while the 8/Middlesex of 167th Brigade (56th Div.) advanced towards the Quadrilateral from the south-east aiming to link up with the 1/Grenadiers, but this was not achieved. The Grenadiers continued alone up the Morval road towards the Quadrilateral. The 3rd Guards Brigade was relieved by the 1st and 2nd Brigades during the following night. At the same time the centre of XIV Corps was taken over by the 6th Division, thereby relieving the left of 56th Division. Patrols had established the absence of the enemy in the trenches south-east of Ginchy which prompted an attack on the Quadrilateral from the south-west. At 6.00am on the 13th September the leading companies of the 2/Sherwood Foresters and 9/Suffolk, both of 71st Brigade, advanced as far as the Leuze Wood/Ginchy road where they seized and secured the sunken road. Further movement forward was not possible against the German machine-guns sited on the crest of the rising ground. Alternatively consolidating their position, then resting a while, they prepared for another assault. At 6.00pm they attacked again with great ardour but they were cut down - it was all too resemblant of the attacks from Serre to La Boisselle on the 1st July, the Foresters and Suffolks falling in their hundreds. They suffered the loss of 21 officers and 500 other ranks. An attempt at night by a company of the 2/Irish Guards to neutralise some German strongpoints near the Quadrilateral came to nothing. The German gunners were ready and took a heavy toll of the Guards. Happier news came from north Ginchy where, just before the Flers/Gueudecourt fork a few hundred meters north of the village, the 2/Grenadiers extended and straightened the new front line.

Lord Cavan, having retaken command of the XIV Corps after his temporary command of the XIII, would have preferred to launch his sector of attack on the general advance planned for the 15th September with possession of the Quadrilateral, Straight Trench and the Triangle but this was not to be. Loop Trench to the east of Leuze Wood was still untaken but Major-General Hull, commander of 56th Division in the area judged it would not hinder his operations in two days time.

During the night of the 10/11th September units of the New Zealand Division took over the left wing of the Jeudwine's 55th Division while Lawford's 41st Division relieved 55th Division's centre and finally, Couper's 14th Divison took over Jeudwine's right wing. Before the relief, the 55th Division

Site of Straight Trench which headed north from the Quadrilateral on the Ginchy/Morval road. It was the scene of a protracted struggle by a number of divisions between 15th and 18th September. (author)

had been engaged in valuable preparatory trench work prior to the coming attack on the 15th. In an attempt to isolate the small cluster of enemy-held trenches stubbornly holding out to the immediate east of Delville Wood, units from the 14th Division dug east from north of Ale Alley but, even with the help of the Guards coming in from Ginchy, the venture failed. Rawlinson agreed to a further attack an hour before zero on the 15th involving a preliminary barrage and the use of two tanks, one from XV and the other from XIV Corps.

The 6th Division faced the Quadrilateral which had been a thorn in the side of the British for some time. The preliminary bombardment on the redoubt had caused little damage, the uncut wire still hidden in the tall grass. It was intended to use all three tanks allotted to the Division but unfortunately, only one arrived at the assembly point, the other two breaking down during the journey forward. The single tank moved forward from near the railway towards the Quadrilateral. Seeing troops in front, it opened fire but the victims were the right of the 9/Norfolk. Further injury was stopped by the courageous action of Capt. A. Crosse of the Norfolks who hurried forward under fire to point out the direction needed and the tank then veered north towards Straight Trench. The attack on the Quadrilateral by the 8/Bedford (16th Brig.) was brought to an abrupt halt by vicious machine-gun fire. The 1/Buffs moved up fifteen minutes later and made a frontal assault but fire from the same source checked their advance. Some Bedford bombers were having difficulty in getting forward from trenches to the south-east of the objective

After its incident with the tank, the 9/Norfolk, accompanied by the 1/Leicester, both of 71st Brigade, after passing through the two battalions holding the Brigade front, advanced on Straight Trench and disappeared from view over the crest of the rising ground. Straight Trench, as the name implies, takes a straight line from the Quadrilateral to the "Triangle", a strongpoint 1,200 meters northeast of Ginchy on the Lesboeufs road. Initial progress was good, the Leicesters securing a strongpoint in the cratered ground but then both battalions came upon a belt of impenetrable wire. Here they came under intense machine-gun fire forcing the men to take cover. The tank, although damaged, was still moving but had to return to refuel.

To the left of the 6th Division, Ponsonby's 2nd Guards Brigade was formed up tightly on the right of the Ginchy/Lesboeufs road. Before zero, the three tanks allotted to the Guards moved up the slope to begin their attack on the Triangle. One broke down at the outset and the other two lost their direction, veering east off their northerly course and were of little use during the attack. Having lost the advantage of the tanks, the leading units of the 3/Grenadiers and 1/Coldstream Guards began their uphill attack. As soon as they crested the ridge, they were felled by the most severe fire from Straight Trench and the Triangle, Lt.-Col. Sergison-Brooke of the Grenadiers being amongst the wounded. In

THE INFANTRY ATTACKS - OBJECTIVES 25 and 26

Three aerial photographs showing the successive phases of the destruction of Ginchy which was finally captured shortly after the last photograph was taken. The top picture was taken on 11th July, the middle 21st July, the bottom on 7th September. (Michelin)

traditional Guards style, the survivors stuck to their task, the fire from Straight Trench on the right forcing the Guards to veer left. The 1/Scots Guards and 2/Irish Guards were in close support. A concealed German strongpoint over the crest was neutralised with the rifle and bayonet. Having caught up with the leading units, the four battalions continued together. They found all three rows of wire well-cut and the quarter-circle of trenches around the Triangle badly damaged. The RFC had also strafed the position. The Guards rushed and secured the northern part of the stronghold, some entering Serpentine Trench which runs north-west from the Triangle. During the course of this bitter fighting, Lance-Sgt. F. McNess of the 1/Scots Guards earned the Victoria Cross for his courageous action and determination while wounded.

In similar style, Pereira's 1st Guards Brigade commenced its attack at zero led by the 2nd and 3/Coldstream Guards but these too, were cut down from fire from the northern section of Pint Trench. Against this curtain of fire, the advance slowed slightly. Suddenly rallying, the Coldstreams rushed forward again; this time nothing would stop them as they overran the positions taking some prisoners, four machine-guns and a trench mortar. The 1/Irish Guards in the second wave were already forward and, joining up with some Coldstreams on the left, continued the attack. By around 7.30am, their objective had been taken. The 2/Grenadiers were engaged in clearing the village of Ginchy. Reorganisation of the mixed battalions began and while the units from the 1st Coldstreams were moving to the right to rejoin the 2nd Brigade, their commander, Lt.-Col. G. Baring, was killed.

Another attack was launched at 8.20am by the 6th Division led by the 9/Suffolk and 2/Sherwood Foresters of the 71st Brigade. Despite a renewed bombardment of the Quadrilateral, fire from this strongpoint and Straight Trench took a heavy toll and no meaningful progress was made. The 1/Buffs, 8/Bedford and units of the 2/York & Lancaster were now moved up ready for the next phase of the attack. They were joined by parties from the 7/Middlesex (167th Brig.) of the 56th Division on the immediate right.

After its heroic struggle the 2nd Guards Brigade formed a defensive flank on the right and bombers from the four battalions successfully assaulted the remainder of the battered enemy trenches around the Triangle which the enemy had fought most desperately to retain. No major forward movement could be undertaken while the 6th Division was still in difficulty in the area of the Quadrilateral and Straight Trench but even so, a mixed party of about 100 Guards advanced a further 850 meters to near the sunken road approximately a 1,000 meters south-west of Lesboeufs(i) Now out of Ginchy, the 2/Grenadiers advanced under fire from the untaken part of Serpentine Trench. Quickly deploying, they charged and entered the trench at the point of the bayonet. The bombers deployed north and south securing a link with the 2nd Guards Brigade on the right and to within a short distance of the Coldstreams on the left. The Guards Division had secured all its first objectives and the 1st Brigade now advanced towards the next objective. Passing through a heavy barrage to the sound of its leader's hunting horn(ii), the 2nd and 3/Coldstreams pushed forward and secured part of the second objective. A party of the 1/Irish Guards on the left flank made contact with the right flank units of XV Corp's 14th Division.

Major-General Feilding, commander of the Guards Division, was anxious to consolidate the high-cost gains. He sent the 4/Grenadiers from 3rd Brigade forward to assist the 2nd in this work. The Grenadiers moved north to reinforce the exposed flank near the Triangle. The 2nd Brigade had been subjected to counter-attacks by bombers and the mixed group which had advanced to the sunken road was forced to withdraw. With the help of the 4/Grenadiers, the 2nd Brigade repulsed all other attacks. A little later the 2/Scots Guards (2nd Guards Brig.) went through the German barrage to support the 1st Guards Brigade but the trenches were so crowded that three companies were ordered to withdraw to the support positions. There being no further attack until the morrow, the Guards now had time to complete their consolidation.

Meanwhile the artillery laid another barrage on the Quadrilateral and Straight Trench but the latter in particular was sheltered on the reverse slope and suffered little damage. This meant the infantry of the 6th Division had to do it the hard way. Assembled in haste north-west of Leuze Wood, half the freshly-arrived 11/Essex (18th Brig.), not even aware of the exact position of the Quadrilateral, advanced at 7.30pm to be caught from fire from its rear by the German guns in Bouleaux Wood. The support of the 2/York & Lancaster (16th Brig.) failed to change the ill-fated attack and the men were withdrawn. The 2/Durham Light Infantry (DLI) sent two companies to the Triangle, now occupied by the Guards. Bombing their way uphill towards the crest, the most they could secure was about a hundred-meter section.

Some time before zero hour the 6/KOYLI (43rd Brig. 14th Div.) was engaged in a bitter contest to secure the Alcohol trenches to the immediate east of Delville Wood. Out of the three tanks allotted to the 14th Division, only tank D1 left its starting point from Pilsen Lane at 5.15am followed by the KOYLI bombers a few minutes later. The advancing tank caused the evacuation of Ale and Hop Alley but its steering gear was damaged by a shell. Continuing alone, the KOYLI's came under machine-gun fire from the direction of Pint Trench, almost all the officers becoming casualties. The Yorkshiremen turned to face the fire and, advancing with bomb and bayonet, silenced the machine-gun crew. The NCO's reorganised the men to take part in the main advance towards Flers, described under Objective 27.

The previous abortive attack on the Quadrilateral by the 6th Division was about to be avenged. At 5.50am on the 18th September, after a most successful preliminary barrage, the 1/KSLI (16th Brig.) advancing close to the now creeping barrage towards the Quadrilateral, stormed the strongpoint. Aided by the 14/DLI (18th Brig.) the Shropshires engaged the enemy in a short but bitter contest forcing him at last from this strongpoint which had resisted all previous attacks. The two battalions now pushed their way into the sunken road on the Morval road clearing dug-outs as they advanced. At the same time the 2/York & Lancaster (16th Brig.) bombed its way on the right flank between the Quadrilateral and Middle Copse. Supported by accurate Stokes mortar fire, this battalion too secured ground which had hitherto defied all former assaults. Units on the right flank linked up with the 56th Division in Middle Copse. With the Quadrilateral now secure, the way was clear for an assault on Straight Trench. An inititial frontal attack by the 1/West Yorks (18th Brig.) was strongly resisted but the battalion bombers worked their way down from the north while the 14/DLI moved north from the sunken road to take the Germans from the rear. At last the German resistance began to weaken and 140 unwounded prisoners surrendered. Seven machine-guns were taken and German dead and wounded lay thick on the disputed ground. Patrols from the 20th Division also captured some twenty prisoners. Work immediately began on forming a new line overlooking the valley in front of Morval (Vallée du Marécage on your IGN map). British artillery dispersed Germans seen to be gathering on the high ground near Morval.

The general attack which commenced on the 15th September and which is known as the Battle of Flers-Courcelette had been in progress for three days when the last of the German defences east of Ginchy were secured and we shall now turn our attention to these events.

(i) They halted at a point only a few meters from where the Guards Cemetery can now be seen on the Sunken Road.
(II) Lt.-Col. J.V. Campbell of the 3/Coldstream Guards was awarded the V.C. for his courageous leadership in this and other actions on the 15th September.

OBJECTIVE 27
FLERS VILLAGE, FLERS TRENCH EAST, SWITCH LINE EAST, GAP TRENCH, BULLS ROAD

The village of Flers was protected to the immediate south by Flers Trench East and a little farther south by the eastern section of the Switch Line. Further defences to the west included Flers Support Trench and Abbey Road while Bulls Road protected entry from the south-east. During the night of the 5th September, preparations were under way to assault Tea Support Trench, situated about half-way between Longueval and Flers. Having moved up from reserve and taken up new positions in the forward line on the 8th September, the 5/South Lancs and 10/King's (166th Brig. 55th Div.) extended the forward positions in preparation for the main attack scheduled for the 15th September.

Although mentioned elsewhere in this guide, it would be perhaps useful to reiterate a few facts about the tanks, especially as the use of tanks at Flers made headlines in the newspapers and the attack "down the High Street of Flers" was to go down in history. The Mark 1 tank came in two versions, the more heavily armoured male tank equipped with two 6 pounder guns and 4 Hotchkiss machine-guns. About a ton heavier than the female tank, it weighed 28 tons and had maximum speed of 3.7 mph and

Objective 27

Flers, Flers Trench East, Switch Line East, Gap Trench, Bulls Road

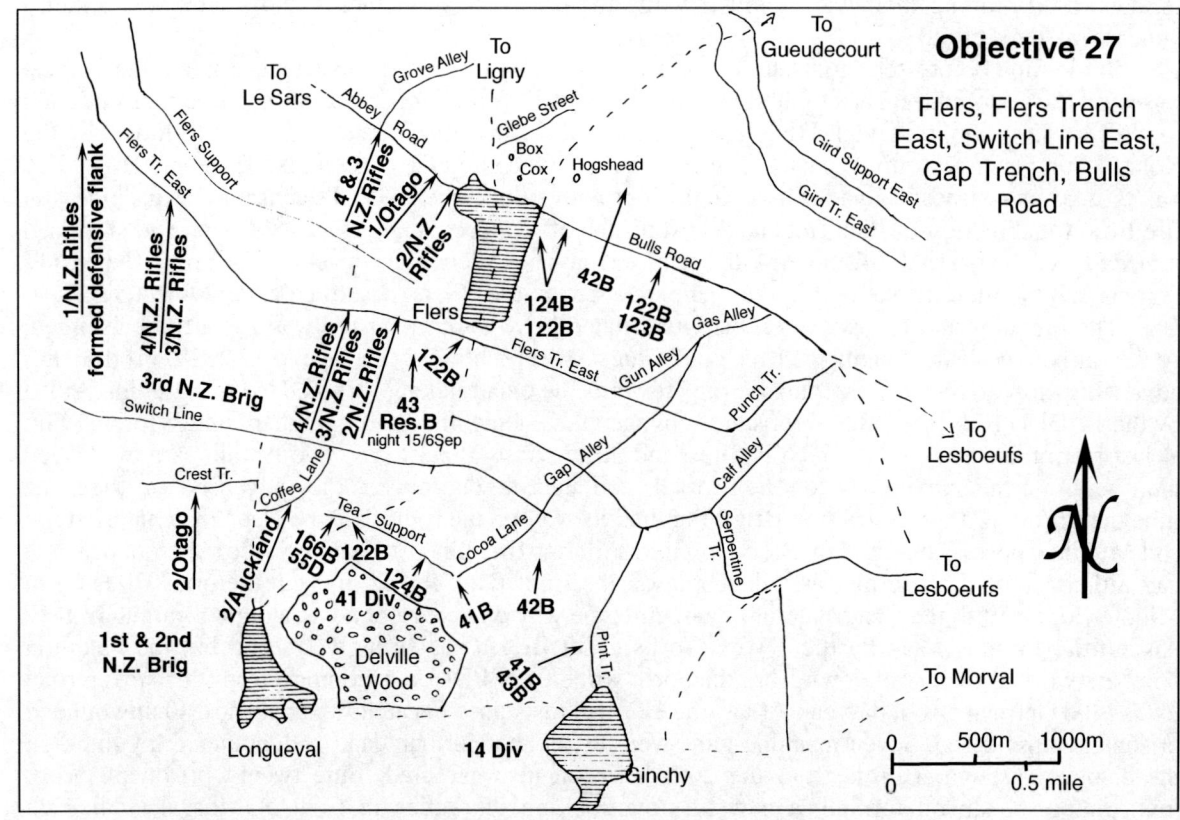

a range of about 23 miles from its 46 gallon tank and 6 cylinder 105 bhp Daimler engine. The female tank was equipped with much more SAA, having a total of six machine-guns, five Vickers and one Hotchkiss. The "Heavy Branch Machine-Gun Corps" was composed of six companies, labelled A to F, of 25 tanks each numbered 1 to 25. Thus Tank D4, was tank number four of D Company.

Major-General Couper's 14th Division formed the right wing of Horne's XV Corps and its 41st Brigade's objectives were the ridge to the south-east of Flers and Flers village. The 42nd Brigade was to pass through the 41st to secure Bulls Road and continue to Gueudecourt. A few minutes before zero, tank D3 commenced its attack from the northern edge of Delville Wood heading for Tea Support Trench and Cocoa Lane. At the junction of the latter-named trenches, the tank was damaged beyond repair by shell fragments. However its start had served a useful purpose. Meanwhile on the right of D3, the 8/Rifle Brigade and 8/KRRC (41st Brig.) advanced under fire from the unsecured Pint Trench sustaining considerable casualties. The men did not falter and continued towards Tea Support Trench clearing German pockets of resistance from the shell craters on the way. Deploying on each flank, both Tea Support Trench and Pint Trench were secured. Most of the Germans were killed or wounded, the remainder surrendering without further resistance. On they went up Cocoa Lane and entered part of the infamous Switch Line facing them. Bombers deployed right and left silencing strongpoints and dug-outs. Two machine-guns and many prisoners were taken. Small parties were sent out to deal with snipers firing from the craters. Consolidation began immediately - the Germans would certainly wish to retake the captured part of the Switch Line, it being one of the longest and well-constructed of the German defence lines. The 6/KOYLI, which had cleared the Alcohol trenches to the immediate east of Delville Wood, had also advanced on the right flank and now joined up with the leading battalions of 41st Brigade. Not having linked up with the 41st Division to the left, the 8/KRRC constructed a defensive flank in the valley south of Flers. The enemy counter-barrage was not long in coming - knowing its precise coordinates, the occupied part of the Switch Line was subjected to a heavy and accurate barrage during which Lt.-Col. W.R. Stewart was wounded. The support battalions, the 7/KRRC and 7/Rifle Brigade left Delville Wood and advanced over the open ground towards the Switch Line. Small pockets of resistance caused some delay and casualties but, pausing only a moment, small parties of the Rifle Brigade moved up Cocoa Lane, Gap Alley and then north-east to the junction of Gap Trench and Punch Trench. Gap Trench, the second objective, was secured with minimal resistance, a link up being made with the Guards division on the right which had veered a little off course. The 7/KRRC also advanced to the second objective, but, like its 8th sister battalion, no contact was made with the 41st Division.

Meanwhile, the 42nd Brigade, some 2,600 meters to the south, began to move foward, the leading companies of the 9/Rifle Brigade and 9/KRRC to the east of Delville Wood while the 5/KSLI, followed later by the 5/OBLI, passed through the wood. At first advancing with few losses to Gap Trench, the Rifle Brigade came under heavy fire as it headed north towards Bulls Road situated on the Flers/Lesboeufs road. The fire was so severe it brought the attack to a halt, the men going to ground. Lt.-Col. T. Morris died from his wounds and almost every other officer was down. The KSLI advanced north from Delville Wood and once having passed through the 41st Brigade, encountered similar fire as it approached the objective. Following in the path of the Rifle Brigade, the 9/KRRC and the 5/OBLI caught up and slightly passed the line of the leading battalion, the KRRC losing its commander at the start of the attack. The losses of the 14th Division had been substantial but they had advanced beyond the divisions on either flank and were thus exposed to fire from three sides.

Major-General Lawford's 41st Division was about to be engaged in its first action on the Somme and Brig.-Gen. Clemson's 124th Brigade on the right, led by the 10/Queen's and 21/KRRC advanced across no man's land at zero towards the Switch Line, its first objective. Moving forward close behind the barrage, they crossed the western end of Tea Support Trench with little difficulty and advanced a further 350 meters to the Switch Line which was also secured by 7.00am. The support battalions, the 26th and 32/Royal Fusiliers had now joined the first waves and the Switch Line was consolidated without delay. The advance continued to the second objective, the western end of Flers Trench, immediately south of Flers. There was more resistance here but the section allotted to them was taken after thirty minutes or so.

Meanwhile, the 15/Hampshire and 18/KRRC of Brig.-Gen. Towsey's 122nd Brigade advanced on the left, their final objective being to make a defensive flank facing north-west across the Gird Lines to the north of Flers. The men had a long way to go to arrive at the Gird Lines but they had the assistance of seven tanks from "D" Company. Infantry and tanks advanced together, the German machine-guns slowing the infantry advance near Tea Support Trench. Only a few minutes after zero hour, Lt.-Col. C. Marten, (18/KRRC) and three other officers were killed by a single shell. Some twenty minutes later, Tea Support was secured, bombers clearing any remaining Germans in the trench. D15 was out of action, its crew killed in front of the Switch Line but the trench was crossed with little difficulty and the elated men and tank crews advanced to their section of Flers Trench to the left of the 124th Brigade. The surprise and fear of the Germans at the sight of these large iron monsters slowly but irrevocably bearing down on them in the trenches can well be imagined. Many fled throwing down their arms and equipment becoming targets for the tank machine-guns, others simply surrendered. However, the tanks were susceptible to mechanical failure and the enemy, once over the initial surprise, put several out of action using shells. D14 ditched south of Flers and D18 was damaged and brought to a temporary halt at Flers Trench. The 11/Royal West Kent and 12/East Surrey (122nd Brig.) had already moved forward and were ready to take part in the attack on Flers. Four tanks were still in action. At 8.20am as the barrage lifted, D16 entered the village and made its way down the main street, guns firing and followed by parties of cheering infantry. A spotter plane reported the following message, "Tank seen in main street Flers going on with large number of troops following it"(i). Such good news was indeed rare and these famous words were soon making the headlines in the newspapers. Meanwhile, the other three tanks, D6, D9 and D17 were making their way up the eastern side of the village crushing and destroying strongpoints on their way, driving the terror-struck Germans before them. Only the most hardened soldier attempted to resist the advance of the new war machine. By 10.00am, it was all over, Flers was secured.

The third objectives lay just beyond Flers, to the east, Bulls Road leading to Lesboeufs while to the west the German line ran through Box and Cox Trenches to the junction of Grove Alley and Abbey Road. A lack of officers and a delay in receiving orders prevented the reorganisation of the units in order to continue to the next objective while some of the men, in a state of euphoria, deemed their duty done with the capture of Flers and when the enemy retaliatory barrage commenced, some parties retired to the relative safety of the trenches just south of the village. Others however, continued the advance as far as the third objective.

Major-General H. Russell's New Zealand Division formed the left wing of XV Corps and Brig.-General Braithwaite's 2nd Brigade, like the divisions to the right, had its share of the Switch Line to secure. For the 2nd Brigade, it was the section to the east of High Wood and the 2/Auckland and 2/Otago were on the move even before zero. Men were falling from the British barrage while to the west, machine-gun fire from High Wood caused heavy casualties. In spite of this, the Otago battalion swept over Crest Trench and on towards the Switch Line. On the right the 2/Auckland passed over Coffee Lane and had secured its section of the Switch Line by 6.50am. While consolidating their gains, the leading units of the 4th Rifles of Fulton's 3rd Brigade passed through the leading battalions until, catching up with the creeping barrage, were obliged to halt their advance for a short while. Followed closely by the 2nd and 3rd Rifles of the same brigade, the attack was launched against Flers Trench and Flers Support just to the north. Four tanks had been scheduled to take part in the attack and although they were assembled before zero hour, the élan of the infantry had carried the New Zealanders well in front of the tanks. The 2nd Rifles seized Flers Trench taking some 85 prisoners and continuing the advance, found Flers Support empty. Further movement northward on the west side of Flers was hampered by machine-gun fire from the north-west of the village. There was an obvious reason for the vacation of Flers Support, the Germans had retired to Abbey Road on the reverse slope where their strongpoints were relatively intact. The 2nd Rifles were involved in a bitter contest using bombers, rifles and Lewis-guns before being able to secure the area at about 11.00am. Still advancing under fire, they eventually dug in just to the north of Flers, the right units in touch with 41st Division. To the left, the 3rd Rifles were held up by uncut wire and waited for the tanks to clear the way. D10 had been damaged by shell-fire at Fat Trench to the south of Flers Trench but D11 and D12 came lumbering up crushing the wire and all other obstacles in their path allowing the 3rd Rifles to proceed north to join its sister battalion. Some hundred prisoners were rounded up. Having been delayed by the halt of the 3rd Rifles while awaiting the tanks, the 1st Rifles task of making a defensive flank was retarded and some units had joined the 2nd Rifles fighting at Abbey Road.

The dying moments of the epic struggle for High Wood were being played out but at this moment of the New Zealand Division's attack, there was no contact with the 47th Division at High Wood. However, reports received at Corps Headquarters were encouraging and there was a certain optimism concerning the prospects for the rest of the day. The 21st Division in Corps reserve was ordered to move up to Pommiers Redoubt, east of Mametz on the Montauban road.

It has been mentioned that the 14th Division had advanced further than the divisions on either flank, creating a potentially dangerous situation and the men had been busy digging in. On the right at 11.20am the 9/Rifle Brigade and 9/KRRC (42nd Brig.) in close support advanced across Bulls Road

The main road through Flers, 1916. (Michelin)

to attack the Gird Trenches, about 700 meters to the north. Reduced to about twenty per cent of their normal strength, the depleted battalions could make no progress against the machine-gun fire from the Gird Lines. The Rifles moved towards Gas Alley to the south of Bulls Road and consolidated positions in that area, linking up with the 5/OBLI of the same brigade. Maj.-Gen. Couper ordered the 43rd Brigade to move up from its reserve position to assist in another attack on the Gird Lines.

The leading battalions of the 124th Brigade were unaware of the success of the tanks in Flers, causing a delay in their preparations for their advance to the third objective. Although Brig.-Gen. Clemson went up personally to organise the next attack, it was mid-afternoon before mixed units of the 10/Queen's and 21/KRRC started towards Bulls Road. These men were reinforced by units of the 23/Middlesex (123rd Reserve Brig.). While entering Bulls Road just east of Flers, Lord Faversham, commanding the 21/KRRC fell mortally wounded. Joining up with parties of the 122nd Brigade, the mixed units assaulted the Gird Trenches only to be repulsed by machine-gun fire.

On hearing of the withdrawal of some of his troops in 122nd Brigade from Flers, Brig.-Gen. Towsey ordered his brigade major to take charge of the situation and, on arrival, the latter instructed the Royal Engineers and any other available troops to advance round the eastern and western perimeters of Flers which was still under fire from the German counter-barrage. This movement was effected and the men joined up on Bulls Road at the northern end of Flers. Joined now by the remainder of 124th Brigade, the Hogshead and Box and Cox Trenches, a little to the north of Bulls Road were secured. These new positions were immediately consolidated and Lewis-gun emplacements established.

The unproven tanks, or perhaps it would be fair to say, the tanks which had had but a minimum of testing and never in anger, had not fared too badly. Tank D6 had reached the fourth objective and engaged a German battery with its six-pounder and after destroying one cannon, the tank set on fire when it received a direct hit. The exploits of D16 have already been described but its glorious day did not end in the High Street of Flers - the tank continued north to the third objective before returning without mishap. D9 reached Box and Cox and was severely damaged while moving up towards Glebe Farm to the north-west of Box and Cox. D17 received two shells and was abandoned to the east of Flers, but was recovered later. It is worth noting the special team spirit needed to operate the new war machine. The heat and noise were infernal, it was almost impossible to make verbal communication. Vision was necessarily limited, the crews thrown about making it difficult to take correct aim with the guns. The landscape bore no resemblance to the maps at headquarters - the terrain on open ground had been battered beyond recognition. A better shot could be made with the tank stationary but that presented a sitting target to the eager German gunners wanting to have the honour of putting a stop to these large monsters. Even on the move, with a speed of under 4mph, they were not a difficult target for the German gunners once they had got over the initial surprise of seeing the new weapon. To abandon a disabled tank was a dangerous operation, nearly all the crew of D9 were killed as they left the tank. Even worse was the fate of the crew whose tank caught fire, trapping the men inside.

With almost all the third objectives secured on this front, Lt.-Gen. Horne ordered a fresh barrage on the Gird Lines and Gueudecourt and a renewal of the advance at 5.00pm. However, this was postponed until the following day in order to give sufficient time to consolidate the new positions. The end of the day's advance

The main road through Flers today. (author)

Bulls Road, Flers, leading to Lesboeufs. Bulls Road Cemetery is on the right. (author)

signalled the start of the counter-barrage and the first of the enemy attacks was checked by the 14th Division. A second counter-attack gained some ground from the Irish Guards some way in front of the second objective. In the early evening the Germans delivered two strong attacks in an attempt to retake Flers but each was checked by Clemson's 124th Brigade. Brig.-Gen. Davidson ordered the remainder of his 123rd Brigade forward to help the 233rd Field Coy. R.E. with the consolidation work. The 10/Royal West Kent took charge of the Switch Line while the 20/DLI occupied the newly dug trenches around the village. The work continued throughout the night under fire, the losses including Lt.-Col. Walmisley-Dresser of the 12/East Surrey, mortally wounded by a shell. The relief began under cover of night, the 42nd Brigade being partially relieved by the 43rd reserve Brigade. The 11/Queens and part of the 23/Middlesex took over 41st Division's front on Bulls Road and further north at Box and Cox Trenches.

A little later, bombers of the 1st Otago (1st N.Z. Brigade) pushed the Germans back up Flers Support Trench, thus securing the western edge of Flers. With Flers now secure, plans were being made to continue the attack north-east towards Gueudecourt (Objective 32), Le Transloy ridge, Lesboeufs and Morval to the east and south east (Objective 30) and finally, the so called "Meteorological" Trenches between Lesboeufs and Gueudecourt (Objective 31). The three objectives have been grouped together and are described later.

(i) Official History, page 323

OBJECTIVE 28
MARTINPUICH, SWITCH LINE WEST

To the left of the 7th Division engaged at High Wood, the 100th Brigade of Landon's 33rd Division was in position between Bazentin-le-Petit and High Wood. At 9.00am on the 15th July units of the 9/HLI and 1/Queen's attacked the Switch Line West, south of Martinpuich. Although preceded by a thirty minute bombardment, the attack failed and the men were withdrawn.

On the left wing of 33rd Division, the 1/Middlesex from Carleton's 98th Brigade, advancing north from Bazentin-le-Petit towards the Switch Line West came under machine-gun fire from both flanks. The Middlesex losses were mounting at an alarming rate and made worse by a German barrage which had the exact range. Following a further abortive attempt later in the day, the attack was called off and the men withdrew.

In accordance with plans made for the attack on the 23rd July, the 1st Division, followed the barrage in the early hours of the 23rd towards the Switch Line West. Brigadier-General Reddie's 1st Brigade came under heavy fire, the area having been lit by German flares, the 10/Gloucester and

Objective 28

Martinpuich, Switch Line West

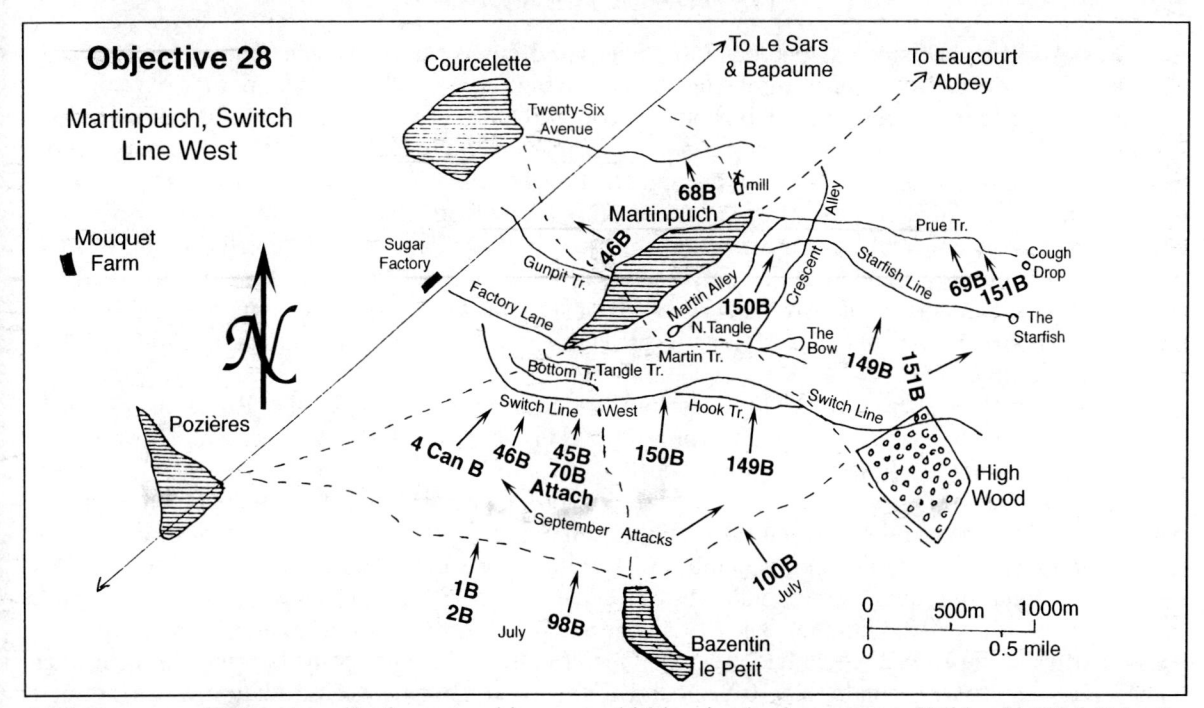

1/Camerons suffering heavily from machine-guns hidden in the long grass. Hubback's 2nd Brigade fared no better when the 2/KRRC and 2/Royal Sussex assaulted the Switch Line. Lt.-Col. Bircham, CO of the Rifles fell while trying to hold a brief entry in the enemy trench.

The immediate priority was the capture of High Wood to the south-east(i) and although a few tentative assaults had been made on the Switch Line West to the south of Martinpuich, the first direct attack on the village formed part of the plans for the general attack on the 15th September. Following its action to the west of High Wood, Major-General Wilkinson's 50th Division moved into position during the night of 9/10th September to the south of Martinpuich ready for the attack on the Switch Line, where the 15th Division on the left of III Corps was already making its preparations and in the evening of the 14th September the leading battalions took up positions in the assembly trenches. Tanks were laboriously making their way forward under the cover of night. The role of the III Corps was to protect the flank of the Fourth Army's main attack on the 15th, the success of which depended on the success of the 47th Division at High Wood and also of the Reserve Army which was about to assault Courcelette.

In the centre of III Corps, Major-Gen. Wilkinson's 50th Division commenced its attack forward of the 47th and 15th Divisions on either flank The 4th and 7/NF (149th Brig.) advanced with the intention of covering the 1,200 meters of ground between the assembly trench and Prue Trench at the third objective as quickly as possible. Crossing over Hook Trench (see section concerning High Wood), the Fusiliers advanced to their second objective, the Starfish Line. Assailed by fire from High Wood, the advance slowed. The 4/NF was diverted to help the 47th Division clear the machine-gun posts on the north-western corner of High Wood while the 5th and 6/NF moved forward to support the attack. The 6/NF worked its way to the right to establish a defensive flank while the 5th and 7/NF fought their way forward to a point in the sunken road just south of the Bow. Taking a short rest, they secured the Bow and parties entered the Starfish Line which runs east from north Martinpuich.

Brig.-Gen. Price's 150th Brigade had two tanks at its disposal, both of which did fine work on this day. Arriving at the first objective, one of the tanks poured enfilade fire down the trench taking a heavy toll. Struck a short while later by two shells, the crew joined the infantry after bringing out the machine-guns. The second tank fared better - driving the terrified Germans before it, the tank silencing three German machine-guns before reaching the eastern approach to Martinpuich. Under constant fire and slightly damaged, it managed to return to base to refill with petrol. The attack was led by the 4/East Yorks, 4th and 5/Green Howards and supported by the 5/DLI. The men had seized the first objective by 7.00am, encountering many Germans fleeing to the British Lines. The advance continued past the second objective but from here the enemy resistance was much more resolute. On the right, the 4/East Yorks right flank was open and the men withdrew to Martin Trench. The Green Howards held the Starfish Line to the junction of Martin Alley which runs south from that point.

Major-Gen. McCracken's 15th Division also had two tanks at its disposal but one was hit before starting and could not participate in the attack. Brig.-Gen. Allgood's 45th Brigade on the right assaulted the southern approaches to Martinpuich. Led by the 11/A&SH and 13/R Scots with the 6/Camerons and 8/York & Lancaster (attached from 70th Brig. 23rd Div.) in support, excellent progress was made, the preliminary barrage obviously having done its work well. Again, a good number of Germans surrendered, the only real resistance coming from the Tangle, just south of Martinpuich. Bombers from the A&SH cleared this obstacle as well as quelling some resistance in the sunken road between Martinpuich and Longueval. The 7/RSF remained in reserve and would take part in the action a little later.

On the left flank Brig.-Gen. Matheson's 46th Brigade, led by the 10/Scottish Rifles and 8/KOSB and the 12/HLI in support made very good progress. Their objective, Factory Lane, which heads west from the south-west corner of Martinpuich, was reached a little after 7.00am. Linking up with the 4th Canadian Brigade of the Reserve Army, the objective was secured and consolidated. The leading battalions had left before the tank but the crew did some fine work in causing the flight of the enemy from Bottom Trench and silencing a number of machine-gun posts. The driver then negotiated his machine to crush dug-outs at the south-western entry of the village. Although damaged, the tank managed to return to refill with fuel. The British barrage lifted from Martinpuich at 9.20am and patrols joined parties of the KOSB which had previously entered the village under the barrage. An hour later entrenching work was begun by the 10/Scottish Rifles of 46th Brigade. Consolidation was continued at the held portions of the Starfish Line but it was imperative to secure the whole of the line. To this effect, 50th Division's 149th Brigade, composed entirely of Northumberland Fusiliers, was ordered to secure the trench along the line of the 50th Division. The increasing volume of fire in the afternoon was such that the men from Northumberland found it impossible to secure their objective and were forced to retire to Hook Trench or the sunken road to the immediate north. It was the intention of 150th Brigade to send out patrols to the north of Martinpuich to link up with the 15th Division but the 150th Brigade had been forced to evacuate the Starfish Line and was consolidating in Martin Alley.

In the early evening the 151st Brigade attacked Prue Trench from its junction with Cough Drop west towards Martinpuich. The 5/Border, 6th and 9/DLI followed a special bombardment and left Hook Trench at 9.45pm. The Border got off to a late start but the Durhams advanced against a curtain of fire, just small parties reaching Prue Trench. Here they were killed or wounded, the suvivors taking cover in the Starfish Line where they were pushed out by German bombers and close-range fire. The Durhams eventually joined the Northumberland Fusiliers in the sunken road. An hour and a quarter later the Borders made their belated attack but could not make much ground and the men dug in.

Although the enemy still held strongpoints to the north, west and east of Martinpuich, the village itself was in control of the 45th Brigade of 15th Division, the 6/Cameron bombers driving out the last remnants of German resistance eventually linking up with the 5/Green Howards of 150th Brigade in Martin Alley. The leading battalions of 46th Brigade constructed a chain of strongpoints to the north-west facing Courcelette. During the night the 12/HLI (46th Brig.) and 9/Yorks & Lancaster (attached from 70th Brig. 23rd Div.) took over the front line linking up with the Canadians in Gunpit Trench.

A further attack on the Starfish Line and Prue Trench was made on the 16th September by Cameron's 151st Brigade (50th Div.). Under heavy fire, units of the 5/Border and 9/DLI reached and entered Prue Trench but the attackers were driven out. The Germans still had control of the Starfish Line and the attack brought no positive result. The 5/DLI of Price's 150th Brigade also attacked Prue Trench from west of Crescent Alley but veered too far west under enemy fire. Later in the day the Durhams made a bombing attack on Prue Trench from Martin Alley, just to the east of Martinpuich but the enemy stubbornly defended the trench and little gain was made. While these attacks were being made to the east of Martinpuich, the 15th Division was engaged in repulsing a German counter-attack. Thereafter, the village and the area to the north-west were targetted all day by the enemy artillery.

Two days later, after being held up by an enemy barrage and wet ground conditions, bombers from the 5/DLI, 4/Green Howards and 5/Green Howards, all of 150th Brigade, worked their way along Starfish Line and Prue Trench to the immediate north towards Crescent Alley. Although gains were made in both trenches, Crescent Alley, which crosses a number of east/west trenches was proving to be an effective block and the 8/DLI (151st Brig.) could make no progress in its attack up Crescent Alley from the south. The 15th Division, still charged with improving the hold on Martinpuich and ground to the north of the village, was again subjected to a heavy enemy barrage. Nevertheless, forward posts were consolidated including the position at Martinpuich mill. The relief of the 15th Division by the 23rd began on the 18th September and was completed the following day.

The next three days were spent bringing up supplies, a tedious and tiring task in the bad conditions. Movement was difficult without having the burden of carrying food and ammunition and pack animals were used as much as possible. The Germans were suffering from the same conditions which limited the possibility of an infantry counter-attack but their artillery was still very active. In these awful conditions the III Corps managed to secure without opposition the rest of its original objectives. Units from the 47th and 50th Divisions began digging a trench between the Starfish and the Bow, the latter being to the immediate east of the southern end of Crescent Alley.

Major-General Babington's 23rd Division relieved the 150th Brigade (50th Div.) near Starfish and Prue Trenches. Crescent Alley was still under German control and an enemy counter-attack was launched against Brig.-Gen. Lambert's 69th Brigade (23rd Div.) driving the men back for a short while. However, the lost ground was quickly regained. The next two days were relatively quiet and night patrols established that Prue and Starfish Trenches had been vacated. The trenches were occupied and consolidated while later patrols reported little enemy activity south-west of Eaucourt Abbey or at the eastern end of Twenty-Sixth Avenue which runs east-north-east from Courcelette.

However, the western part of Twenty-Sixth Avenue north-west of Martinpuich was still to be cleared and the 10/NF (68th Brig. 23rd Div.) had the assistance of two tanks to complete the operation. One ditched at the start while the second attracted the attention of the German artillery which, coupled with deadly machine-gun fire, prevented the Fusiliers from getting forward. A subsequent bombing attack starting from the mill between Martinpuich and Twenty-Sixth Avenue produced no better result.

At least, the village of Martinpuich was now secured. Destrement Farm to the north and Le Sars and Eaucourt Abbey to the north-east were to be assaulted at a later date and are described under Objective 36. Before an attack could be mounted against Courcelette (Objective 29) from the south-east, Twenty-Sixth Avenue would have to be secured.

(i) See Objective 23

OBJECTIVE 29
COURCELETTE, REGINA TRENCH EAST, DESIRE TRENCH AND TRENCHES NORTH TOWARDS GRANDCOURT

Following the capture of Pozières, patrols had been sent north-east towards Courcelette but no direct attack had been made on that village before the 15th September. On the extreme right of the Reserve Army's front, Major-General Turner's 2nd Canadian Division had its 4th Brigade (Brig.-Gen. Rennie) on the right wing and the 6th Brigade (Brig.-Gen. Ketchen) on the left. The attack was to be supported by tanks but the infantry was under strict orders not to wait for the tanks and to take full advantage of any gains made. Two tanks set off from the Windmill just north-east of Pozières, one machine each side of the main Bapaume road while the third was to attack the Sugar Factory situated 1,000 meters north east of Pozières on the main road. The first two tanks were then to turn right on to Factory Lane which leads to Martinpuich. From a point approximately 900 meters west-north-west of the Windmill, three more tanks were to make their way up Sugar Trench until they were opposite the Sugar Factory, whence they would wheel right to assist in its capture. Three hours prior to zero, German bombers attacked the Canadian positions calling for quick action from the 4th Brigade. The attack was checked in time by the 18th, 19th and 20th Ontario Battalions. Infantry and tanks set off together, one tank was quickly out of action with damaged steering gear, its crew trying desperately to get it back to base. The other two machines made their way to the Sugar Factory. The leading waves of 4th Brigade, the 18th and 20th Ontario Battalions, advanced from south of the main road behind the creeping barrage and

Objective 29

Courcelette, Regina Trench East, Desire Trench and trenches towards Grandcourt
key
CB - Canadian Brigade
B - British Brigade

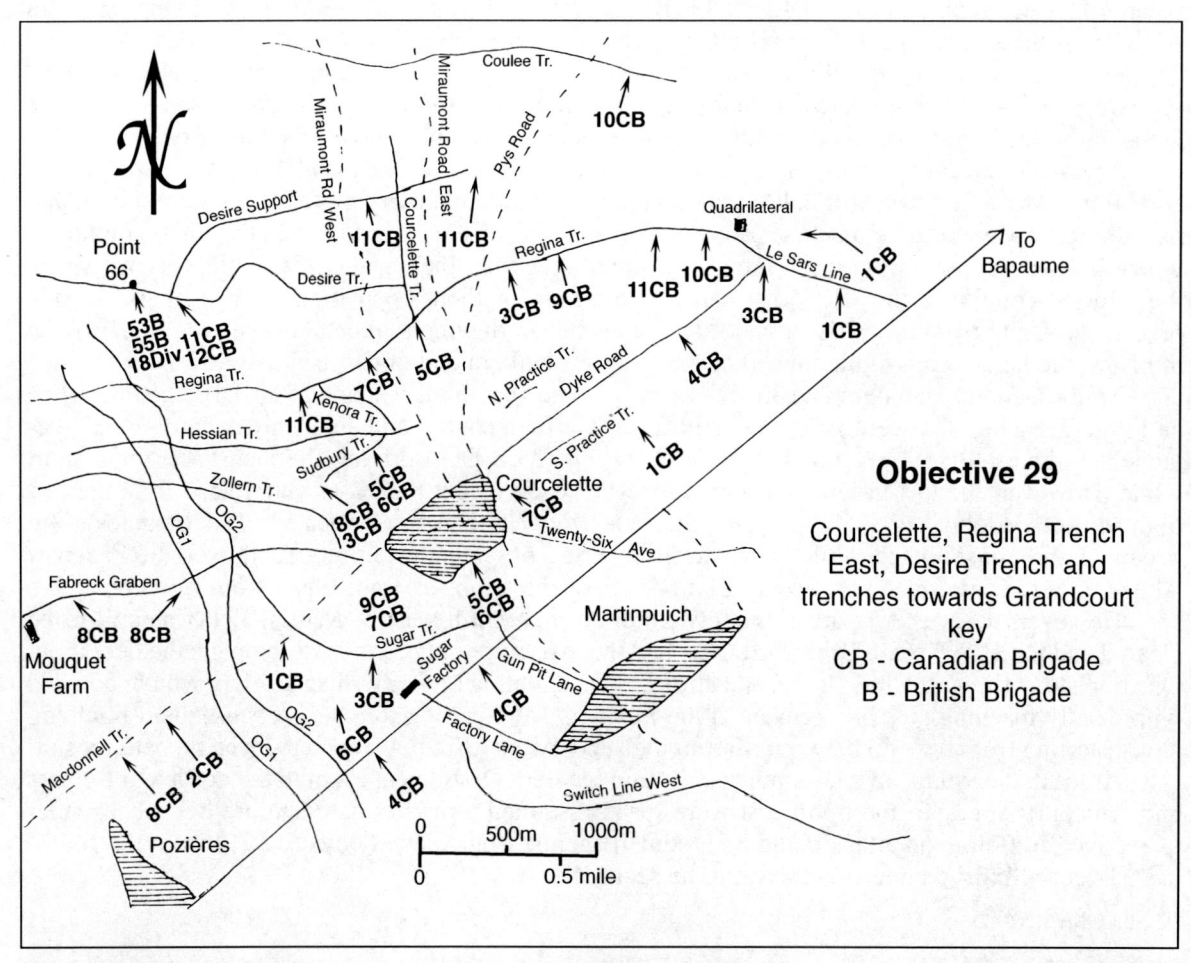

got quickly forward, aided by covering machine-gun fire. Against some stiff resistance, the Germans were pushed out of their front line trench and the Canadians had reached Factory Lane at about 7.00am. The British barrage and machine-guns had accounted for many dead and wounded found in the trench. A number of German snipers and machine-gun crews who refused to surrender were killed with rifle or bayonet. The 21st Bn. had now joined the fray and with units from the 20th, cleared the Sugar Factory taking well over a hundred prisoners. Finding a deep dug-out beneath the factory, the Canadians added another twenty or so prisoners to their total, including six officers. All this was achieved before the arrival of the tanks.

On the left, the 27th and 28th Battalions of 6th Brigade advanced well from the north of the Bapaume road, not quite as quickly as the 4th Brigade, but the first objective was reached about 7.30am. From hereon, enemy fire caused heavy losses. The 28th Bn. had to neutralise a strongpoint at the junction of the line of the first objective with Courcelette Track. Once this was accomplished, the battalion bombed its way along MacDonnell Trench securing an enemy strongpoint having a dominant view over the Canadian line. The 31st Bn. followed, its role being to "mop up" and carry the necessary supplies forward. Scattered over a wide front to ensure their task was completed correctly, the men were exposed to fire of all kinds. Of the tanks engaged with the 6th Brigade, one had laid 400 meters of telephone cable from a rear-mounted drum on its way to the rendez-vous and, on arrival, did sound work with its Hotchkiss guns. The other two tanks finally ditched in MacDonnell Trench but not before being put to good use. The crews laboured for hours in a vain effort to get their machines freed.

The 4th Brigade was already establishing Lewis-gun posts beyond Gunpit Trench and was now within striking distance of Courcelette. On the opposite wing, the 6th Brigade was digging in, a few parties managing to reach the western edge of Courcelette. Major-General L. Lipsett's 3rd Canadian Division was in position to the left of the 2nd, its role being to protect the left flank of the attack on Courcelette. The 5th Canadian Mounted Rifles(i) of 8th Brigade, advanced from south of Mouquet

Farm towards Fabeck Graben. With few casualties and accounting for many of the enemy, they reached their objective east of Mouquet Farm and constructed a bombing block in their section of trench just south of Fabeck Graben. The 1st CMR, quite close to Mouquet Farm, made an abortive raid on the original German 2nd line while other parties from the same battalion made a small raid on the farm itself, taking some prisoners.

Major-General Turner instructed his 2nd Division to extend the forward positions from Gunpit Trench preparatory to an attack on the southern part of Courcelette. This was done without delay and preparations began, the 5th Brigade for the attack on Courcelette and the 3rd Division for its assault on Fabeck Graben.

The 22nd Bn. (French Canadians) and the 25th Bn. (Nova Scotia Rifles), both of 5th Brigade, advanced on Courcelette behind the creeping barrage at 6.15pm. Courcelette was stormed and taken with little difficulty and the men set about clearing the ruins and particularly, the underground cellars. The 26th Bn. (New Brunwick) came forward to help with the "mopping up" and many prisoners were escorted to the rear. Strongpoints were constructed at the quarry and cemetery on the east side of the village and a further toll was taken of the fleeing Germans.

The 7th Brigade of 3rd Division started its attack from Sugar Trench and Princess Patricia's Canadian Light Infantry (PPCLI) was somewhat disorientated by the lunar-type landscape, receiving fire from isolated pockets of resistance established in the many shell craters. Some Germans in dugouts in MacDonnell Trench were forced to surrender and the PPCLI continued its advance. Although still losing heavily, they joined up with the 5th Brigade at the eastern section of Fabeck Graben. The Germans still held the junction of Zollern Graben and Fabeck Graben and successfully resisted a Canadian attempt to force an entry. The remainder of the PPCLI linked up with the 42nd Bn. (Royal Highlanders of Canada) well to the west of the troublesome junction. On the far left, the 4th CMR of 8th Brigade extended under heavy fire the hold on Fabeck Graben in the direction of Mouquet Farm. Some twenty prisoners and two machine-guns were taken, the CMR then establishing a bombing block to protect the flank. In the dimming light the 49th Bn. (Edmonton) was ordered to reach a position from which an attack could be launched against Zollern Graben East but machine-gun fire prevented any appreciable movement. However, two companies advanced past Fabeck Graben to the quarry which they took and held. The Germans mounted a number of counter-attacks but these were all repulsed. Meanwhile the 3rd Canadian Pioneers had been busy working on forward communication trenches while the 5th Coy. Canadian Engineers further south had fortified the Sugar Factory and reinstalled a water supply from the well within the factory perimeter.

On the 16th September bombers of the PPCLI and 49th Edmonton Battalions (7th Brig.) successfully bombed their way forward and closed the troublesome gap in Faben Graben taking a good number of prisoners(ii) On the following day, the 5th Brigade attacked the German defences to the east

Courcelette Château before the War (J. Bommeleÿn)

of Courcelette but the positions were bravely defended. A bitter bombing contest ensued but the enemy could not be dislodged. During the night of the 17/18th September, Maj.-Gen. Currie's 1st Canadian Division took over the sector and immediately set about pushing out forward posts towards North and South Practice Trenches. A strong counter-attack was launched against the 4th Bn. (Toronto), the Germans forcing an entry into the north-eastern edge of the village. It took over two hours for the men from Toronto to eject the Germans but their respite was short lived. The German policy of retaking lost ground was vigorously pursued, the 4th Battalion having to repulse another attack a few hours later with bombs, Lewis-gun and rifle fire.

A further attack on Zollern Graben East was made at 5.00am on the 20th September by the 3rd Canadian Division. The 58th Bn. (9th Brig.) bombed its way from the east while the 43rd (Cameron Highlanders) of the same brigade attacked from the south. The Highlanders caught the Germans by surprise and a good section of the trench was secured. Recovering quickly, the Germans again showed stern determination resulting in close-quarter fighting. Little by little the Canadians gained ground but the struggle had taken its toll. Apart from their casualties, the two battalions were physically spent and the newly arrived German 26th Regiment, covered by a smoke screen, counter-attacked with rifle grenades and retook most of the Canadian gains.

During the evening of the 22nd September, a further attack was made on some of the trenches east of the village - this time with little opposition. While most of Zollern Graben East was still under German control, an entry was made at the junction of Zollern and Fabeck Graben. However, the Canadians still held strong forward positions in Fabeck Graben itself.

The eastern approach to Courcelette was protected by Twenty-Sixth Avenue which was entered by units of the 50th British Division by way of Spence Trench, south of Le Sars. The 23rd Division, on the left of the 50th, linked up with the Canadians in Twenty-Sixth Avenue at its western end near the Bapaume Road.

At 12.35pm on the 26th September, zero hour for the general advance on Thiepval and Mouquet Farm, the Canadian 6th Brigade at Courcelette advanced to the north of the village. The leading battalions, the 29th (Vancouver) and 31st (Alberta), covered by extensive machine-gun fire and a shrapnel barrage, moved forward. An immediate counter-barrage concentrated on the forward positions and village. The two tanks accompanying the advance were quickly out of action, one ditching while making its way to the front and the other literally blown apart with its own ammunition after receiving a direct hit. The 28th Bn. (North West) on the east of the village was pinned down in its assembly trenches by accurate enemy fire. However, the 29th succeeded in its forward movement, quickly seizing the German front line with little opposition, the occupants fleeing north towards Miraumont. On the left, the 31st Bn. was caught by machine-gun and rifle fire and only platoons from the right company reached the German front line adjacent to the 29th where an entry was made and a Lewis-gun emplacement established. The position of the remainder of the 31st Bn. was precarious to say the least with nearly all officers down and men pinned down in shell holes in front of the objective. A fresh British bombardment was quickly arranged and, under cover of dark, the 31st Bn., aided by a company of the 27th (Winnipeg), left the shelter of the craters and assaulted the German front line, this time occupying it between Miraumont Road East and Courcelette Trench. Quickly consolidating the entry, they were able to repel two German counter-attacks.

To the north-west of Courcelette lay Staufen Riegel, renamed Regina Trench by the Canadians, which formed the final objective. It was a long trench, in the Thiepval sector, known as Stuff Trench, then passing over the Grandcourt road, Miraumont Road West, Courcelette Trench, Miraumont Road East, Pys Road before turning south to cross the main Bapaume road to join the Le Sars line. North of Courcelette, the trench lay out of sight on the reverse slope of the ridge. This being the case, it was not known if the wire had been cut by the barrage. Accordingly, patrols were to go forward to establish the possibility of an attack on Regina Trench.

The leading companies of the 14th Royal Montreal Regiment and 15th Bn. (48th Highlanders) had advanced sufficiently to clear the German barrage which was laid as soon as the attack was spotted. Although clear of the barrage, the men from Montreal suffered heavy casualties from machine-gun and rifle fire. The survivors pressed on and secured Sudbury Trench, their first objective to the north-west of Courcelette, taking some forty prisoners. Pausing briefly, the 14th Bn. then continued its attack, securing the eastern section of Kenora Trench which runs north-west from

Sudbury Trench to join Regina Trench. The Highlanders on the left could not keep abreast of the 14th. With its flank in the air, the 14th was subjected to attacks by German bombers but these were checked with grim determination by the Montreal men. Eventually, the tired battalion was reinforced by a company of the 16th (Canadian Scottish) Bn. Meanwhile the Highlanders were still under fire from Regina Trench and were fighting desperately to get forward. With great resolve but with mounting casualties, including many officers, the Highlanders advanced yard by yard to the western end of Sudbury Trench. After a brief pause, the advance continued, the Highlanders having to clear hidden machine-gun nests and sniper positions over the cratered ground. By mid-afternoon, they were at the second objective just north of Sudbury Trench and eventually linked up with the 14th Battalion on the right, contact also being made with the 2nd Brigade on the left. Another company of the Canadian Scottish reinforced the line during the night. The Highlanders had had a hard day and were eventually to hear that its headquarters had been struck by a shell, igniting a store of petrol, killing Lt.-Col. Buchanan and many of his staff.

The Canadians were subjected to a long and heavy barrage which continued through the night of the 26/27th September until about noon on the 27th. Rain fell in the afternoon. On the right flank the 28th Bn. was still in touch with the III Corps of Rawlinson's Fourth Army while its left linked with the 29th Bn. in the former German training ground named North and South Practice Trenches. The 31st Bn. (Alberta) secured the remainder of the previous day's objective and later, with units of the Winnipeg Battalion, sent patrols forward to consolidate the position in Regina Trench.

The situation of the 14th Battalion in Kenora Trench was precarious. Having casualties of over 60 per cent, it had lost touch with the 48th Highlanders (15th Bn.) on the left. Kenora Trench becoming untenable, the 14th Battalion withdrew with its wounded to a freshly-dug support trench some 150 meters to the rear. Reinforced a little later by the 16th Bn., Kenora Trench was reoccupied with little loss. A hostile barrage harassed the men while strengthening the defence of the trench, the Canadians encountering much difficulty in checking a bombing counter-attack. The trench was then raked by machine-gun fire while some two hundred Germans were preparing another counter-attack. The decision was taken to make a further withdrawal to the support trench. The weakened 3rd Brigade was due to be relieved by the 5th Brigade of the 2nd Canadian Division but Lt.-Gen. Byng ordered the 14th Battalion to renew the attack, insisting that Kenora Trench be secured before the relief. As the relief of other battalions was under way, Lt.-Col. Clark delivered his attack at 2.00am on the 28th through the rain and slush with 75 weary men who, against the barrage and enemy machine-guns stood no chance of success.

East of the secured North and South Practice Trenches, patrols were harassed by enemy shell fire. The situation regarding Regina Trench was still unclear. Reports had been received that the enemy had vacated Regina Trench north of Courcelette but these reports were not confirmed. Reconnaissance by a cavalry patrol towards Destremont Farm was halted by machine-gun fire

At 7.00am on the 28th September, the 26th Bn. (New Brunswick), having passed through the 31st, advanced up Courcelette Trench where it came under heavy fire from Regina Trench. There was no doubt now that the reports about Regina Trench were incorrect. A further forward movement was attempted in mid-afternoon following a fresh barrage but little gain was made. The 26th made yet a further attack at 8.30pm but this too, was repulsed. The battalion, with casualties of around two hundred, then retired to the line of the first objective where they relieved the 31st Alberta. Meanwhile the 5th Brigade had relieved 3rd, the 25th Nova Scotia Rifles taking over the front of Lt.-Col. Clark's 14th Royal Montreal Battalion.

The newly arrived 5th Brigade engaged its 24th and 25th Battalions. Bombers of the 24th endeavoured to work their way in from the left while other companies made a frontal attack. Both movements were brought to a halt by the enemy barrage and uncut wire.

The 19th and 21st Battalions of the 4th Canadian Brigade on the extreme right flank strengthened their defences north of the Bapaume road towards the Practice trenches, the 19th in touch with the left flank of the Fourth Army. It was uncomfortable work under the continuous hostile bombardment and heavy rain which fell for most of the day. The new line ran west from Destremont Farm (see Objective 36), then north-west to cross Dyke Road, passing to the right of the Practice Trenches before turning south-west and finally west crossing Miraumont Road East, Courcelette Trench and Miraumont Road West. Meanwhile the 21st Battalion, under the same bombardment, moved north from the Practice trenches.

The question of Regina Trench was forever in the minds of the commanders. The awful weather would not permit useful aeroplane reconnaissance and so patrols from the 5th Brigade were sent out to establish the position at Regina Trench with orders to retire if the trench was still strongly defended. Met by heavy fire on their approach, the patrols withdrew as arranged. During the night the 26th Battalion was relieved by the 22nd (Canadien Français), attached from 4th Brigade.

The following day, the weather had improved and, although dull, there was no rain. A proposal for another attack on Regina Trench East was received with little enthusiasm by Brig.-Gen. A.H. Macdonnell, whose 5th Brigade was down to under 1,200 fighting men. They were tired and had sustained a good number of casualties from friendly fire. MacDonnell requested a fresh bombardment during which the range should be carefully verified. While waiting for the revised zero hour on the 1st October, the French Canadians dug and established an advanced line across Courcelette Trench.

On the 1st October the 20th and 18th Battalions of the 4th Canadian Brigade straightened the line of the Brigade front under periodical fire and linked up on the right with 23rd Division of III Corps. Regina Trench was assaulted by the 22nd Bn. (Canadien Français, 5th Brig) on the west side of Miraumont Road East. The leading waves were cut down by machine-gun fire while trying to cut a way through the wire while the rear waves were caught by the hostile barrage on their way forward. The casualties were very severe and the few French Canadians who succeeded in reaching the trench were not seen again. Kenora Trench which runs south-east from Regina Trench was assaulted by the Nova Scotia Rifles, also of the 5th Brigade. In spite of heavy frontal and enfilade fire, an entry was forced and the Rifles dug in. The Victoria Rifles advanced on the left making good progress until the advance of the 5th CMR of 8th Brigade (3rd Can. Div.), next on the left, was halted by a strong counter attack leaving the Victoria's flank in the air. A bitter struggle ensued but, by the end of the day, the biggest part of Kenora Trench was secured along with outposts in the sunken road and in Courcelette Trench on the right.

The 4th and 5th CMR of 8th Brigade (3rd Can. Div.) attacked Regina Trench further to the left. Here also, the Rifles met uncut wire and were sprayed by machine-gun fire while clearing a way through but units from both battalions forced an entry. The intrusion was strongly contested by the defenders and German and Canadian bombers were engaged in a bitter contest. The right of the 5th CMR began to give way as mentioned above but the left, along with the 4th CMR held on until the early hours of the morning when, with heavy losses and almost out of bombs, the men withdrew.

It rained at some time during the next six days and a renewal of the attack was not practicable. The wounded needed to be brought back and supplies moved up. The ground conditions deteriorated daily and any movement demanded a huge physical effort. While plans were being made for the next attack on Regina Trench and the Quadrilateral at its eastern end, the artillery continued its bombardment of Regina Trench and the uncut wire before it. During the intervening period, the 2nd Canadian Division was relieved by the 3rd whose left flank was eventually taken over by Maj.-Gen. Bainbridge's 25th Division in II Corps sector. The right of the Canadian Corps was taken over by the 1st and 3rd Brigades. Much preparatory work was done prior to the next attack on the 8th October. Assembly trenches and advanced posts were dug, often under fire. Patrols reported the wire to be reasonably cut but that the enemy was continuously filling the gaps with loose wire.

At 4.50am, zero hour on the 8th, the 1st Brigade advanced as far as Le Sars line with little difficulty. The 4th Battalion, crossing Dyke Road(iii), approaching the Quadrilateral from the south-east, found the wire uncut. Veering to the left to find a way through, the left of the 4th Battalion linked with the 3rd. Clear of the wire, an entry was forced into the forward trench of Le Sars line where the enemy was cleared by the Canadian bombers. A counter-attack in the early afternoon was checked by accurate artillery fire. Just as a direct attack on the Quadrilateral was about to be launched, the Germans counter-attacked in large numbers. Heavy losses were incurred on both sides during the bitter close-quarter bombing struggle for supremacy. During such bombing encounters, the victor is usually the one with the most bombs and now, the Canadians, out of bombs with no possibility of being supplied, with great gaps in their ranks, were forced to withdraw to their starting line. During the night, the survivors of the 4th Battalion dug a forward trench within 50 meters of Le Sars line, linking up with 23rd Division.

The 3rd Canadian Brigade attacked Regina Trench from the east of the Pys road. On the right the

16th Bn. (Canadian Scottish) passed through the wire with some difficulty(iv), forcing an entry in Regina Trench but, due to the withdrawal of the 3rd Battalion, the Canadian Scottish right flank was in the air, the men having to withdraw against accurate enfilade fire. The left battalion, the Royal Highlanders could not penetrate the wire and after taking heavy casualties, they too, had to withdraw.

Astride the Pys road, the 3rd Canadian Division assaulted its section of Regina Trench. The attack was carried with the utmost resolution but was destined to fail against a superior defence manned by stubborn defenders. Small parties of the 58th Bn. (9th Brig.), although held up at the wire, made a lodgement in the trench but were not seen again. In the centre, the Cameron Highlanders (43rd Bn.) of the same brigade, forced an entry just west of Miraumont Road East but were driven out. The Royal Canadian Regiment of 7th Brigade assaulted the trench from west of Miraumont Road West. It succeeded in gaining an entry and bombers immediately worked both up Miraumont Road and west along Regina Trench. The first signs of a permanent hold looked very possible but such hopes were dashed when yet another strong counter-attack was launched. Although contested with great determination, the Canadians were pushed back and lost all their gains. Newly-laid wire before Kenora Trench prevented entry by the right battalion, the 49th Edmonton which lost heavily from machine-gun fire. Repeated efforts by the battalion bombers to bomb up Kenora Trench failed to secure a permanent hold.

On the 10th October, Maj.-Gen. Watson's 4th Canadian Division began to relieve the 3rd. Rain during the next few days caused postponements in the action. But at least, the problem of uncut wire before Regina and Kenora Trenches was at last resolved by accurate artillery fire and at 12.06pm on the 21st October the 4th Canadian Division advanced behind the barrage. The 87th and 102nd battalions enjoyed the assistance of machine-gun fire, the infantry taking the defenders by surprise. The trench facing the attackers was secured, the 87th Battalion establishing advance posts north of Regina Trench and a defensive flank on the right to the east of Pys road. After crossing Desire Trench, the 102nd entered and secured Regina Trench immediately east of Courcelette Trench and also established an advanced post on Miraumont Road East. Contact was made with the 18th Division on the left of Courcelette Trench. Next on the left, the 53rd Brigade of 18th Division attacked with the 10/Essex and 8/Norfolk. The Essex, after a sharp encounter, seized its section of Regina Trench. The Norfolk on the left was equally successful, securing its objectives, its bombers, assisted by parties of the 11/Lancs Fusiliers from 25th Division, winning a bombing encounter with the enemy.

On the 25th October the 44th Battalion of the 10th Canadian Brigade endeavoured to extend the hold on Regina Trench as far as Farmer Road but the mud, counter-barrage and machine-gun fire brought the attack to a halt. A long rainy spell put an end to hostilities and no further attack was possible until the 9th November. Following a preliminary barrage the 4th Canadian Division made a further attack on the eastern extremity of Regina Trench. Leading at midnight on the 10/11th November from advanced positions, the 46th Bn. (South Saskatchewan) and 47th (British Columbia), supported by a company of the 102nd Battalion leaped into the trench before the Germans realised what was happening. Advance posts were established without delay. Numerous counter-attacks were successfully repulsed and four machine-guns and eighty-seven prisoners were taken.

The next week was spent strengthening Regina Trench, now almost entirely secured, pushing out advance posts and preparing new assembly trenches for the next attack. Little rain fell between the 9th and 16th November but the saturated ground benefitted little, the nature of the soil and subsoil apparently holding a large percentage of the water. Fanshawe's V Corps had been heavily engaged on the 13th November north of the Ancre while the left of Jacob's II Corps had attacked the Thiepval Ridge. It was now necessary for the 4th Canadian Division on the right of II Corps to continue its attack north of Courcelette. The next objective was Desire Trench and its support trench, appropriately called Desire Support. The former was approximately 500 meters north of Regina Trench, while Desire Support branched first north and then turned east off Regina Trench at a point just east of the Miraumont/Pozières road. Further to the north, Grandcourt Trench and its continuation, Coulee Trench, protected Grandcourt, Miraumont and Pys.

On the extreme right, one company of the 46th Battalion and two of the 50th (Brig.-Gen. Hughes' 10th Brigade) covered the eastern flank, adjacent to the left of Rawlinson's Fourth Army. The attack began on the 18th November at 6.10am, after the first snow of the winter, and now in driving sleet, the 46th (S. Saskatchewan) was met by deadly machine-gun fire and had to go to ground, eventually withdrawing to the rear. The 50th Bn. (Calgary) pushed forward with little opposition capturing about a hundred prisoners but, losing contact with the 11th Brigade on the left, dug in and consolidated just south of the eastern end of Desire Support Trench. Many casualties were sustained from front and flank fire while endeavouring to dig in, the Calgary battalion having to retire to Regina Trench.

Next on the left, Brig.-Gen. Odlum's 11th Brigade attacked Desire Support from east of the Miraumont Road East road with two companies each of the 54th (Kootenay) and 75th (Mississauga) Battalions. In the driving sleet both battalions veered to the right but seized part of Desire Support, taking a good number of prisoners. The 87th (Canadian Grenadier Guards) attacked with 'A' and 'D' companies, passing between the two roads leading to Miraumont. Crossing Desire Support the Guards, assisted by the 38th Bn. (Ottowa, attached from 12th Brig.) pushed out strong patrols to the north and entered Grandcourt Trench near its junction with Coulee Trench. While 'C' and 'B' companies moved quickly forward to clear any remaining Germans in Desire Support, the Guards, seeing Germans approaching from the east, prepared to receive the expected counter-attack. But the barrage, the cold, the determination of the Canadian attack had taken its toll, the Germans threw down their arms and surrendered. The British barrage now targetted and dispersed German troops seen to the south of Pys. The 44th Battalion and 47th (Brit. Columbia) of 10th Brigade were sent up to extend the hold on Desire Support. In the mid-afternoon, the enemy still held Desire Support a litttle each side of the Pys road while the Canadians had secured a fair length on the front of the 11th Brigade. Much to their displeasure the Guards and Ottowa were later ordered to withdraw from their hold in Grandcourt Trench. Having consolidated their gain, they were very reluctant to abandon it. Watson was well pleased with the results of his 4th Division's attack, almost all of the objectives had been taken and had a total of 620 prisoners to its credit. The Germans had suffered heavy losses from the barrage and the attacking infantry.

On the left of the Canadian Division, Brig.-Gen. Price's 55th Brigade of Maxse's 18th Division was assembled for its attack on Desire Trench. All four battalions were engaged, moving well forward of Regina Trench before zero and thus escaping any harm from the hostile barrage ranged on Regina. The 8/East Surrey advanced in fine style and had soon secured its section of Desire Trench and linked with the Canadians on the right while consolidating the new position north of the objective. In the centre the 7/Royal West Kent also secured its section of Desire Trench but there was a gap between the East Yorks and West Kents which was eventually closed by the battalion bombers. Divisional headquarters had no news of the success or failure of the other two battalions, the 7/Queen's and 7/Buffs and all seven runners sent forward to make contact became casualties. Later reports confirmed that the Queen's had been heavily hit by shell fire and that fire from snipers and machine-guns from specially prepared positions in the shell craters using a new technique of cutting firing slots, had continuously harassed the Queen's as it endeavoured to clear the dug-outs at the junction of Desire Trench and Stump Road. The same fire had brought the advance of the Buffs to a halt and only a few men reached the objective. However, help was on the way. Bombers of the West Kents attacked west down Desire Trench while riflemen of the same battalion advanced over the open ground astride the trench. Twenty-five Germans were killled, the rest fled to the north and the Buffs were now able to get forward to the objective. Contact between three of the four battalions was now established. Due to the failed attack on the 19th Division on the left, the flank of the Queen's was in the air, the fire so dense from Stump Road and Grandcourt Trench that the two leading companies were almost annihilated. Now aware that the left flank could not be held, Maxse ordered the evacuation of Desire Trench west of point 66 (see map) while the gains to the east should be immediately consolidated. The 92nd Field Coy. RE helped with the consolidation and dug a line back to Regina Trench.

Ground conditions had now deteriorated to such a degree that all hope of further offensive action towards Miraumont or Pys had to be abandoned.

(i) CMR, always dismounted and engaged as an infantry battalion on the Somme
(ii) During this action, Private J.C. Kerr of the 49th Edmonton Battalion, although wounded, ran along the parapet killing several Germans at close range. Awarded the VC, he survived the war and died in 1963 at the age of 76.
(iii) A sunken road or perhaps better described as a ravine, the depression being about twelve meters deep.
(iv) While his 16th Battalion was held up at the wire, Piper J. Richardson continued to play under heavy fire and later displayed great courage during bombing work. He was reported missing and was awarded the Victoria Cross posthumously.

OBJECTIVES 30, 31 & 32
GUEUDECOURT & LESBOEUFS VILLAGES, "METEOROLOGICAL" TRENCHES NORTH-WEST OF LESBOEUFS, GIRD LINES EAST, MORVAL & TRANSLOY RIDGE

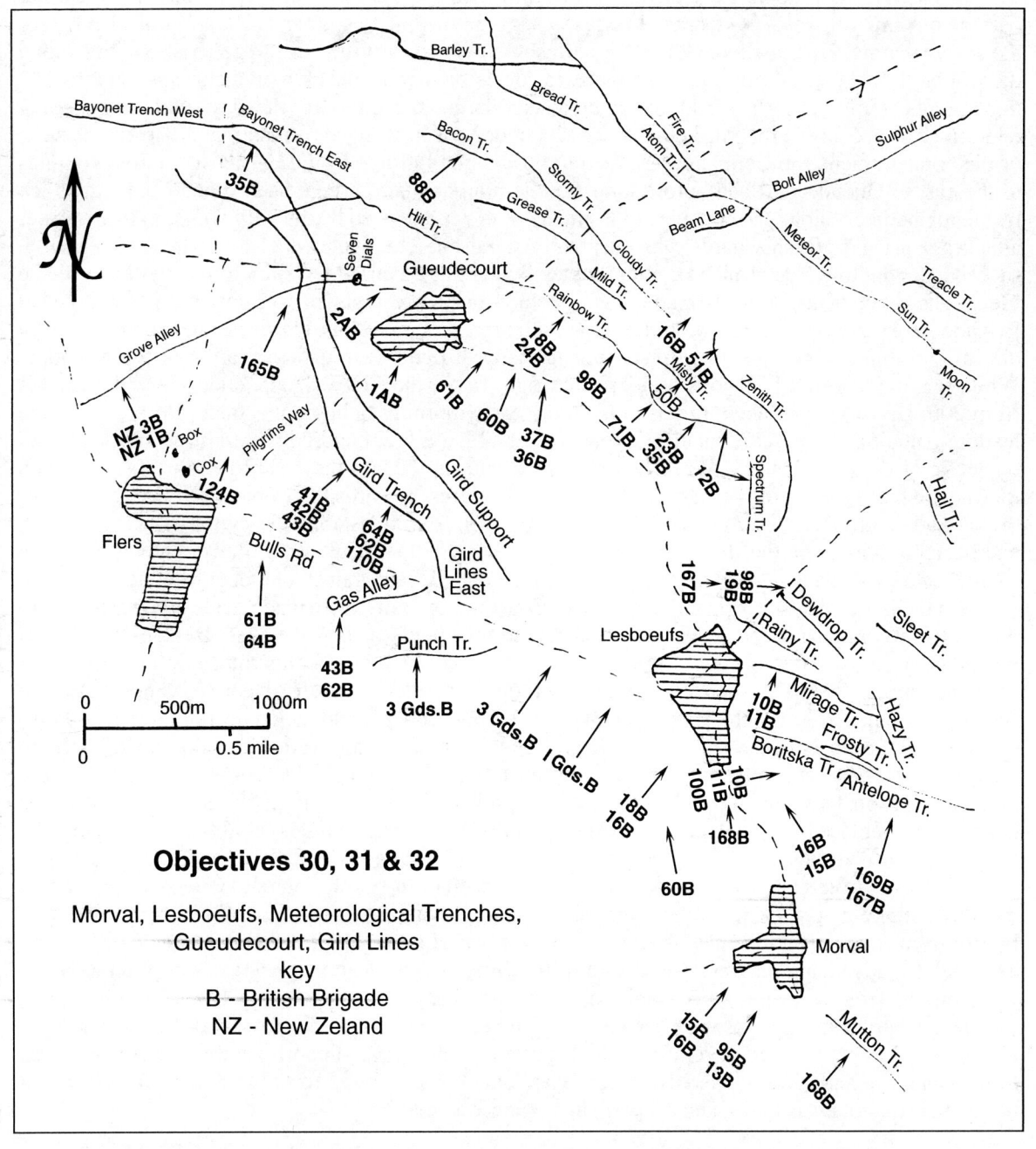

At 10.45am on the 15th September, following the attack on Flers, the 42nd Brigade (14th Div.) along with units of Major-General Lawford's 41st Division were at the northern edge of the village and within striking distance of Gird Trench to the south of Gueudecourt. Uncertain as to the outcome of the divisional attacks on either flank, there was some hesitation as to whether the attack should be continued. Lawford received orders to advance towards Gueudecourt and protect his flank as best he

could. Mixed units from all three brigades advanced but were brought to a halt by fire from Gird Trench and no further forward movement could be made on the 15th.

The role of the New Zealand 3rd Brigade was to establish a defensive flank to the north-west of Flers and at 11.30am, half the 1st Rifles, covered by machine-gun fire from its sister battalion the 2nd, assaulted Grove Alley which runs north-east from Abbey Road. The centre and left made entries into Grove Alley but the right units suffered heavy casualties from German positions on the ridge. It is now known the Germans were under orders to retake the lost parts of the Gird Lines at all costs and the New Zealanders could see a large force of Germans coming in their direction. Being too few in number to hold the trench, they withdrew to the third objective to prepare for the counter-attack. Tank assistance provided to the New Zealand Division did not, apart from the initial impact, amount to much practical help. D12 was hit and had to be abandoned. D8 could do little with its 6 pounder mechanism damaged while D11 was ordered to take up position on the Ligny road which it did and awaited further orders.

In the early afternoon, Lt.-General Horne informed the 14th and 41st Divisions that a further barrage on Gird Trench and Gueudecourt was to begin immediately and that an attack was planned for the early evening. By this time, General Rawlinson had been informed of the general situation along the whole of the Fourth Army's front and took the decision that the attack would continue on the 16th, the remainder of the 15th would be spent consolidating the gains of the day, including linking operations, especially at the third objective - this under cover of the aforementioned barrage. In spite of the bombardment, the Germans were seen to be arriving in force from Le Transloy, only two miles to the east of Gueudecourt and before long, the German counter barrage had commenced. An attack by enemy bombers along the Gueudecourt/Ginchy road was checked by the 14th Division while a later attack against the 1/Irish Guards was eventually repulsed. The depleted 124th Brigade (41st Div.) succeeded in halting a determined counter-attack by the 6th Bavarian Division in its attempt to retake Flers. The relief of the 42nd Brigade was effected during the night in wet conditions by the 43rd Brigade preparatory to the attack on the 16th. Although no rain fell during the 16th, the overcast sky did little to improve ground conditions, causing many difficulties in the evacuation of the wounded. While the 56th and 6th Divisions protected the right flank, the 61st Brigade, attached to the Guards from 20th Division, had taken up positions in the early morning to begin its attack. Before dawn the leading battalions, the 7/DCLI and 7/SLI were assembled a few hundred meters in front of the first objective of the previous day. While awaiting zero hour at 9.25am they had to endure heavy trench mortar fire but once out of the trenches, made good progress towards their objective taking a number of prisoners from the *52nd Reserve Division*. Strongpoints were established by the 7/KOYLI which had been brought up on the right flank while the 12/King's supported the left flank. The whole of the weakened 61st Brigade had been committed but held its position against a number of counter-attacks.

On the left of the 61st Brigade, Brig.-Gen. Corkran's 3rd Guards Brigade made a late start due to difficulties in organising its attack formation. Some four hours behind the 61st Brigade, the Guards moved forward without artillery support in the face of accurate enemy machine-gun fire. The 1/Grenadier and 1/Welsh advanced grimly to within 250 meters of their objective where they were forced to dig-in. The left of the Welsh Guards was in Punch Trench and facing north rather than north-east towards Lesboeufs. The 3rd Guards Brigade had been much involved in carrying up supplies on the 15th September as well as being involved in the actual fighting. The courageous attacks of the Guards Division between the 10th and 16th September had cost over 4,900 casualties(i) and the division was replaced in heavy rain by the remainder of Maj.-Gen. Douglas Smith's 20th Division, the 59th and 60th Brigades.

Lt.-Gen. Horne's XV Corps objectives were a section of the Gird Lines and Gueudecourt village, eventually linking up with III Corps beyond the village. On the right, the 6/SLI (43rd Brig.) was halted by fire from enemy positions in Gas Alley. Entry was made into what was thought to be Gird Trench. The 10/DLI, advancing on the western side of the Ginchy/Gueudecourt road was caught in a hail of fire from the front and right flank from hidden enemy positions in the sloping ground leading down to Gueudecourt. The Durhams were forced to take cover in shell holes while the 6/KOYLI and 6/DCLI endeavoured to get forward to reinforce the leading battalion. The support battalions were subjected to the same fire and suffered heavily. A combined attack was ordered for 7.00pm but this too served to increase the total casualties and no permanent gain was made.

The 41st Division's attack was made by the 64th Brigade, attached from 21st Division. After experiencing much difficulty in getting forward in the rain and dark of the previous night, the commanders of the leading battalions being unaware of the front line position, the 15/DLI and 9/KOYLI advanced a long way behind the barrage line. Coming immediately under shrapnel and machine-gun fire, their losses mounted at an alarming rate. The close support of the 10/KOYLI and 1/East Yorks did not achieve entry into Gird Trench and although some prisoners were taken in the many shell craters, Gird Trench remained inviolate. Tank D14 caught up with the KOYLI's but was put out of action by a shell. The survivors retired and reformed at Bulls Road, east of Flers.

Meanwhile the 1st N.Z. Brigade, after repulsing a German attack from the Ligny Road, attacked and secured its section of Grove Alley. The check of the 64th Brigade on the right prevented further movement by the New Zealand Brigade. The 1st Canterbury secured the gain and dug a trench back to Box and Cox Trenches. Tank D11 had played an important role in the check of the German attack earlier in the day and had subsequently advanced a further 300 meters before being put out of action by an enemy shell.

On the 17th September while improving and consolidating the line, the 60th Brigade (20th Div.) repelled a frontal attack south of the Ginchy/Lesboeufs road with Lewis-gun and rifle fire. Enemy bombers were working down either flank but were checked by the 12/Rifle Brigade and 12/KRRC. Heavy rain began to fall in the early evening when Shute's 59th Brigade received orders to complete the capture of the third objective. In a position which offered practically no advantages, now deteriorating rapidly in the heavy rain, the depleted 11/KRRC with the 10th and 11/Rifle Brigades, having insufficient artillery support(ii) and with both flanks in the air, stood little chance of success and the advance was quickly brought to a halt. Encouraged by the abortive attack, the German bombers advanced down Gas Alley which runs south-west from Gird Trench to the south of Gueudecourt. The attack was repulsed with some difficulty and the decision was taken not to make a further attack on the third objective until the whole line had been straightened and improved. The unpleasant and dangerous task of digging in under fire and coping with the bad ground conditions continued for many hours.

On the 21st September the 3rd Guards Brigade had relieved the weakened 20th Division while Major-Gen. Campbell's 21st Division was taking over the positions of the 14th. An early attempt to bomb up Gas Alley towards Gird Trench by the 1/Lincoln (62nd Brig. 21st Div.) on the 20th was unsuccessful. Two days later the 4/Grenadier (3rd Guards Brig.) made a similar attack but little ground was gained.

After several days of rain, blue skies appeared again over the battlefield from the 22nd September and at 7.00am on the 24th, the preliminary bombardment began in preparation for the asssaults on Morval, Lesboeufs, Gueudecourt and Gird Lines, collectively known at the Battle of Morval. At 12.35pm, zero hour on the 25th, the men moved forward behind the creeping barrage on the first of a planned three-phase attack. The 1/East Surrey of Brig.-Gen. Gordon-Lennox's 95th Brigade on the extreme right of 5th Division came under heavy fire from hidden German positions to the north of Combles railway while the 1/Devon to its left was held up for a while in front of a German strongpoint

The ruins of the church at Morval in 1918.
(A. Delmotte)

The church at Morval was rebuilt in the 1920s but it was soon in need of costly repairs which accumulated to such an extent that the Mayor of Morval, Monsieur Alfred Delmotte, in conjunction with the authorities, took the sad decision to have the church demolished. A photographer took this amazing picture on 22nd September 1987. (courtesy of La Voix du Nord, Lille 27th September 1987)

on the Ginchy/Morval road. The Devons veered to the north and then attacked the strongpoint from that direction. On the immediate left of 95th Brigade, Brig.-Gen. Turner's 15th Brigade, led by the 1/Norfolk, stormed and took its first objective, taking a heavy toll of the defenders and sending over a hundred prisoners to the rear. The advantages of a narrower front were now becoming evident, the Norfolks being assisted by bombers of the 1/Buffs (16th Brig.) as well as Stokes-mortar fire from the right-hand brigade (the 16th) of 6th Division. Immediately north of the 6th Division, Brig.-Gen. C. Pereira's 1st Guards Brigade was held up by three lines of uncut wire. The casualties in the ranks of the 2/Grenadier mounted rapidly while hacking a way through the wire but in traditional Guards style, continued to cut a passage through under covering fire from bombers and marksmen from the other companies. The Grenadiers then rushed forward and seized the German trench. The 1/Irish Guards of the same brigade and 2/Scots of the 3rd Guards Brigade advanced successfully crushing the weakened opposition before them. On the extreme left of the Guards frontage, the 4/Grenadier of 3rd Brigade, in the face of heavy fire, stormed the German defenders of a trench in front of the Gird Lines at the point of the bayonet, killing over a hundred before arriving at the first objective. The Germans retained control of Gird Trench and the northern end of Gas Alley but the first phase of the attack had generally gone well.

At 1.35pm promptly, the Guards, with the 5th and 6th Divisions, commenced the second phase of the attack. The East Surrey battalion had achieved its objective and consolidated its gains. The 1/Devon (95th Brig.) and 1/Bedford, the latter passing through the Norfolks engaged in the first phase, quickly secured its section of the sunken road, crushing all opposition before them and taking a good number of prisoners. The 6th Division enjoyed equal success, the 1/Buffs which had earlier helped the Norfolks, now advanced with the 2/DLI and 11/Essex of 18th Brigade gaining their objectives with few losses, the resistance of the Germans facing them seemingly much reduced. The 1st Guards Brigade, not to be outdone, set about clearing the many dug-outs in the sunken road and bombers of the 2/Grenadier and 1/Irish Guards did sterling work. Similar good work was executed by the 2/Scots Guards of 3rd Brigade which reached its second objective. However, the divisions of Horne's XV Corps to the immediate north of the Guards Division had failed to take the Gird Trenches with the result that the 4/Grenadier, having previously secured the trench immediately in front of Gird Trench now had its flank in the air. It was decided that part of the battalion would stay and face north against possible counter-attacks from the Gird Trench while the remainder joined the Scots Guards in its advance to the next objective.

A little after 2.30pm began the third and last phase, the attack on Morval and Lesboeufs. The 5th Division sent in the 12/Gloucester (95th Brig.) and 2/KOSB (attached from 13th Brig.) to assault the southern half of Morval village. Original plans were for the village to be secured in four stages, but the objective was taken in less than ninety minutes. The main reason for this was the excellent results obtained by the artillery and the Gloucesters and Scottish Borderers met little resistance. The 1/Cheshire (15th Brig.), after passing through the Bedfords, had secured the northern part of Morval by mid-afternoon.

Before the demolition of the church, the bells were taken down and later reinstalled on the village green (author)

The 6th Division advanced with the 2/York & Lancaster, closely followed by the 1/KSLI, both of 16th Brigade. Passing through the Buffs, the battalions advanced east of the Morval/Lesboeufs road meeting little opposition. On the left flank the 1/West Yorks of 18th Brigade successfully assaulted the southern entry into Lesboeufs, again, virtually unopposed. Meanwhile the 2/Grenadier and 1/Irish Guards had secured the rest of the village with relative ease. The two Coldstream Guard battalions of the 1st Brigade were moved forward in support. The 1/Grenadier (3rd Guards Brig.) secured the final objective north of Lesboeufs and one company was left astride the Gueudecourt road while the remainder dug in some distance to the east of the road. The attack had been most successful and the lack of the usual stern enemy resistance was encouraging. Germans were seen hurrying away to the north. As is always the case, some hardened soldiers clung to their posts and snipers had to be dealt with(iii). The Germans laid a barrage on Morval and Lesboeufs but only a minimum number of British troops were kept in reasonably secure posts. The high ground on the east and north-east of Morval was occupied and in the early evening the 16/Royal Warwicks of 15th Brigade dug in on the east side of the village while the 2/York & Lancaster established strongpoints in the general direction of Lesboeufs. Both the 5th and Guards Divisions had the facility of the use of tanks. The 5th Division's machines followed up the attack towards Morval - two ditched and the third was sent back. The Guards kept their three tanks in reserve near Ginchy.

On the right flank of Horne's XV Corps, the 64th Brigade of Campbell's 21st Division was to assault the Gird Trench facing its sector and then secure the village of Gueudecourt. The leading battalions, the 10/KOYLI and 1/East Yorks were held up, like the Guards previously, by uncut wire and German machine-guns took a heavy toll while efforts were being made to cut a passage through. The survivors took cover in shell craters, eventually making their way back after sunset. The German range on the 1/Lincoln (attached from 62nd Brig.) was so accurate that the battalion was halted at the very outset on its own front line and only the right company was able to get forward. These men joined the 4/Grenadiers of XIV Corps in an attempt to secure the junction of Gas Alley and Gird Trench. The German defenders beat off the attack and no entry was made. At least the Lincolns and Guards retained the link between the XIV and XV Corps.

The leading companies of the 8th and 9/Leicester of Brig.-Gen. Hessey's 110th Brigade (attached to 21st Division) attacked the Gird Trench from east of Flers. Initially held up by the German barrage, they pressed on, taking Goat Trench to the south of Gird Trench. Although the British artillery had succeeded in cutting the wire to a fair degree, the attackers came under machine-gun fire from Gird Trench causing considerable loss in both battalions. The C.O. of the 9/Leicester, Lt.-Col. Haig went forward up Pilgrims Way which runs from north-east Flers to south Gueudecourt to see what was happening. It was obvious support was needed and so Haig sent a runner to order up the remaining companies but he never arrived with the message. Pinned down by enemy fire, no further movement on the right was possible and a defensive flank was dug in Watling Street, part of the sunken road south of Gueudecourt. On the left, the 8/Leicester gained entry into the long Gird Trench and linked up with 55th Division. Consolidation began immediately. Reports that the 8/Leicester had entered Gueudecourt led Headquarters to believe that the village had been taken, but this was not the case.

Next to the 21st Division, Jeudwine's 55th, facing its section of the Gird Lines, engaged its 165th Brigade, composed entirely of battalions of the King's Liverpool Regiment. Following the barrage, the 6th, 7th and 9/King's advanced with great determination and, in just over half an hour, had secured the Gird Trench at its junction with the sunken road from Gueudecourt to Factory Corner. Sappers from the 1/West Lancs R.E. blew in Gird Trench while the 6/King's blocked entry from the north. Grove Alley was successfully assaulted with bomb and bayonet by the 9/King's. An hour later the 7/King's along with parties from the 6th gained entry into the sunken road towards Gueudecourt, eventually linking up with the 110th Brigade.

The following day the 5th Division made some progress down Mutton Trench between Morval and Frégicourt as well as making an entry into Thunder Trench to the east of Morval. German machine-gun fire from Sailly-Saillisel prevented the complete capture of the trench but the lodgement was useful. During the night the 5th Division was relieved by the 20th and 6th Divisions, the latter moving its front a little to the south.

Patrols of the Guards Division in front of Lesboeufs were subjected to sniping and machine-gun fire during the 26th. A strong German counter-attack was launched in the late morning between Morval and Gueudecourt but artillery fire took a heavy toll, the survivors fleeing without arms towards Le Transloy.

The Leicesters were now to have a second attack on Gueudecourt and bombers from the 7th Battalion (110th Brig.) followed the allotted tank firing as it advanced up Pilgrim's Way towards Gird Trench. Two companies from the same battalion were in close support. The combination of tank, bombers and infantry worked to perfection, the Germans being pushed ever closer to the waiting Guards Division. The Leicester bombers dealt death to those who sought cover in the shell craters while Lewis guns fired with equal accuracy on those who fled across the open ground. Following a short barrage on Gird Trench, a call to the RFC to clear the remaining Germans in Gird Trench was executed with courage and efficiency. The Germans had nowhere to go, machine-gunned from the air, bombed from the ground, all open ground covered with Lewis gun and rifle fire, they died or surrendered. The assault had been completely successful and Horne's disappointment with the failure to secure the Gird Trench on the 25th must have changed to great satisfaction on receiving the news. The Official History records the surrender of 362 men and 8 officers against the total of five British casualties. German dead lay thick on the ground.

The 15/DLI (64th Brig. 21st Div.) occupied the Gird Trenches while the jubilant tank crew, now close to Gueudecourt, returned to refill. Reports that the ground was suitable for cavalry use brought about the forward movement of the 19th Lancers of the Sialkot Cavalry Brigade. Leaving Mametz, the cavalry passed east of Flers, then turned towards Gueudecourt. A patrol reached the village but had to retire against the heavy shelling. The remainder dismounted and entered the village from the south-west while dismounted troops of the South Irish Horse opened fire with Hotchkiss guns and rifles to the north and north-east. The 6/Leicester, also of 110th Brigade, had sent patrols into Gueudecourt and the remaining companies now moved slowly into the village. About 5.00pm a large number of Germans were seen advancing from the direction of Le Transloy through the high corn but an effective barrage brought any further movement to a halt. The Leicesters made their way through the village and dug in on the northern edge. The cavalry was then withdrawn.

In the early evening the 15/DLI moved up from Gird Trench and, with the 10/KOYLI and part of its sister battalion, the 9th (all of 64th Brig.), took up positions a little way short of the final objective, the Gueudecourt/Le Transloy road. The 12/NF (62nd Brig.) took over the front and pushed the position forward to the Lesboeufs road where contact with the Guards was made. During the night the 10/Green Howards of the same brigade relieved the left sector.

Although the Gird Trench was secured before Gueudecourt, the remaining section of this long trench still needed to be taken and orders were issued to the 55th and New Zealand Divisions to take the Gird Trench from the secured part to Factory Corner and from there, the remainder of Goose Alley. Zero hour was set for 2.15pm the next day, the 27th. This action is described under Objective 36.

During the morning of the 27th September, a further assault was made on Mutton Trench and although some progress was made, the trench was not completely secured. The British artillery was also very active during the 27th, being ranged on roads, trenches and any observed infantry or supply movement. This prompted a retaliatory barrage on Morval, Lesboeufs and Gueudecourt during the night of the 27/28th September. At this time the right wing of the XIV Corps was handed over to the French and the 6th, 20th and 56th Divisions were withdrawn a short distance to the north.

Heavy rain fell on the 29th September creating conditions which were to worsen as autumn approached. In XIV Corps sector, the 6th Division and the Guards occupied trenches to the north of Lesboeufs before being relieved by Maj.-Gen. Hull's 56th Division which now formed the right hand division of XIV Corps. After a postponement of a few days, the general attack restarted on the 7th October, preceded by a preliminary barrage. The 168th Brigade on the right was to advance east and keep in touch with the French on the right, but the 14/London (London Scottish) line of advance being slightly different from that of the French, contact was quickly lost. However, the London Scottish secured the southern section of "The Gun Pits" to the east of Lesboeufs as well as the southern end of Hazy Trench, one of many trenches carrying meteorological names situated between Lesboeufs and Gueudecourt(iv) A few minutes later the 4/London were checked by fire from the northern part of the Gun Pits and veered to the right. The London Rangers began its attack north-east from Lesboeufs four minutes after zero, supported by Stokes mortar fire, the objective being too close for artillery support. The mortars had not silenced the guns in Dewdrop Trench and the Rangers attack was checked. The 1/London and 7/Middlesex of 167th Brigade on the left attacked east from north of Lesboeufs. The Middlesex was quickly engaged in bitter close combat but eventually secured the southern end of Rainbow Trench. Fire from Spectrum Trench halted the attack of the 1/London. However, bombers on the left wing entered the northern end of Spectrum linking up with the Middlesex. The 4/London and London Scottish repulsed a counter-attack but a second attack at night forced a withdrawal.

North of the 56th Division, Maj.-Gen. Douglas-Smith's 20th Division had assembled immediately south-east of Gueudecourt, the 61st Brigade on the left and the 60th on the right. The 6/OBLI and 12/Rifle Brigade secured their section of Rainbow with little difficulty, in contrast with the 7/Middlesex which had been engaged in a hard contest to secure the southern part of the trench. It appears the Germans facing the 60th Brigade lacked that sternness of resistance normally so prevalent and fled east making easy targets for the attackers. The attack continued to Misty Trench where a link was made with the Middlesex on the right and with the 61st Brigade on the left. The 7/KOYLI and 12/King's led the 61st Brigade's attack, firstly having to deal with a good number of the enemy wishing to surrender. Now quickly crossing Rainbow, they secured the south-eastern part of Cloudy Trench. The 37th Brigade of 12th Division had not come up, leaving the flank of the 12/King's in the air. Consequently, consolidation was made in Cloudy and west along Shine Trench towards the Gueudecourt/ Beaulencourt road. A little over 300 meters of Rainbow south-east of the road still remained under German control. A counter-attack was made from the Beaulencourt road but was checked with little difficulty.

On the right of XV Corps, the 36th and 37th Brigades of Maj.-Gen. Scott's 12th Division were

These dry and dusty fields were part of the area containing the 'Meteorological' Trenches where, in the autumn of 1916 both British and German barrages turned the whole sector into a sea of clinging mud and filth. (author)

subjected to a barrage and raked by enemy machine-gun fire before zero hour. Gueudecourt was also shelled which hampered the advance of the 37th Brigade on the right. The 6/Buffs advanced through the barrage and reached Rainbow but could not hold, sustaining many casualties. The left battalion, the 6/Royal West Kents was halted by fire from Hilt Trench. The Royal Fusiliers of 36th Brigade fared no better, both battalions, the 8th and 9th, being cut down by fire from Bayonet Trench East and the few Fusiliers who reached the trench were not seen again.

The Lesboeufs road out of Morval where the unusual Calvary can be seen, (author)

Rain fell during the morning of the 8th October and the wounded were brought back in very difficult conditions, the stretcher-bearers accomplishing Herculean labours. The men were withdrawn from most of the previous day's gains on the 56th Division front in order to allow a thorough bombardment of the area. Following the barrage the London Rifle Brigade reached Hazy but no contact was made with the French. The 9/London (169th Brig.) and 3/London (167th Brig.) were halted by machine-guns mounted in advance positions in the shell craters It was painfully obvious nothing more was going to be achieved in the Meteorological Trenches on the 8th and a general withdrawal to the original line began before midnight. The enemy promptly occupied the vacated Rainy Trench.

General Rawlinson ordered a renewal of the attack at the very earliest moment in order to secure the untaken objectives of the 7th October but time was needed to make preparations and particularly to get supplies forward. It was necessary also to relieve units which were well below strength. Both these priorities were hindered considerably by the ground conditions. Although there was to be no rain of any consequence during the next six days, the thousands of craters formed over the whole front had turned the ground into a sticky mess, making it difficult even to walk. Many craters were partly filled with water. Conditions do not improve rapidly in autumn. In the mud and sludge, between the 8th and 11th, Lambton's 4th Division relieved the 56th and Ross's 6th Division replaced the 20th.

A German soldier on duty in front of his sentry box at Gueudecourt. The picture probably dates from 1915. (J. Bommeleÿen)

The next attack was scheduled for 2.05pm on the 12th October. The 4th Division on the extreme right of XIV Corps assaulted with the 1/R Warwick of 10th Brigade. The Warwick advanced east of Lesboeufs, the right getting forward as far as Antelope Trench to the south of Hazy and dug in. A link was at last made with the French on the right. A counter-attack some hours later was successfully checked. The 1/Royal Irish Fusiliers (10th Brig.) on the left was stopped before Rainy and Dewdrop. To prepare the way for the 12th Brigade, Stokes mortars bombarded Spectrum and the 2/Lancs

Fusiliers secured a hold on the northern section. The mortar barrage did not silence the enemy defences in the southern part of Spectrum, the 2/DWR on the right taking heavy losses before forcing an entry and linking up with the Fusiliers. Pausing a while to recover and regroup, both battalions advanced towards the southern end of Zenith, about 200 meters up the slope where the enemy was engaged at close quarters but after a protracted struggle, the men withdrew to Spectrum Trench.

On the left of 4th Division, the 16th Brigade of 6th Division attacked the northern part of Zenith but the leading battalion, the 2/York & Lancaster could not force an entry. Meanwhile the 9/Suffolk of 71st Brigade in the centre which had taken over the previous gain at the junction of Misty and Cloudy, and was not required to advance further at the moment and the Sufollk strengthened the defences in its line. The 18th Brigade on the left flank to the north east of Gueudecourt had attacked Mild and Cloudy trenches, the 1/West Yorks making a determined frontal attack, supported by a bombing attack from the right. The enemy defences were too strong, the Yorkshiremen sustaining heavy losses. On the left, the 14/DLI of the same brigade fared better, securing the remainder of Rainbow from where its bombers attacked enemy dug-outs to the north-east in the sunken road in the direction of Beaulencourt. Parties of the Durhams linked up with the West Yorkshires while on the northern side of the Beaulencourt road and a link was made with the Newfoundland Regiment of 88th Brigade. (attached to 12th Div. from the 29th)

The Newfoundlanders and 1/Essex made a vigourous attack on part of Hilt Trench, a continuation of Rainbow, forcing an entry. Parties of the Essex continued, entering Grease Trench, a continuation of Mild. The Essex was well forward, but its hold on Grease was quickly becoming untenable with its left flank in the air and the Battalion was ordered to fall back to join the Newfoundlanders in Hilt Trench. Good progress had been made up Hilt by Newfoundland bombers and, following the failure of the 35th Brigade on the left, a barricade was constructed at the start of Bayonet Trench East. The unfortunate 35th Brigade was faced with uncut wire where the courageous efforts of the 7/Norfolk and 7/Suffolk to cut a way through were halted by enemy fire. Taking cover in the water-logged shell holes until nightfall, the survivors made their way back to the starting line.

Some small gains in the 'Meteorological' trenches had been made but the attack of III Corps on the left of Rawlinson's Fourth Army had failed. Rawlinson ordered the capture of Zenith, Mild and the unsecured part of Cloudy trenches, at night if necessary, before resumption of the general attack provisionally set for the 18th October.

The 10th Brigade of 4th Division took up the challenge. On the 14th October at 6.30pm the 2/Seaforths secured the gun-pits south of Dewdrop and, surprising the enemy, entered Rainy Trench. A subsequent counter-attack forced the Highlanders to withdraw. Meanwhile the 2/Royal Dublin Fusiliers, on the right, could not secure the Gun Pits before Hazy. Bombing attacks down Spectrum on two consecutive nights by the 1/King's Own (12th Brig.) failed to make any progress.

To the north, the 6th Division was engaged north-east of Gueudecourt, the 2/Sherwood Foresters (71st Brig.) securing and holding a strongpoint before Cloudy. On the left, the 11/Essex (18th Brig.) crossed Mild to reach the sunken road but were driven back by German bombers.

On the 18th October the 4th Division engaged the 11th Brigade to attack Frosty, Hazy, Rainy and Dewdrop trenches. The 1/Rifle Brigade reached the gun-pits assaulted by the Dublin Fusiliers on the 14th but fire from Hazy forced a withdrawal. The 1/East Lancs suffered a similar fate being halted in front of Dewdrop by fire from hidden machine-guns. Just to the north, the 1/King's Own of 12th Brigade was engaged in a bitter struggle to secure Spectrum and some 70 meters were eventually secured towards Dewdrop. The next day, the 1/SLI (11th Brig.) occupied Frosty which was found to be empty, the Somersets consolidating its defence and repulsing a night counter-attack.

The 9/Norfolk of the 71st Brigade (6th Div.), under artillery fire in the assembly trenches, was impatient to get started. Progress was very slow through the mud, the creeping barrage now well in advance of the Norfolks. Even so, the north-western part of Mild at its junction with Grease Trench was secured and held, the Norfolk repulsing a subsequent counter-attack.

On the right of XV Corps, the 88th Brigade, still attached to the 12th Division, attacked Grease Trench with the 2/Hampshire and 4/Worcester. The Hampshire, in touch with Norfolks of 6th Division, successfully secured Grease with relatively few casualties.The Worcester enjoyed a similar result, securing and holding Hilt Trench before moving north to Grease. The two battalions then attempted to continue the advance towards Stormy but sustained heavy casualties from accurate fire,

The crossroads at the centre of Gueudecourt. the Canadian Memorial 1000m towards Beaulencourt commemorates the action of the Newfoundland Regiment at Hilt and Rainbow Trenches in October. (author)

the men then retiring to their former gains. The 9/Essex of 35th Brigade on the left succeeded in finding a gap in the wire before Bayonet Trench East and made an entry. German bombers hurried down from the north-west and forced the Essex out of the trench. The 4/Worcester constructed a barricade in Hilt Trench to protect its left flank.

All movement was painfully slow through the flooded trenches and water-filled craters, particularly at the southern extremity of the 4th Division's front and further rain (4 mm) fell on the 18th and the same quantity was to fall during the following day. The only possible way for the infantry to take its objectives was to get to close grips with the enemy and fight with bomb and bayonet. Some way had to be found to keep the Germans down while the infantry got forward. Some serious thought needed to be given to the use of artillery, mortars and machine-guns.

No further offensive was possible before the 23rd October and even then, ground conditions had changed little. The British artillery ranged on Le Transloy village and cemetery and at zero hour 2.30pm, having already been postponed for one hour because of the mist which was slow in clearing, the 11th Brigade on the right of 4th Division moved to attack, led by the 1/Hampshire with the 2/Royal Dublin Fusiliers (attached from 10th Brig.). The Hampshires on the right, in touch with the French, were caught by fire from their objective, Boritska Trench, just over a thousand meters east of Lesboeufs, as well as from machine-gun nests set up in shell-holes. The 1/Rifle Brigade (11th Brig.) moved up in support and posts were established north of the gun pits. As darkness fell, a link was made with the Fusiliers(v) which had made a frontal attack on the gun pits and secured a strongpoint slightly to the north-east. The 1/Royal Warwick, also attached from 10th Brigade, moving forward to pass through the leading battalion, linked up with the Dublin Fusiliers and the two battalions were fighting their way forward at close-quarters. Fire from the flanks prevented further progress. The 12th Brigade on the left had similar problems. The 2/Essex came under fire from Dewdrop Trench and the few who managed to reach the objective were overwhelmed. However, the 1/King's Own on the left, entered the southern part of Spectrum but, even with the help of bombers of the 2/DWR, was unable to make further progress towards Zenith and Orion Trenches.

Maj.-Gen. Hudson's 8th Division, having replaced the 6th, attacked Zenith with the 2/Scottish Rifles and 2/Middlesex of 23rd Brigade. The Scottish Rifles crossed Zenith and entered Orion just ahead but the hold was untenable against the hostile barrage. The Middlesex bombers on the left were unable to make any progress at the centre of Zenith. The 25th Brigade on the left attacked the northern section of Zenith with the 2/Lincoln but fire from the objective halted the attack. Just a few parties linked with the Middlesex which, as stated, had made little progress. The 2/Rifle Brigade also failed to make progress against the salient formed at the junction of Zenith and Eclipse Trenches. However a series of posts was established a little to the south-east of the salient. The left of the 8th Division front was taken over by the 24th Brigade facing Mild with Cloudy immediately to the rear. The 2/East Lancs stormed and succeeded in securing most of Mild Trench, constructing a barricade at each end and also taking some 50 prisoners. The barricades facilitated the checking of a number of bombing counter-attacks. In the pouring rain the 2/Royal Berks and 1/Royal Irish Rifles of 25th Brigade laboriously made their way forward in a further attempt at Zenith Trench. It stood little chance of success, the waiting Germans had the advance covered, the two battalions advancing only 70 meters before being checked. The relief of the 4th Division, with casualties of over 4,000 during October, began the same night, Maj.-Gen. Pinney's 33rd Division taking over the front. The proposed attack on Le Transloy on the 24th October was postponed until the 26th, then the 31st and, with no improvement in ground conditions, the attack was postponed again until the 5th November.

On the 28th October, Rainy and Dewdrop were assaulted at 6.00am by the 1/Middlesex and 4/King's (98th Brig. 33rd Div). The Middlesex advanced through machine-gun fire, forcing an entry into Rainy, leaving a few "moppers up" to consolidate the trench, the Middlesex continued to Dewdrop, again forcing an entry. The bombers then worked down the trench pushing the Germans in front of them and by 9.30am, the southern part of Dewdrop was secure. Sixty-two men and two machine-guns were captured. On the left, the King's assaulted the northern section of Dewdrop, its bombers working north and securing the whole trench. Almost a hundred and fifty prisoners were taken. At 5.45am on the 29th, the 6/Scottish Rifles of 19th Brigade, passing over Dewdrop and north of the gun pits were checked by machine-gun fire from shell craters. Veering right, some of the Rifles entered the north-western section of Boritska Trench where they engaged the enemy with the bayonet but no permanent gain could be made.

The next attack was on the 1st November when the 9/HLI and 2/Worcester (100th Brig.) sustained casualties from a barrage while awaiting zero hour. They advanced at 3.30pm through the slime and mud towards Boritska. The sheer physical effort of getting forward brought the attackers to the point of exhaustion and, when coming under machine-gun fire from the cemetery at Le Transloy, the attack failed. The south-eastern part of Boritska was successfully assaulted by the French and, the following day, had secured entry into Tranchée de Conté, north-east of Boritska.

Two sculptures carved from the stumps of Giant Red Gum trees at the Lake Entrance, East Victoria. Australia to commemorate the action of Australian Soldiers and Nurses in the Great War. (D. Heaney)

Maj.-Gen. Roberton's 17th Division took over the front of the 8th at the end of October and engaged its 51st Brigade on the 2nd November on an attack on northern part of Zenith Trench. Parties of the 7/Border assaulted at 5.30pm, storming the trench with bomb and bayonet, killing a good number of Germans and sending some prisoners to the rear. After checking a counter-attack and with the salient now secure, a barricade was erected 150 meters up Eclipse Trench, the 7/Lincoln of the same brigade moving up to reinforce the Borders a little later. A strong counter-attack on the following day was repulsed by the Lincoln. There were still a few isolated pockets of German resistance in Zenith and these were eventually cleared by the Lincoln with the help of bombers of the 7/Green Howards of 50th Brigade.

In the afternoon of the 3rd, the 1/Queen's of 100th Brigade (33rd Div.) endeavoured to force an entry into Boritska but the attempt was unsuccessful. However, two days later, the 2/Worcester of the same brigade seized and secured Boritska and Mirage a few meters to the north-east. This was achieved by attacking from the French front and bombing up both trenches. The 16/KRRC, also of 100th Brigade stormed and secured Hazy by a frontal assault and linked up with the Worcester. Problems were encountered by the 2/RWF (19th Brig.) while trying to get forward on the Le Transloy road due to the failure of the right of the 17th Division in Eclipse Trench and the Fusiliers could make little progress. The 7/East Yorks and 7/Green Howards of 50th Brigade tried again at night to get forward from the secured part of Eclipse, but to no avail.

The Ist Anzac Corps, commanded by Lt.-Gen. Birdwood had replaced XV Corps in the centre of Rawlinson's Fourth Army and the 1st Australian Division relieved the 29th. Assembling in the rain, bombers of the 3rd NSW (1st Brig.) went forward at 12.30am on the 5th November in an attempt to enter the German front line north of Gueudecourt and then bomb north and south. The bombers working south were checked by machine-gun fire but towards the north, the NSW inflicted heavy loss on the Germans and finally linking with other troops which had advanced up the sunken road on the right. The 1/NSW (1st Brig) advanced from the line of the 2nd Brigade and was in immediate trouble from fire from Hilt Trench. A second assault and a subsequent bombing attack produced no better result. The troops complained they could not get forward through the mud and the creeping barrage was soon well ahead of them. The failed attack meant that the left flank of the 3rd NSW was in the air and the gains had to be abandoned. Meanwhile, the 2nd Brigade had assaulted Hilt from the south but could make no progress against the enemy, the men eventually retiring to the starting line south-west of Hilt Trench.

Apart from some minor actions, mainly by patrols and an irregular bombardment of the German positions, there was no further action in this sector in 1916.

(i) Official History, Vol. 2, page 350
(ii) A severe shortage of ammunition, particularly of 18 pounder, was the cause of an order to reduce the expenditure of ammunition to ensure sufficient supply for the preliminary barrage planned in a few days time.
(iii) The Victoria Cross was awarded to Private T. A. Jones of the 1/Cheshire for his bold action in silencing a German sniper, then accounting for two more who fired on him while showing a white flag, and then rounding up over a hundred prisoners from dug-outs. He survived the war and died in 1956.
(iv) Named collectively by the author for convenience as the "Meteorological Trenches". There being a large number of trenches in this group, the word "Trench" has occasionally been omitted to avoid repetition.
(v) During this action, Sgt. R. Downie of the 2/Royal Dublin Fusiliers attacked single-handed a machine-gun nest killing its detachment. He was awarded the Victoria Cross and, surviving the War, died in 1968 at the age of 74.

OBJECTIVE 33
COMBLES, BOULEAUX WOOD, LEUZE WOOD, WEDGE WOOD, FALFEMONT FARM, ANGLE WOOD

On the right of XIV Corps the 5th Division, commanded by Maj.-Gen. Stephens had taken up position at the end of August in the Maltzhorn Farm area in preparation for an assault on Falfemont Farm and Leuze Wood. A few days earlier the French First Corps had secured Angle Wood and the German 2nd position to the south-east of Falfemont Farm but fire from the farm halted further progress.

The British bombardment commenced on the 2nd September, the 13th Brigade taking up its positions facing Falfemont Farm and Leuze Wood. The 2/KOSB of 13th Brigade advanced between Savernake Wood, objective of the French, and Falfemont Farm. The French advance, in unison with the British, had been brought to a halt under enemy fire leaving the right flank of the Borderers wide open. Assailed by frontal and enfilade fire, nearly three hundred Scots Borderers lay dead or wounded. An enemy barrage then poured fire on to the support positions. At noon, a second attack was made by the 14th and 15/Royal Warwicks of the same brigade. The 15/Warwick on the right was caught in the same manner as the Borderers by enfilade fire and could make no progress. On the left, the 14/Warwick managed to enter a trench just south of Wedge Wood. Two machine-guns and some prisoners were taken.

Brig.-Gen. M.N. Turner's 15th Brigade moved up quickly to take up the challenge of Falfemont Farm while another barrage was laid on the German positions. At 6.30pm the 1/Cheshire with units of the 16/Royal Warwick in close support, attacked on the right wing. The barrage had not silenced the enemy strongpoints and the leading companies suffered heavy casualties. Better fortune awaited the 1/Bedford on the left, whose leading companies took possession of Wedge Wood . A link up was made on the left with 95th Brigade which was now in a position to make an assault on Leuze Wood to the east but permission was refused on the grounds that the unsecured Falfemont Farm could cause serious problems. However, at dusk, patrols of the 7/SLI (61st Brig.) were sent out towards Leuze Wood. These were not challenged.

Rawlinson ordered the attack to be continued on the following day, the 5th Division to secure Falfemont Farm and Leuze Wood including Valley Trench, near to the Wood. The 1/Norfolk (15th Brig.), now in touch with the French, advanced on Falfemont Farm but the French on the right were held up and did not move. The Norfolks were caught by fire from Combles Ravine, although a party on the left entered the farm only to be bombed out. The 1/Cheshire crept up under cover of the rising ground while the 1/Bedford approached from Wedge Wood bombing its way down the trench. Supported by the 16/Royal Warwick, the Norfolks made another abortive attack on the farm. The

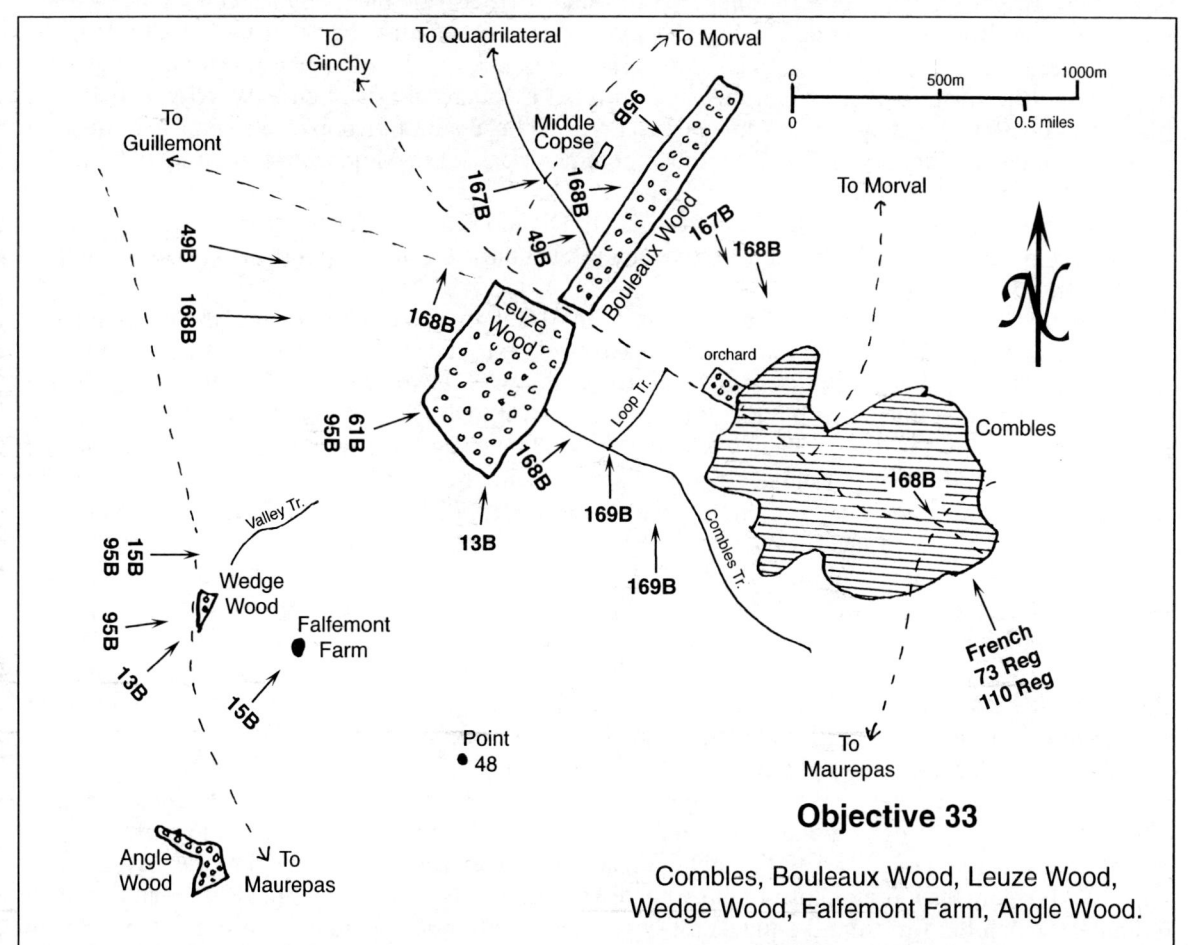

Objective 33

Combles, Bouleaux Wood, Leuze Wood, Wedge Wood, Falfemont Farm, Angle Wood.

Warwicks were then instructed to sap forward after dark while the Bedfords and Cheshires, sited on top of the ridge, opposed the approach of any German reinforcements. In heavy rain the Norfolks, using to full advantage the work of the Warwick sappers, successfully attacked and secured Falfemont Farm. Patrols were then sent south-east towards Point 48.

At the same time the 1/East Surrey (95th Brig.) had seized Valley Trench, having approached from the Ginchy road without opposition. Two companies of the 1/Devon, also of 95th Brigade, passed through the leading battalion and reached the edge of Leuze Wood at 7.30pm, linking up with patrols from the 7/SLI (61st Brig.) After a short delay while waiting for the barrage to clear the wood, the Devons entered without opposition, consolidating a line just inside while the Somersets established posts in a north-westerly direction to join the Ginchy/Maurepas road south of Guillemont. The northern section of Leuze Wood was not yet secured.

At 8.30am, two companies of the 16/Royal Warwick established a line along the ravine to link up with the 1/Devon on the edge of Leuze Wood. The Warwicks were then relieved by the 7/Royal Irish Fusiliers from 49th Brigade, the latter being detailed to attack Combles Trench, just west of the village of Combles. The Irishmen were soon in difficulty, coming upon uncut wire, completely hidden in the high corn, the German machine-guns ready and waiting. The attack failed as did a subsequent assault at 7.30pm. The Devons now set about clearing the remainder of Leuze Wood and a German trench running through the wood was quickly overrun.

The newly arrived 49th Brigade (16th Div.) established a continuous trench along the road leading from Guillemont to Leuze Wood and the 7/Royal Inniskilling Fusiliers linked up with the 5th Division at Leuze Wood. Another relief was effected on the 6th when the 168th Brigade of 56th Division replaced the 7/Royal Irish Fusiliers. The Kensingtons (13/London) formed up in Leuze Wood at night to attack down Combles Trench but their preparations were interrupted by a German barrage on the wood. During the night the 8/Royal Irish Fusiliers (49th Brig.) relieved the Devons on the front of the 95th Brigade and by early morning, the Fusiliers were astride the Combles/Ginchy road, some units entering Bouleaux Wood. During the evening the remaining battalions of 168th Brigade gradually relieved the front of the 5th Division. The London Scottish (14/London) were in the process of relieving the Irish Fusiliers when they had to repel a counter-attack. Leuze Wood was held, the London Scottish taking a number of prisoners. There had been a lot of troop movements over the past two days and at 1.35am on the 7th September, command passed to Major-General Hull, commander of the 56th (London) Division.

On the extreme left, patrols from the 16th (Irish) Division came under fire from the Quadrilateral, a thousand meters east of Ginchy and any movement over the crest of the spur provoked an immediate spate of fire from the north.

The 169th Brigade relieved the right flank of the 168th on the 8th September, the London Rifle Brigade taking over the southern part of Leuze Wood close to the French in Combles Ravine west of the village. The remaining battalions of 168th Brigade took over the road running west from Leuze Wood to Guillemont, thus relieving the 49th Brigade of 16th Division. At 11.30am the Rifle Brigade bombed its way down Combles Trench which runs south-east from the eastern edge of Leuze Wood. Some gain was made and consolidated before a strong counter-attack by German bombers forced the men to retire. At zero hour, 4.45pm on the 9th, the LRB advanced against Loop Trench which leaves the Combles/Ginchy road and heads south-west to the southern tip of Leuze Wood. The Londoners were checked by machine-gun and artillery fire. Reinforced by units from the 2/London and later by the 16/London, the attack was renewed the following day. At the same time, the Queen's Victoria Rifles (9/London) had entered and secured the southern section of Bouleaux Wood and extended north-west towards the Quadrilateral. The 168th Brigade, to the west of Leuze Wood, advanced almost due north, the Royal Fusiliers (4/London) on the right quickly reaching the Leuze Wood/Guillemont road. The London Rangers (12/London) encountered machine-gun fire on their approach to the Leuze Wood/Ginchy road, only the right company managing to reach the road. The Fusiliers reached their ultimate objective, the enemy trench running south-east from the Quadrilateral to the southern end of Bouleaux Wood. Although having lost heavily, a link-up with the 9/London enabled them to dig in. Patrols were sent towards Morval, the few Germans encountered being dispersed by Lewis-gun fire.

At 7.00am in the morning mist of the 10th, the Westminsters (16/London) made their attack down Combles Trench but fire from the untaken Loop Trench to the immediate north brought the advance to a halt. Following a Stokes mortar barrage and aided by the 2/London, the Westminsters tried again. After an initial gain of some 100 meters, they were obliged to withdraw under the enemy fire.

The next four days saw great preparations being made along a wide front ready for the attack on the 15th September and the 56th Division was assembled on the extreme right flank of the Fourth Army, forming the right wing of Cavan's XIV Corps. The immediate task of the Brig.-Gen. Coke's 169th Brigade was the capture of Loop Trench. Forward trenches had been dug during the night by the 2/London south of Leuze Wood and south of Combles Trench. A tank was to assist in the attack and the new-fangled machine lumbered forward at 6.00am followed twenty minutes later by the leading companies of the 2/London advancing close to the creeping barrage with the right heading for the junction of Combles and Loop Trenches. Any uncut wire had been destroyed by the tank and Combles Trench was secured with little loss. Moral was boosted by the sight of Germans recoiling from the advancing metal monster. Loop Trench branched off Combles Trench to the right and the Londoners, now supported by the two remaining companies, attacked north but enfilade machine-gun fire checked further progress. Assisted by bombers of the London Rifle Brigade (5/London), the 2/London moved down Combles Trench and north up Loop. Immobilised by a shell in the afternoon, the tank nevertheless held off the enemy with its Vickers machine-guns. Receiving a direct hit later in the afternoon, the machine set on fire and the crew successfully abandoned it. Steady progress was being made up Loop Trench but the *28th Reserve Regiment* defended it well and at the end of the day the Londoners were still 80 meters from the "T" junction with the sunken road at the northern end of the trench.

Brig.-Gen. Freeth's 167th Brigade was to clear Bouleaux Wood which would give a dominating position over Combles a few hundred meters to the south-east. Twenty minutes prior to the infantry attack, two tanks moved forward, one breaking a track but the second moved towards Middle Copse about half-way between the Quadrilateral and Bouleaux Wood, drawing fire as it went. The 1/London advanced at zero and were immediately held up by uncut wire. The right hand section of the German trench within the wood was fully manned and, untouched by the British barrage, enemy fire caused heavy losses in the ranks of the Londoners. The left flank of the 1/London managed to enter the trench and progressed north towards Middle Copse. The tank eventually ditched and was attacked by German bombers, no doubt very curious to have a good look at the revolutionary machine. Being as yet unaware of the outcome of the 1/London's attack, the 7/Middlesex was sent forward. The men were met by the same fire which had diverted the advance of the 1/London and, in spite of a second attempt, Bouleaux Wood remained the property of the Germans. The Middlesex and London men linked in the vicinity of Middle Copse.

Orders had been issued for a new bombardment of the German defences east of Ginchy and orders were sent to delay the planned afternoon attack by the 6th and 56th Divisions. Unfortunately the 8/Middlesex (167th Brig. 56th Div.) did not receive the order in time and made its attack at 1.40pm losing heavily in the process. The troops were then withdrawn to Leuze Wood. In the late afternoon, Middle Copse had been occupied by the 1/London and 7/Middlesex.

At 5.50am on the 18th September the 169th Brigade (56th Div.) on the extreme right flank of the Fourth Army with the French on its immediate right, launched an attack on the sunken road to the west of Combles. The 16/London, (Queen's Westminster Rifles) was halted by machine-gun fire during its advance towards the Loop. On the right, the London Rifle Brigade bombers gained some ground. The now atrocious ground conditions in this area prevented the 167th Brigade from making its attack to secure the remainder of Leuze Wood.

Further fighting was impossible and so the 56th Division began digging assembly trenches as close as possible to the western edge of Bouleaux Wood, still held by the Germans. There was further rain over the next few days and the next attack was planned for the 25th September. After an eighteen-hour barrage the 4/London and 14/London Scottish of the 168th Brigade to the immediate south of 95th Brig. of the 6th Division, began their advance. The 4/London cleared the enemy from forward positions in shell craters in front of Bouleaux Wood and then successfully secured the northern part of the wood. The London Scottish seized the first trench within the wood but then came under the same fire as the 1/East Surrey of 95th Brigade. The Surreys rushed the position and with the help of the London Scottish, crushed the last of the enemy resistance in the trench and dug-outs. Eighty prisoners and four machine-guns were taken. The advance then continued, the London Scottish securing a newly discovered trench from where good observation was enjoyed over the Combles-Morval valley.

Intelligence gained from a German officer indicated that the enemy was about to evacuate Combles and during the night of the 25/26th September, parties of the London Scottish (168th Brig.) advanced to within 500 meters of the village. Patrols through Bouleaux Wood by the 4/London of the same brigade and 1/London (167th) confirmed the withdrawal of all Germans from the wood. German flares were seen going up from the west of Combles and this transpired to be the signal for the withdrawal from that sector. The orchard was reached unopposed by the 1/London and other parties from the same battalion were in the village itself and had made contact with the French. The 5/London (LRB) of 169th Brigade was advancing south-east down Combles Trench towards the southern end of Combles. Linking up with the French at 4.15am, they handed over 200 Germans collected on the way. The French 110th and 73rd Regiments had also put patrols into Combles during the night. Some desultory fighting took place but resistance was soon quelled and many prisoners were taken. At 7.00am, the London Scottish had linked up with the French - a happy occasion for the Allies. Machine-guns were turned on the Germans as they fled to the north-east. It was necessary to search the village for booby-traps and hidden Germans in the cellars and this continued throughout the day. At the same time, along the Combles/Morval road the men dug in facing east. Frégicourt had now been secured by the French and the 168th Brigade (56th Div.) was ordered to keep in touch with the French along the former German 3rd Line, now renamed "Mutton Trench". But, at the moment, the trench was still in enemy hands. An attack by the 12/London (Rangers) supported by two tanks had to be postponed when both machines ditched.

Falfemont Farm, Combles, Leuze and Bouleaux Woods had been secured and the subsequent offensive towards Morval, Lesboeufs and Gueudecourt has been described in the grouped Objectives 30, 31 and 32.

OBJECTIVE 34
THIEPVAL VILLAGE & RIDGE, SCHWABEN REDOUBT, ST. PIERRE DIVION, STUFF REDOUBT, REGINA TRENCH WEST, GRANDCOURT,

In the evening of the 1st July, the 32nd Division was holding on to a lodgement in the Leipzig Salient, 1,200 meters south-south-west of Thiepval, while the 36th (Ulster) Division had been forced to abandon its entry into the Schwaben Redoubt. It being impossible to supply or reinforce the Ulstermen in the redoubt, the men were withdrawn to the German first line secured earlier in the day. Between hostile bombardments, both divisions were subjected to German bombing attacks which were successfully repelled. The casualties in the 36th Division were in excess of 5,100 and yet there were still some men holding on desperately to some small gains. The 36th Division sent up mixed units from the three brigades to supply and reinforce the men until they could be relieved. This was effected during the night of the 3/4th July by the 49th Division.

A proposed attack on Thiepval was reduced in scale, the original objectives of a line from the Leipzig Redoubt to the Wonderwork, 800 meters south of Thiepval, being scaled down. The change in plans caused some confusion and, in the end, two companies of the 15/HLI (14th Brig.) made two unsuccessful assaults on the German front line on the 3rd July. The 75th Brig. (attached from 25th Div.) fared no better. The 11/Cheshire and 2/South Lancs were cut down in front of unbroken wire, the 8/Border also having difficulty in getting within striking distance of the front line. No further supplies of bombs could be brought up and the men were instructed to return as best they could. The casualties for the abortive assault were almost 1,100 all ranks. The 32nd Division was withdrawn during the night of 3/4th July, the 75th Brigade remaining in the line and the 7th and 74th Brigades of 25th Division arriving later.

THE INFANTRY ATTACKS - OBJECTIVE 34

The 4th July brought a German barrage and bombing counter-attacks but these were successfully repelled. Two attempts at night by bombers of 49th Division to neutralise some machine-gun posts in front of St. Pierre Divion produced no satisfactory result. The next day, the 1/Wiltshire (7th Brig.) forced an entry into Hindenburg Trench, just north of the Leipzig Salient. Three days later the 49th Division, which was to have taken part in the assault on Ovillers, was itself subjected to a severe bombardment in the early hours of the morning followed by an assault by German bombers anxious to eject the British units from their lodgement north of Thiepval. Using a new type of grenade, the Germans took a high toll of the 4/KOYLI. Reinforced later by the 5/York & Lancaster, the latter retaliated with Mills bombs, but after a prolonged struggle, both battalions were obliged to withdraw to their original line. The enemy now flung itself from three sides at the Leipzig Salient but was thrown back by the 1/Wiltshire aided by an accurate British barrage. Not only was the counter-attack

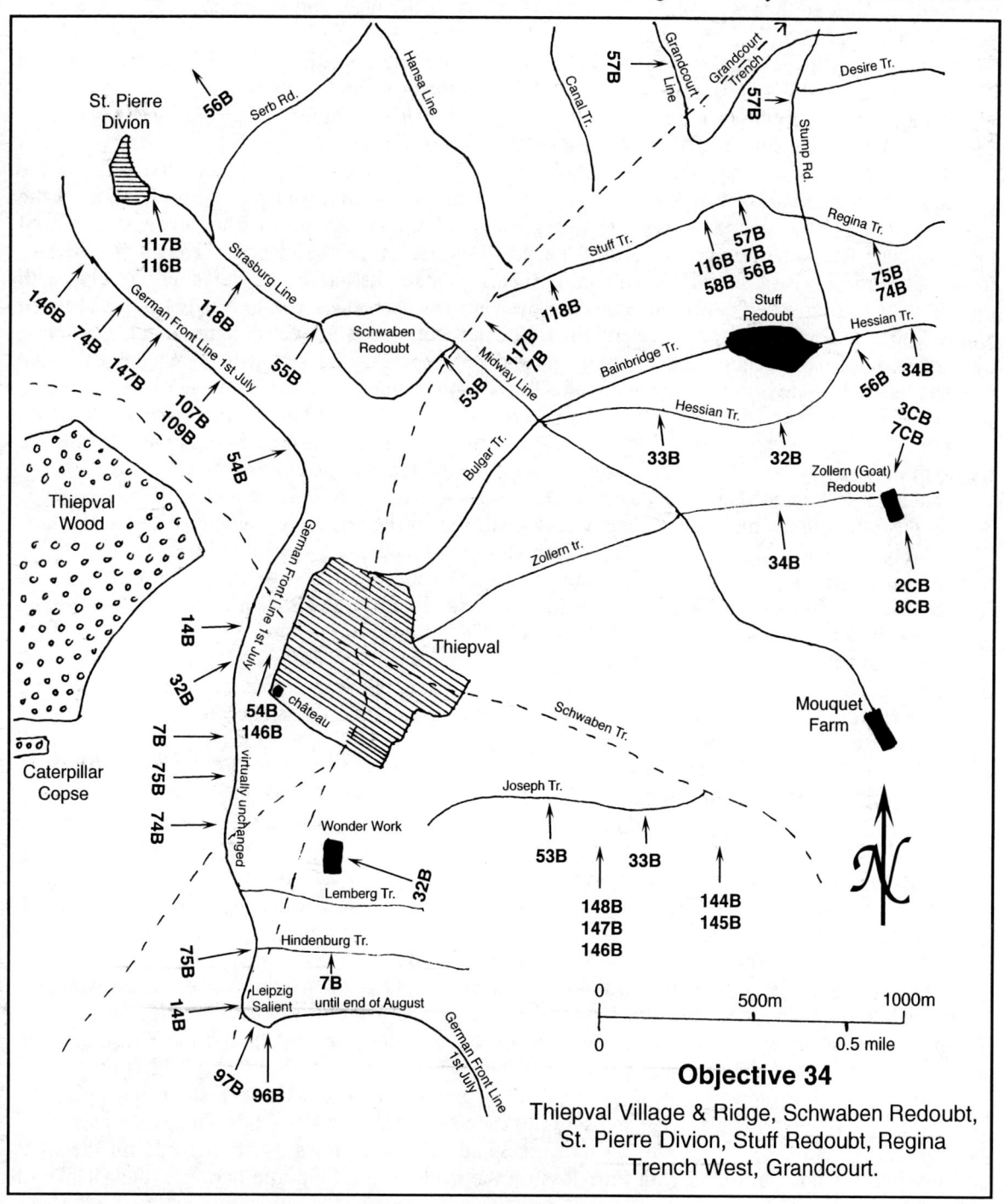

Objective 34

Thiepval Village & Ridge, Schwaben Redoubt, St. Pierre Divion, Stuff Redoubt, Regina Trench West, Grandcourt.

contained, but the hold in the salient was improved with the help of the 3/Worcester.

The main action on the Somme was to the south and east by Rawlinson's Fourth Army whereas in the Thiepval sector, small engagements designed to harass the enemy were taking place fairly regularly to improve the line or take the line forward. The Germans did not neglect Thiepval during the general advance of the Fourth Army on the 14th July. High explosive and gas were used periodically. On the 15th July, Perceval's 49th Division repulsed an attack on the Leipzig Salient using bombs and flame-throwers.

The British artillery on General Gough's front engaged in sporadic bombardments of the enemy positions to retain German forces in the Thiepval area, thereby preventing reinforcements being sent to Rawlinson's Fourth Army front or, indeed, elsewhere. The 49th Division had been engaged in essential work in straightening the line, improving roads, trenches and communications as well as bringing up supplies and giving assistance to other units in the area. On the 12th August the 8/West Yorks of 146th Brigade attacked the original German front line opposite the Nab. The objective was to destroy a barricade in the enemy trench and although the blockade was not destroyed, a lodgement was made in the enemy front line trench. The 49th Division had been in the Thiepval sector since the 1st July and was relieved on the 19th August by the 25th Division.

On the 21st August at 6.00pm the 144th Brigade (48th Div.), having replaced the 145th in the Mouquet Farm sector, attacked further to the west with the 4/Gloucester to extend the gain in the Leipzig Salient. Close behind the barrage, the Germans were taken by surprise and many surrendered to the Gloucesters, now in touch with its sister battalion, the 6th, in Hindenburg Trench. To the left, the 1/Wiltshire (7th Brig. 25th Div.) entered Lemberg Trench, just north of Hindenburg Trench, with the aid of a "pipe pusher" which demolished the German barricade. The two battalions had taken almost 200 prisoners during the day. Early the following morning, a German counter-attack was halted by Stokes mortars. A second enemy attack in the evening was checked with heavy losses in the German ranks. In spite of this, they mounted a further three counter-attacks in an endeavour to retake the lost position but the Gloucesters held firm, each attack being repulsed with bombs and Lewis-gun fire. On the 23rd, two companies of the 1/Buckingham (145th Brig.) sustained heavy casualties while trying to move up Nab Valley in an attempt to improve the forward positions.

Thiepval still blocked any advance north towards the Schwaben Redoubt and its associated ridge and on the 24th August the artillery laid a heavy barrage on the objectives, aided by trench mortars. The 3/Worcester and 1/Wiltshire (7th Brig. 25th Div.) advanced so close behind the barrage that they surprised the Germans and most of Hindenburg Trench was captured with bomb and bayonet with little resistance. However, units of the Wiltshire were held up on the left flank where no man's land was a little wider, the Germans aware of the attack. The left of the Wiltshire, forcing its way forward, entered the trench. Consolidation work under enemy fire was undertaken without delay. The German barrage intensified in the early evening of the 25th August and enemy troops could be seen forming up in preparation for a further counter-attack. A call was made for a British barrage which turned out to be most effective, disrupting for the time being, the counter-attack. The strongly-held western flank of Hindenburg Trench resisted the assault of the 8/LNL in the evening of the 26th.

Over half an inch of rain fell on the 25th and 26th August turning the chalky ground into a sea of white mud. In these trying conditions the 25th Division relieved the 48th which was to be transferred north of the Ancre. The next two days saw little action, the men being engaged in getting rid of the water in the trenches and bringing the wounded to the rear. The last attack during August was made on the 28th by the 8/South Lancs (75th Brig.) against the left flank of Hindenburg Trench. Forward movement of any kind was strenuous and physically draining and the Lancashires were halted by machine-gun fire as they approached their objective.

General Gough, in coordination with his divisional commanders had, in spite of the bad weather, been making preparations for the attack on the 3rd September. Before any assault could be made on Thiepval and the Schwaben Redoubt, it was necessary to secure St. Divion and the Strasburg Line which ran from the village to the Schwaben Redoubt. The 146th and 147th Brigades of Major-General Perceval's 49th Division were assembled on the Thiepval/Hamel road. The British artillery placed a heavy barrage on the Strasburg Line and south to the Schwaben Redoubt while Thiepval was shelled with ammonal and gas. At 5.13am on the 3rd September, Brereton's 147th Brigade on the right advanced across no man's land with little loss on the heels of the field-gun barrage, the 4/Duke of

Wellington (DWR) seizing a good part of the front line and defending the gain with bomb and rifle. Amid the smoke and morning mist, the 5/DWR lost direction, sustaining heavy casualties when coming upon uncut wire. The result was that the position called Pope's Nose(i) was not taken. Hostile fire came from the Strasburg Line, the Schwaben Redoubt and unsecured parts of the enemy front line. The 146th Brigade on the left flank made its attack between the Pope's Nose and the river. The right hand battalion, the 6/West Yorks was caught by enfilade fire from the Pope's Nose which prevented entry into the front line. Close to the river, its sister battalion, the 5th, was in trouble from the start, being pinned down by fire from the Strasburg Line. The Yorkshiremen could not get forward and they made their way back as best they could. The 4/DWR fought hard to retain its entry in the German trench but, out of bombs and with nearly all the officers down, had to withdraw. The two brigades had lost over 1,200 men. With poor visibility, runners had to be used to report the unsatisfactory outcome of the attack. Many of these men fell while trying to negotiate the German counter-barrage. The British artillery barrage had been well prepared and executed and when news reached H.Q. that the attack had failed, the staff expressed its disappointment with the result. Amongst the leading companies were many tired and untrained replacements who did not, according to reports of the action, display the same determination as the NCO's and seasoned soldiers.

Meanwhile, Bainbridge's 25th Division of II Corps advanced north from Hindenburg Trench towards the Wonder Work. On the right were the 2/South Lancs (75th Brig.) and 3/Worcester (attached from 7th Brig.) while 1/Wiltshire (7th Brig.) prepared its attack on the left flank. All three battalions made a determined advance against an equally determined defence by the 5th Foot Guard Reg. With mounting losses, they forced an entry into the German front line where their position became untenable against the heavy and accurate artillery and machine-gun fire. Unable to consolidate their entry, withdrawal was inevitable. Amongst the dead were Lt.-Col. W.B. Gibbs (3/Worcs.) and Lt.-Col. H.T. Cotton (2/South Lancs), Major S.S. Ogilvie (1/Wilts) being wounded.

Lt.-General Woollcombe's 11th Division, having been transferred to the Western Front after its participation in the Gallipoli Campaign and a short stay in Egypt, took over the right wing of the Jacob's II Corps on the 7th September. Prior to its arrival, the 49th Division on the left had been subjected to a two-hour bombardment with H.E. and gas, the latter affecting nearly a hundred men. Without doubt, the Germans were expecting another attack by the 49th Division and it was Jacob's intention to undertake minor attacks to keep the enemy guessing. Brig.-Gen. Price's 32nd Brigade (11th Div.) took up its assembly positions in Hindenburg Trench while the barrage prepared the way. At 6.30pm on the 14th September, half the 8/DWR and 9/West Yorks advanced north close to the barrage, seizing the front line with little opposition. Continuing forward, they captured the Wonder Work as well as Hohenzollern Trench to the west as far as the Thiepval road and east for some 250 meters. Just to the west of the Thiepval road, units from the 6/Green Howards of the same Brigade made progress towards Thiepval. The action had been short and decisive with few casualties but while the men were digging in, the Germans laid a heavy and accurate barrage on the known positions and during this period the casualties rose to over 700 all ranks.

While the Fourth Army to the east made its major attack on the 15th September, Gough's Reserve Army artillery laid a harassing fire on the German positions in II Corps sector. After some two months of minor but sometimes fierce engagements designed to keep the enemy guessing as to the intentions of Jacob's II Corps, plans were now in hand to make a major attack in the Thiepval area towards the end of September. While the plans were being finalised, and in order to keep up pressure on the German positions, two battalions of the 7th Canadian Brigade attacked on the 16th up Zollern Graben West towards Zollern Redoubt, sometimes referred to as Goat Redoubt. The artillery had caused little damage to the objective and the German machine-guns took a heavy toll before the Canadians were brought to a halt. A planned attack on the redoubt from the east by the 9th Brigade had to be postponed and the Brigade relieved the 7th which was now well under strength.

On the left wing of II Corps the 49th Division had also been harassing the enemy with counter-barrage work and raiding parties. The original German front line had changed little in the two and a half months since the attack on the 1st July but some progress was made towards Thiepval between the 16th and 22nd September, the 7/Duke of Wellington of 147th Brigade (Brig.-Gen. Lewis) taking a number of prisoners.

The plans for the major attack on Thiepval, Schwaben and Zollern Redoubts and Mouquet

Farm(ii) were now completed and zero hour was fixed for 12.35pm on the 26th September. The preliminary bombardment by eight hundred pieces began on the 23rd. Artillery observation was hindered by morning and evenings mists, almost inevitable after temperatures had soared to between 66 and 73 degrees Fahrenheit between the 23rd and 25th September. A most successful lacrymatory barrage was laid on the German trench mortar positions on the 24th. While waiting for zero hour, Maj.-Gen. Maxse's 18th Division relieved the 49th apart from the 146th Brigade which was attached to the 18th Division. The preparation of assembly trenches attracted occasional fire from the German artillery but caused no serious delay, the troops taking up their positions before dawn on the 26th.

On the extreme left of the Canadian front the 2nd Brigade attacked on the east side of the Pozières/Grandcourt road which represented the Corps boundary. Its path was directly under observation from Zollern and Stuff Redoubts which were to be assaulted by the 11th Division of II Corps. The 5th Bn, (Western Cavalry) and the 8th, (90th Rifles), each attached with a company of the 10th Bn, started their advance. The leading waves were ahead of the counter-barrage but the support companies were caught in the German barrage. Enfilade fire from Hessian Trench took a heavy toll in the ranks of the 8th Battalion. However, the Germans in Zollern Trench were taken by surprise, the objective being secured apart from the eastern extremity, about a hundred prisoners being captured and sent to the rear. With the help of the 10th Bn, the hold was extended a little on both flanks. Leaving parties to consolidate, the remainder continued the advance north towards the next objective, Hessian Trench. After a brief encounter, the trench was penetrated and the entry consolidated. Meanwhile, the 5th Battalion's right flank, linked with 3rd Brigade, got well forward but the left was raked by accurate fire from the German positions on the II Corps sector. Reinforcements from the 10th Bn. attempting to support the hold in Zollern Trench were caught in a murderous fire, the survivors erecting a barricade to prevent entry by the Germans from Zollern Redoubt, a few hundred meters to the west. Fire from Stuff Trench and Stuff Redoubt rendered the hold untenable at the western end of the gain in Hessian Trench and the 8th Bn. moved east along the trench where it continued to strengthen the defences. The pioneers of the 3rd Brigade were brought up from reserve to form a defensive flank south from Hessian Trench to Zollern Trench. They were assisted by the 7th Bn., the 1st British Columbia, which had also been brought up in the late evening. Lt.-Gen. Byng, commander of the Canadian Corps was reasonably pleased with the day's results. The right wing and centre had achieved their objectives and only on the left wing was there some disquiet. Gough reassured Byng that the 11th Division would restore the situation.

Turning now to the attack of the II Corps, the 34th Brigade of Lt.-Gen. Woollcombe's 11th Division formed the right wing with the Canadians on the immediate right. The Brigade's first task was to relieve the pressure on the Canadian left wing described above. Starting from Mouquet Farm where they had previously bombed the known cellar exits, the leading companies of the 8/NF and 9/Lancs Fusiliers advanced north taking their first objective south of Zollern Redoubt with few casualties. The 5/Dorset of the same Brigade was caught by the enemy barrage while moving up to support the first wave, all four company commanders becoming casualties. After a pre-arranged halt, the first waves continued but were met with accurate fire from Stuff and Zollern Redoubts and Hessian

Thiepval Château before the War. (J. Bommeleÿn)

Trench. In spite of alarming casualties the Northumberland Fusiliers, heads bent against the murderous fire, continued on and on, and were finally engaged in the most desperate struggle at the southern entry of Zollern Redoubt. They fought on until they were almost all killed or wounded. With only one officer and some fifty Fusiliers standing, they dug in. The Lancashire Fusiliers had also been severely mauled, the right company being almost wiped out by fire from Zollern Redoubt. Isolated parties from both Fusilier battalions had taken refuge in shell craters. The centre and left were heavily engaged in the Midway Line which joins Mouquet Farm to the Schwaben Redoubt where six enemy trench mortars caused horrendous casualties, only one officer and a few men reaching Zollern Trench. Although entry was made, both flanks were in the air, the Canadians having been pushed out. Parties of the Dorsets north of Mouquet Farm were urgently sent foward to reinforce the two Fusilier battalions. A combination of smoke from the barrage, the lack of movement by troops pinned down by enemy fire, made it difficult for observers to gain any accurate information as to how the battle was progressing. The reports tended to be somewhat over optimistic and orders were given to consolidate the Zollern Redoubt which was still brimming with German machine-guns. The Fusiliers, also uncertain of what was happening, cautiously moved out to reconnoitre the ground. An attempt was made to unite the scattered parties but this was hampered by machine-gun fire from Zollern Redoubt and Mouquet Farm. Many small groups in the shell craters were pinned down for hours. The men had assaulted with great determination and Brig.-Gen. Hill could correctly report that his 34th Brigade could have done no more.

Attacking from Nab Valley the task facing Brig.-Gen. Erskine's 33rd Brigade was a little less daunting, its front reducing from 1000 meters at the start to 500 meters at the final objective. The narrower front would allow the use of less troops with reinforcements readily on hand if needed. The 9/Sherwood Foresters and 6/Border led the attack, the Foresters almost too impetuously, being caught in the creeping barrage. Joseph Trench which runs east from south Thiepval was taken with little loss once the two German machine-guns had been silenced. The Borders stayed to mop up and consolidate the line while the Foresters continued north, crossing Schwaben Trench and disappearing out of sight over the crest. Stubborn enemy resistance in the dug-outs in Midway Line was finally overcome by the Forester bombers who, within half an hour had secured their section of Zollern Trench taking almost 200 prisoners and three machine guns. Reinforced an hour later by the last company, the Foresters continued towards the final objective, Hessian Trench. The trench was assaulted by all four companies and finally captured except for a length of about 250 meters on the right flank north of Zollern Redoubt from where machine-gun fire opposed its occupation. Some more prisoners were sent to the rear and the Foresters were now in touch with 18th Division's 53rd Brigade on the left. Moving up in support, the 7/South Stafford (33rd Brig.) was engaged in tidying up the communcation lines to the advanced front and was also engaged in mopping up isolated pockets in Midway Line. The Staffords reinforced the Foresters in the early evening and together, the battalions checked a counter-attack. The 33rd Brigade had certainly enjoyed a good day, having achieved over ninety per cent of its objectives. Most of the 600 casualties were sustained by the Foresters, but these were fortunately mostly wounded cases. Only three battalions had been used although the fourth, the 6/Lincoln had supplied a good number of men for carrying parties.

Brig.-Gen. Higginson's 53rd Brigade (18th Div.) was formed up on the left of the 33rd Brigade. This Brigade too was to start its advance from Nab Valley but included in its objectives were the eastern edge of Thiepval and on the opposite flank, the north-western section of the Midway Line which leads to the Schwaben Redoubt. At 12.35pm the 8/Suffolk and 10/Essex were out of the assembly trenches before the preliminary bombardment had lifted and dashed forward right on the heels of the creeping barrage. With great verve, the Suffolks swept over the German front line leaving the platoons of the following 8/Norfolk to mop up any resistance. Many Germans surrendered without arms. Within a few minutes Schwaben Trench which runs east from Thiepval was also overrun, the enemy resistance giving way under the fierce attack. On the left the Essex found numbers of dispirited Germans wishing to surrender on the St. Pierre Divion road. The *153rd Regiment* and *77th Reserve Regiment* had both sustained very heavy casualties and it is possible the appearance of the tank assigned to the attack may have put an end to the normally stubborn resistance of the defenders of Thiepval. The tank later ditched in Schwaben Trench. The Suffolks and Essex now pushed on to Zollern Trench which they occupied, again with little loss, the Suffolk linking up with the Foresters of 33rd Brigade. There were still many obstacles facing the successful advance of the leading battalions and German resistance towards the final objective would be up to its usual standard. After covering about 250 meters, the Suffolks came under accurate frontal and enfilade fire, forcing them to take cover. A similar fate awaited the Essex which came under fire from the north-west corner of Thiepval as well as from Martin's Lane and Bulgar Trench, the latter linking to Hessian Trench. An attack by bombers up the two communication trenches achieved only a small gain. Other units of the 8/Norfolk had now moved up and joined the "moppers up" in Schwaben Trench.

The well-fortified western edge of Thiepval, the original German front line and the Schwaben Redoubt faced Brig.-Gen. Shoubridge's 54th Brigade on the left flank of 18th Division. The 49th Division laid a creeping barrage to accompany the assault while two tanks were on hand in Caterpillar Copse. The leading companies of the 11/Royal Fusiliers and 12/Middlesex were well forward from their assembly trenches at zero and the former was quickly engaged in close combat with the enemy. The Germans had had almost three months to improve their defences and the Fusiliers found themselves in a bitter struggle using every available type of weapon - bombs, Lewis-guns, rifle, machine-guns and bayonet. Although the bombardment had damaged the original German front line trenches, there were many deep dug-outs, old and new, still intact and clearance was taking longer than planned, the creeping barrage already distancing itself from the leading waves. Another company of the Fusiliers was moved up to help in the "mopping up". Meanwhile, the right of the Middlesex was at first getting forward reasonably well through the centre of the village but the left was held up by fire from the ruins of Thiepval château. One of the tanks moved forward crushing everything in its path, but ditched shortly after while the second tank, ditching on the way forward, could take no part. The Middlesex now moved forward again but clearance of the dug-outs and pockets of resistance was slow and another company of Fusiliers moved up in support. The Germans had by this time laid a barrage on the British assembly trenches and the 6/Northampton sustained some casualties while preparing to move forward. The leading wave of the Fusiliers was now in the centre of the village, its numbers much reduced while the Middlesex struggled to get forward to the north-western corner of Thiepval. A number of machine-gun nests were still intact and snipers were taking a toll of both battalions. Help was needed to assure success and a defensive barrage was laid to the north of Thiepval, ranging east and west of the civil cemetery. The German retaliatory barrage took a heavy toll of the 6/Northampton while moving forward(iii) but the survivors were immediately sent forward to support the Royal Fusiliers and Middlesex. At dusk the mixed units were reorganised by Lt.-Col. Maxwell, the sole surviving battalion commander, the new line on the right which was just north of the village being consolidated while further efforts were made in the north-western corner to silence isolated pockets of resistance. The Germans holding out in the village church would not surrender and the struggle continued until the enemy was overwhelmed. With the Germans still in control of north-west Thiepval and considering the exhausted state of the Middlesex and Fusiliers, there was no possibility of any further forward movement towards the Schwaben Redoubt that evening. The day's fighting had revealed many acts of heroism by individuals and small groups, each dug-out, each stronghold, represented a separate target, each demanding imaginative and enterprising action and in this, the 54th Brigade was not lacking. The defenders of Thiepval were, for the most part, not inclined to surrender and the close-quarter fighting was decided solely by the death of one of the antagonists. Numerous courageous acts went unrecognised but two examples which were recorded are Privates Edwards and Ryder of the 12/Middlesex battalion, the former bombing a machine-gun which was holding up the advance and the latter clearing a German trench singled-handed with his Lewis-gun. Both men were awarded the Victoria Cross. The three battalions had total casualties of 840 which represented over a third of the total officers and men engaged. Like the Canadians on the right, Maxse's 18th Division had achieved most but not all of its objectives. North-west Thiepval and the German defences to the north were still intact with no sign of any weakening of the enemy resistance.

Orders had been issued for a continuation of the attack by the 11th Division at 10.am on the 27th and in the early morning, Brig-Gen. Hill sent officers' patrols to ascertain the situation on the front of his 34th Brigade. It was established that Zollern Redoubt had been evacuated which enabled the 8/NF to be withdrawn. West of the redoubt, Zollern Trench was occupied for about 700 meters by the 9/Lancs Fusiliers and parties of the Dorsets. East of the redoubt, the 11/Manchester was in touch with 7th Canadian Battalion.

At 10.00am the Manchester's advance on Stuff Redoubt and Hessian Trench was quickly checked by machine-gun fire and the 9/West Yorks and 6/Green Howards from Brig.-Gen. Price's 32nd Brigade were brought forward, initially to attack at once, but a postponement was ordered to coordinate the attack. Arriving first, the West Yorks, unaware of the postponement and unfamiliar with the terrain, surged forward with great determination across the cratered and littered ground and, against enemy fire, gained an entry into the southern section of Stuff Redoubt. An hour later, the Green Howards drove the Germans out of Hessian Trench to the left of Stuff Redoubt. Catching the occupants by surprise, they overwhelmed the enemy, taking eighty prisoners and two machine-guns. Valuable gains had now been made in Hessian Trench, the Green Howards now in touch with the 33rd Brigade on the left while its right flank linked up with the West Yorkshires. Bombing is one of the best ways of securing trenches and throughout the attack the West Yorkshiremen and Green Howards were supplied with bombs and supplies by the Manchester and Dorset, the former making some gain up the track leading north-east from Zollern Redoubt.

The Chapel of the Count de Bréda, Thiepval. (J. Bommeleÿn)

Meanwhile, on 33rd Brigade's front, the 6/Border began to reinforce the 9/Sherwood Foresters and 7/South Stafford in Zollern and Hessian Trenches. While the left flank was pushing east to join up with the 34th Brigade, the right was engaged in repelling a counter-attack. At 3.00pm the Border successfully attacked the last untaken part of Hessian Trench, taking over sixty prisoners and two machine-guns and linking up later with the Green Howards. The Borders now took over the front and consolidation of the line was undertaken under a hostile barrage.

Returning now to the attack by the 18th Division, the 53rd Brigade had progressed well with its consolidation work and in the morning of the 27th the 10/Essex bombed up Bulgar Trench, assisted by Stokes mortars. Although the enemy was not fully cleared from the trench, some 50 meters were secured. Next on the left, the front of 54th Brigade was taken over by the 7/Bedford which, supported by the 5/West Yorks (attached from 146th Brig. 49th Div.) had the unenviable task of securing the heavily fortified north-west corner of Thiepval. Advancing without artillery support in the hope of surprising the enemy, two companies of the Bedford assaulted with bomb and bayonet. While the left made good progress from the start, the right was held up by accurate fire from a machine-gun, the advance losing its impetus. The situation was reversed by the prompt and courageous action of 2/Lt. T.E. Adlam, an accomplished bomb-thrower, who led a small group across the open ground, silencing the machine-gun and receiving a bad leg wound in the process(iv). Seventy prisoners were sent to the rear and the advance continued past the church where the Bedfords dug in, now in touch with 53rd Brigade. The leading battalions from the previous day were now gradually withdrawn, the Royal Fusiliers, Northampton and Middlesex having a well-earned respite from the action. The 5/West Yorks then moved up to support the Bedford.

With Stuff Redoubt still under German control apart from a small gain in the extreme southern section, it was decided not to attack the remainder of the day's objectives including the final one, Schwaben Redoubt, until the next day. The 74th Brigade (25th Div.) relieved the 146th during the night. But meanwhile, the Green Howards and West Yorkshires in the southern tip of Stuff Redoubt, bombed their way up the eastern edge, securing most of the northern section of the redoubt. Their number greatly reduced and short of bombs, they lost ground against the inevitable counter-attacks.

At 1.00pm on the 28th September began the first major attack against the Schwaben Redoubt since the 1st July. Brig.-Gen. Higginson's 53rd Brigade having assembled under fire, facing north from Zollern Trench. The 8/Suffolk on the right launched its attack on part of Bulgar Trench and the Midway Line. Initially held up by fierce enemy resistance, the Suffolk fought its way forward reaching the eastern side of Schwaben Redoubt by 2.30pm, having taken a number of prisoners on the way. At the same time the 7/Queen's, attached from 55th Brigade, started its attack. Unfamiliar with the ground, the battalion veered to the left and engaged the enemy in Martin Trench. Heavy fire rained down on the Suffolk from the southern face of Schwaben Redoubt. Two companies of the 7/Bedford (54th Brig.) followed on the heels of the creeping barrage, advancing north. Arriving at the northern edge of the cemetery, the right company suffered very heavy casualties from the redoubt. Joined by the 5/West Yorks (attached from 146th Brig.) the Suffolks were engaged in a bitter struggle against strongpoints in the original German front line. Well-prepared for the attack, ensconced in their deep dug-outs and protected strongpoints, the Germans defended their positions with great tenacity. With equal courage, the attackers, in spite of heavy casualties, stuck to their task. The Queen's had now separated from the other battalions and had gained an entry into the southern face of Schwaben Redoubt, taking some fifty prisoners. The south-west corner was also secured, improving the situation of the Bedford and West Yorks. Now starting to advance up the western edge of the redoubt, the Queen's met with stiff opposition and to hold its entry, constructed a barricade and then turned its attention again to the southern face. The redoubt was a veritable warren of trenches and dug-outs, each presenting particular difficulties to the attacker. Bomb, rifle and bayonet were used in the clearance and by 5.00pm the Queen's had secured the whole of the southern face of the redoubt, linking up with the Suffolk on the right. Having cleared the western approach to the redoubt, the Bedford and West Yorks threw themselves at the western face, not only clearing it but securing part of a trench heading north-west from the redoubt. The 11/Lancs Fusiliers of 74th Brigade (25th Div.) had also moved forward and gained further entry into the German front line system and, bombing its way east, linked up with 54th Brigade. The light was failing rapidly, the men were exhausted and the remaining companies and 8/Norfolk, the "moppers up", moved up to help with consolidation of the day's gains. Periodical bombing raids continued through the night although, in the main, they were of short duration - everyone was ready for some rest. The 7/Royal West Kent (Queen's Own) began to take over the front of the 54th Brigade.

It has been mentioned that the 7th Canadian Bn. (1st British Columbia) had been brought forward on the right to help establish the defensive flank. This completed, Columbian patrols pushed west along Hessian Trench towards Stuff Redoubt without meeting any opposition. However, while waiting for the remainder of the company to come forward, the enemy reoccupied the western end of Hessian Trench but he was engaged and pushed back to Stuff Redoubt and Stuff Trench, a few hundred meters to the north. The Columbians' efforts were not yet at an end, having to check a further strong counter-attack and the trench was disputed for some time. Eventually, the Columbians erected a barricade at the corps boundary. This helped to repulse a number of counter-attacks. Under constant machine-gun fire from Stuff Redoubt, little more could be done for the time being. A link up was, however, made with 11th Division in Zollern Trench a little to the south. In the late evening the 2nd Brigade front was taken over by the 8th of Maj.-Gen. Lipsett's 3rd Canadian Division and arrangements were made for a combined attack on Hessian Trench with the 11th Division on the 29th September.

Almost half an of inch of rain fell during 29th, making conditions difficult for the combined attack but at noon the 2nd Canadian Mounted Rifles and units of the 32nd Brigade (11th Div.) covered by fire from two Stokes mortars, advanced north from Zollern Trench, forcing an entry into Hessian. The tenure was hotly disputed, first by a barrage, then by a counter-attack during which some ground was ceded to the enemy. A persistant attack by Canadian bombers recovered the lost ground and the Brigades sent over sixty prisoners to the rear. A fresh barrage was ordered and three companies of the 6/York & Lancaster made a frontal assault on Hessian Trench, securing nearly the whole length with the help of the Canadians. About 200 meters was still held by the enemy east of Stuff Redoubt. Meanwhile the West Yorkshire and Green Howards were still attempting to force the northern face of the same redoubt. After securing most of the edge, now out of bombs, they were again forced back.

No more reinforcements were available from the 32nd Brigade. Accordingly, the 7/South Stafford of 33rd Brigade was ordered forward.

On the front of the 18th Division, the 8/Suffolk occupying Midway Line was relieved by the 6/Royal Berkshire. It was obvious the Queen's Battalion was in no condition to continue the attack, almost all the officers and NCO's were casualties and the Battalion was relieved later by the 8/East Surrey of the same Brigade. In the early evening the 7/Royal West Kent took over the area between the original German front line and the western face of Schwaben Redoubt. The relief of troops under fire in this sector was fraught with difficulties. The Germans were determined to regain their losses - this had been the policy since the start of the campaign. Always preceded by a barrage, enemy bombers then attacked the British positions, resulting in a fierce bombing struggle, the most adept and courageous bombers usually winning the encounter. This close-quarter work with bomb and bayonet was one of the most dangerous aspects of infantry work, during which all opposing sensations were experienced - fear-excitement, indecision-confidence, despondency-elation, despair- hope, what is called today a surge of adrenaline. Gains by both sides were made and lost, and at the end of the day, the line had changed little. Losses were very heavy on both sides. In the evening Brig.-Gen. Jackson took over command of the 18th Division's front.

Ground conditions were still bad but at least there was no rain on the 30th. Hessian Trench was secure apart from a few isolated pockets of resistance. The final clearance was undertaken from three directions. Units of the 6/York & Lancaster (32 Brig.), advancing from the east, the 7/South Staffords (33rd Brig.) from north of Zollern Trench and Captain White's Yorkshiremen from the southern face of Stuff Redoubt. Against this three-dimensional attack, not even the renowned resistance of the hardened German soldier could resist. Fighting was fierce but gradually, the enemy was overwhelmed and the long-disputed trench at last secured. The Germans still held the northern sector of Stuff Redoubt. After dusk, Maj.-Gen. Bainbridge's 25th Division relieved the 32nd and 33rd Brigades.

Brig.-Gen. Jackson was soon busy organising affairs on the front of his 55th Brigade. His own 8/East Surrey was in trouble, having been driven from the southern face of Schwaben Redoubt during one of the lightening counter-attacks. The Germans had caught the Surrey while settling in after its arrival but the Battalion later regained entry at the point of the bayonet. On the western face, the 7/Royal West Kent was forced back and could not recover the position. In the afternoon, the remaining companies of the East Surrey, covered by artillery fire, assaulted the northern face, fighting their way through the trenches until the whole of the northern face was secured. The West Kents, supported by the 7/Buffs attempted to regain the entry into the western face but the Germans could not be dislodged. The Surrey men, still out of breath after their tremendous advance, were now the recipients of the inevitable counter-attack. Attacked from the west, the weary men gave their all to retain their hold but against superior numbers, short of bombs, and at the point of exhaustion, they pulled out of the redoubt into Stuff Trench. Here, they dug in and held on.

The newly-arrived 25th Division was engaged in the arduous work of consolidation. Ground conditions prevented any immediate attack on the untaken northern part of Stuff Redoubt. Meanwhile, the 18th Division continued its efforts to secure the whole of Schwaben Redoubt, the 55th Brigade making a further assault on the 1st October. Before dawn the following day, the enemy made a desperate attempt to recapture the redoubt which commanded advantageous observation over a wide area. The struggle, mainly between bombers, continued throughout the day. After a pause on the 3rd, the close-quarter fighting continued on the 4th without a positive result. At 10.00am on the 5th, bombers of the 8/Norfolk (53rd Brig.) attacked the unsecured part of the redoubt from both flanks. Movement was so slow in the mud that the Norfolks could not get to close grips with the enemy and the attack failed.

Maj.-Gen. Cuthbert's 39th Division relieved the 18th. The 16th and 17/Sherwood Foresters of 117th Brigade were quickly engaged in repulsing a number of counter-attacks using *Flammenwerfer* on the 7th and 8th October. Before dawn on the 9th, the 16/Sherwood Foresters, unable to use the mud-filled trenches, attacked the northern edge of Schwaben Redoubt just over the crest of the ridge over open ground, hoping to surprise the enemy. The right of the Foresters gained entry but unfortunately, the Germans were already standing to arms and the Foresters were dislodged. Eight hours later the same sector was successfully stormed by the 10/Cheshire of the 7th Brigade (25th Div.), many Germans being killed or taken prisoner. Advanced posts were established without delay and two separate counter-attacks were successfully checked. The potential loss of the Schwaben Redoubt was of great significance to the enemy and a number of determined counter-attacks were launched between the 11th and 14th October. On the 12th, the 8/Loyal North Lancs drove back the enemy several times. During the afternoon of the 14th, the same Battalion advanced to the north-west, securing a strong point called "The Mounds" which enjoyed a good view of Grandcourt. Over a hundred prisoners were taken. Meanwhile, the 117th and 118th Brigades of 39th Division assaulted the last remnants of German resistance in the redoubt. The 5/Black Watch and 1/Cambridge (118th Brig.) along with the 17/KRRC, advancing over open ground, stormed the enemy defences but he was not ready to cede ownership without a hard fight and the struggle continued for eight hours. By then, both sides were exhausted but the German resistance was now at breaking point and a hundred and fifty soliders of the *II.110th Reserve Regiment* surrendered. The Schwaben Redoubt was first attacked by the valiant Ulstermen of the 36th Division on the 1st July and now, three and half months later, for the first time, the whole of the complex system of deep trenches and inter-connected dug-outs was secured. It was not necessary to be a good tactician for the exuberant troops to appreciate the advantages of the redoubt which dominated much of the Somme battleground. Since the first of July, every single attack in this sector had been made uphill and, for the most part, facing the rising sun. Little wonder the events of the 14th October created a feeling of general contentment. While consolidation continued, the 6/Cheshire of 118th Brigade extended on the left of the redoubt. No-one was surprised the next day when the Germans mounted three counter-attacks but all were successfully repulsed in spite of the use of *Flammenwerfer.*

Rain fell at some time during the next few days and the plans to continue the attack had to be postponed. There was still hard work to accomplish and on the 21st October, a fine but cold day, the II Corps launched its attack, having the support of over two hundred artillery pieces of all types. While waiting for zero hour, set for 12.06pm, the Germans made a counter-attack on the northern face of Schwaben Redoubt gaining entry at two points. Their hold was short-lived. Attacked in their turn by the 17/KRRC (117th Brig. 39th Div.) and part of the 14/Hampshire, the intruders were expelled.

The 74th Brigade on the right of Bainbridge's 25th Division attacked in line with the 11/Lancs Fusiliers, 9/Loyal North Lancs and 13/Cheshire, surprising the enemy and, brushing aside the slight opposition, secured all its objectives. On the left the 75th Brigade attacked west of the Pozières/Miraumont road with the 8/Border, assisted by a company of the 11/Cheshire, plus the 8th and 2/South Lancs. The men pressed forward eagerly, the Borders too close to the creeping barrage causing some casualties. The artillery had prepared the way forward to a high degree of efficiency and the men advanced rapidly up Stump Road, the bombers clearing enemy dug-outs on the way. The advance was halted as they caught up with the barrage.

Next on the left, the 116th Brigade of Cuthbert's 39th Division faced Stuff Trench. On the right the 13/Royal Sussex got forward after being delayed in a bombing encounter while on the left the 11/Royal Sussex and 14/Hampshire stormed and secured Stuff Trench. The 16/Rifle Brigade plus one company from the 16th and 17/Sherwood Foresters (117th Brig.) made an attack on the area near the Pope's Nose(i) in the vicinity of the original German front line and some ground was gained. Further ground was gained on the 29th October by the 39th Division in the same area.

Almost all Thiepval ridge was secured as well as Stuff Trench in less than thirty minutes. A hostile barrage did not stop consolidation from proceeding. Germans were still being rounded up and by 5.00pm on the 22nd 16 officers and 1,041 other ranks had been taken. The 25th Division was relieved by the 19th on the 22nd and the 7/East Lancs (56th Brig.) was quickly in action halting a counter-attack on Stuff Redoubt. The Lancashiremen took a heavy toll of the attackers as well as taking over forty prisoners.

Then it rained, making further action impossible and the next attack, after a number of postponements, was fixed for 5.45am on the 13th November following a period of four days with virtually no rain. Changes had been made in the delivery of the artillery barrage with a repetitive pattern for several days before the 13th and which remained unchanged on the day of the attack.

At zero hour Brig.-Gen. Rowley's 56th Brigade (19th Div) attacked with the 7/Loyal North Lancs and 7/East Lancs from Stuff Trench. The East Lancs Battalion was to protect the flank of the 39th Division on the left flank of II Corps. A pre-dawn mist reduced visibility to about thirty meters allowing the men to get well forward before zero. Eight machine-guns had been set up in no man's land to cover the advance. The attackers caught the Germans by surprise and they were quickly overwhelmed. The North Lancs overran its objective but was brought back into line a little later. The East Lancs quickly secured its objective on the right, the left doing so a little later after being disorientated in the mist. Dug-outs were cleared in Lucky Way and some prisoners sent back. The advance progressed well, the 18 pounders still firing a hundred meters in front of the main barrage. The 5/SWB pioneer battalion was sent forward to help with the consolidation under cover of the mist. The work became more dangerous as the German machine-gunners obtained better vision in the dissipating mist. A company of the 7/King's Own of the same brigade moved up to support the North Lancs and advanced posts were constructed. On the right flank of the 19th Division, the 6/Wiltshire (58th Brig.) was unable to force an entry into Stump Road which ran north to Grandcourt and formed the divisional boundary. The 19th Division's attack had gone well apart from the refusal at Stump Road. Casualties were low and over 150 prisoners were taken. It would seem that the combination of artillery and infantry supported by machine-guns must have demoralised the Germans to some extent. The retaliatory barrage was not up to its usual standard and there was no counter-attack on the 19th Division front.

Maj.-Gen. Cuthbert's 39th Division on the left of the 19th formed the left flank of II Corps and Brig.-Gen. Finch-Hatton's 118th Brigade had taken up positions just north of the Schwaben Redoubt. On the right the 1/Hertford, close behind the barrage, advanced towards its objective, the Hansa Line. Resistance was minimal, the Hertford taking 150 prisoners and 4 machine-guns. Securing the new line, parties then turned left into Mill Lane meeting up with the right company of the 1/Cambridge. The left of the Cambridge had lost its way in the mist and, as it cleared, came under fire from the Strasburg Line sustaining some casualties but, with the Hertford, cleared the whole of Mill Trench. This allowed the Cambridge to advance up the track from Mill Trench and seize the ruins of Beaucourt mill and the station crossing. At the mill, a link was made with the Hood Battalion and the HAC of the 63rd Royal Navy Division which formed the right flank of Gough's Fifth Army(v). The objectives of the 6/Cheshire were the Strasburg Line and the village of St. Pierre Divion. The Cheshire could not keep up with the barrage, having great difficulty negotiating the maze of trenches. Losing direction, the Battalion was engaged in spasmodic fighting and as the mist cleared, came under machine-gun and sniping fire. Although gains were made, neither objective was secured. The 5/Black Watch had the task of clearing the original front line and support trenches as far as the Ancre but encountered many difficulties including an alarming number of officer casualties. The bad visibility impaired correct assessment of the attack by Forward Observation Officers (FOO).

The attack on the left flank was led by the 16/Sherwood Foresters of Brig.-Gen. Oldman's 117th Brigade. Advancing up Mill Lane at 6.15am, the Foresters surprised the Germans. One of the companies cleared the strongpoints on top of the bank while bombers from two other companies set about clearing the dug-outs with interconnecting galleries in the twelve-foot-high river bank. The fourth company was called forward and helped with the clearance and consolidation. Some determined Germans would not come out of the bunkers and if a bomb did not bring them out, a gas grenade was used. The advance continued north towards St. Pierre Divion, the Foresters now joined by units from the Black Watch and Cheshire. This fortified small village which had caused so much damage on the 1st July was, on the 13th November, captured with little difficulty by the mixed force, one of the trophies being the capture of a battalion staff found in the cellars.

The 14/Hampshire (116th Brig.) was sent forward to assist in the clearance of the Strasburg Line. Orders came from Corps headquarters that all gains must be consolidated and strengthened sufficiently to withstand the expected counter-attack. The 118th Brigade undertook most of this work but was quickly harassed by a hostile barrage. Royal Engineers of 227th Field Coy and the divisional pioneers, the 13/Gloucester set to work repairing the road from St. Pierre Divion to Hamel. In the early evening a group of Germans with machine-guns came unexpectedly on the Hertford line but after a brief struggle, the intruders were captured or killed. A subsequent enemy bombing attack was also repulsed.

Two of the three tanks available for the attack were soon out of action, one becoming stuck in the mud and the other having mechanical trouble. The third reached the German front line where it had a series of adventures. Firstly, its fire beat off some Germans trying to approach with bombs, then crashed through a dug-out while driving to the support line and wound up on its side. The machine was attacked by a group of Germans and a message by pigeon was sent for help but the Black Watch arrived on the scene, driving the Germans away.

The casualties in the attack of the II Corps were under a thousand and most of these being wounded while German prisoners totalled 27 officers and 1,300 other ranks. Accurate casualty figures for the Germans are not known, but their loss in dead and wounded was very substantial.

The only activity on this front on the 14th November was a dawn raid in the 19th Division sector by the 9/Welsh (58th Brig.) and 7/South Lancs (56th Brig.), the objectives being the German strongpoints between Stump Road and Lucky Way.to the south of Grandcourt. The South Lancashire battalion was checked by uncut wire but the Welsh succeeded in bombing some of the dug-outs before being forced back. During the subsequent pause in hostilities the 19th Division began the relief of the 39th.

The immediate objective of the 19th Division was to secure the Grandcourt Line which runs south from the western end of the village. The first winter snow had fallen during the night of the 17/18th November and at 6.10am, the snow having turned to driving sleet, the 57th Brigade, led by two leading companies of the 8/North Stafford advanced, heads bowed and disappeared from sight. There was no news from the Stafford for some time but the Battalion had, in fact, entered the enemy trenches west of Stump Road where, counter-attacked in force, many were taken prisoner including its commanding officer, Lt.-Col. C.L. Anderson. Later in the day small parties managed to withdraw and some seventy men filtered back down Battery Valley. On the left of the Stafford, the 10/Royal Warwick was at first disorientated in the sleet but this corrected, continued the advance. The right came across uncut wire and took heavy casualties while clearing a way through. The assault progressed better on the left when bombers forced an entry into the Grandcourt Line, linking up with the 8/Gloucester. The Gloucester had made a rapid advance, in spite of the sleet and fire, and had stormed the German trenches over a wide front, entering the south-western corner of Grandcourt. The remaining companies of the Gloucester and Warwick moved forward to complete the clearance of occupied shell holes, dug-outs, gun positions and trenches. They completed the dangerous but exhilarating "mopping up" process in the front and support lines but sustained heavy casualties as they moved north, looking for the main German Line.

Two companies of the 7/South Lancs of Rowley's 56th Brigade moved up to the St. Pierre Divion/Grandcourt road to effect an entry from the west. One platoon was detached to make a link via

The village of Thiepval had been totally destroyed. This photograph conveys the utter desolation of a former peaceful village which the people of Albert and neighbouring villages came to visit on a Sunday to admire the beautiful view from the high ridge. (Michelin)

the railway with the V Corps on the eastern extremity of Beaucourt. The main force made its way over open ground which was reasonably firm, capturing some Germans at the western entry into Grandcourt. Bombers pushing towards the centre of the village linked up with the Gloucesters of 57th Brigade. The left, having crossed the Ancre, followed the railway towards the northern edge of Grandcourt where strongpoints were established. At the same time the 7/East Lancs of the same Brigade was pushing forward in its attack to secure Baillescourt Farm. One company was in touch with the South Lancs on the western edge of Grandcourt, but both units were held up by machine-gun fire from unsecured parts of the village(vi). The other company advanced along the railway embankment where it halted under friendly fire.

In an attempt to improve the open left of the 18th Division to the east, the 9/Cheshire, attached to 57th Brigade from the 58th, attacked Desire Trench west of Stump Road. Rapidly losing direction and unfamiliar with the terrain, the Cheshire could not find the trench. At the end of the day the 56th Brigade took over the line taken on the 18th November and completed the consolidation under a hostile barrage.

After a cold night the 7/South Lancs repulsed an enemy counter-attack at the western corner of Grandcourt. Maj.-Gen. Bridges was concerned that the lines gained by his Division could not be held and ordered a line to be constructed up Battery Valley, west of the Grandcourt Line. The divisional pioneers, the 5/SWB, assisted by the 24/Manchester, pioneer battalion of the freshly arrived 7th Division, undertook the work On completion, the East and South Lancs, Gloucester and Warwick withdrew to the new trench.

Although certain parts of the front in this sector drained well, others remained a sea of mud. No further offensive was possible in this sector in 1916.

(i)   The Pope's Nose was situated just a few meters from the Ulster Tower, a memorial to the 36th (Ulster) Division, situated on the Thiepval/Hamel road.
(ii) The Australians were the first to attack Mouquet Farm after taking Pozières and the farm has, therefore, been included in Objective 22
(iii) Colonel G.E. Ripley was mortally wounded at this time.
(iv) A 2/Lieut. in the 7/Bedford, he was subsequently awarded the Victoria Cross. Surving the war, he died in 1975 aged 81.
(v)   The Reserve Army was renamed the Fifth Army, General Gough remaining in command.
(vi) Lt.-Col. T.G.J. Torrie was mortally wounded.

OBJECTIVE 35
BATTLE OF THE ANCRE
BEAUMONT HAMEL, FRANKFURT & MUNICH TRENCHES, BEAUCOURT-SUR-ANCRE, REDAN RIDGE EAST & NORTH TO SERRE

It had been reasonably quiet north of the Ancre since the failed attacks on the 1st July giving both sides the opportunity to reorganise. The German defences had been put to a severe test and had withstood the initial assault but improvements were made during the lull in the fighting. Enemy bombardments including gas, though usually not prolonged, harassed the bringing up of supplies to the dumps, the movement of troops, construction of railways, etc. and on the 18th July, the Germans laid a particularly heavy barrage on Englebelmer, some four kilometers west of Hamel.

There was no major offensive in this sector during the remainder of July or during the month of August but infantry patrols and raids were carried out fairly regularly. General improvements were made to the roads, trenches and communication system. The weather deteriorated at the end of August making life uncomfortable for those involved in trenching and supply work. Meanwhile, the British artillery had sporadically targeted the enemy wire and front line positions and had laid particularly heavy barrages to coincide with major attacks to the south and east by Rawlinson's Fourth Army in order to confuse the enemy as to the direction of attack. Major-General Fanshawe's 48th Division arrived north of the Ancre at the end of August.

Objective 35
The Battle of the Ancre

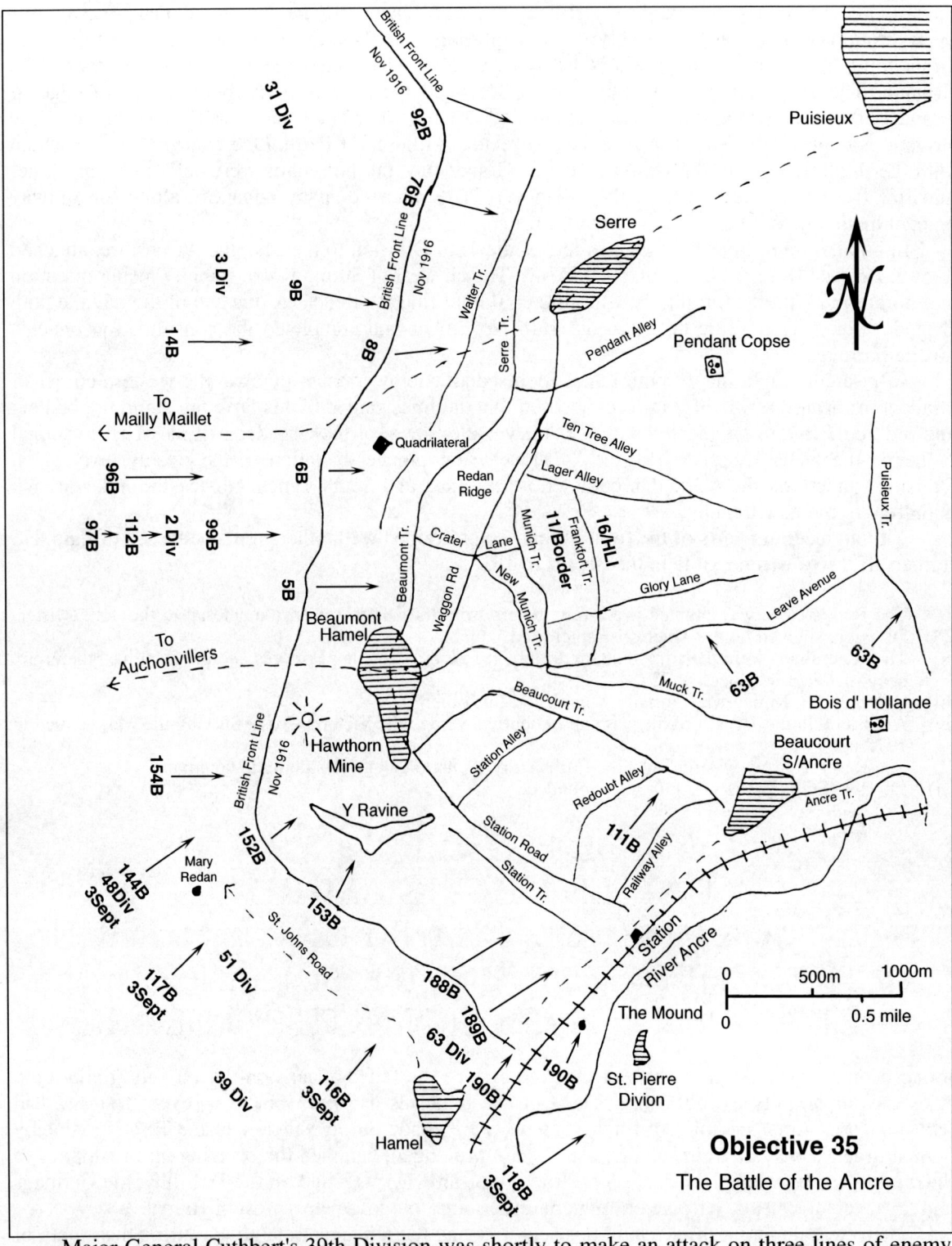

Major-General Cuthbert's 39th Division was shortly to make an attack on three lines of enemy trenches situated on the spur north of Hamel. On the left Brig.-Gen. Oldman's 117th Brigade was assembled just to the south-east of Mary Redan while Hornby's 116th Brigade covered a frontage east from north of Beaumont Hamel as far as the railway. The Division would then work its way up the Ancre valley to protect the left flank of 49th Division during its attack on St. Pierre Divion. At 5.10am on the 3rd September the 11/Royal Sussex (116th Brig.) secured the German front line with little loss and continued its advance past the support line, the Brigade's Lewis-guns taking a heavy toll of the

German's standing on the parapet to fire - no doubt they were thinking this would be a repetition of the 1st July when it was just a question of how many British troops they could kill in the least possible time. The 14/Hampshire of the same Brigade sustained initial heavy casualties and although securing parts of the enemy front line, could make no further progress. The extreme left battalion of 117th Brigade, the 17/Sherwood Foresters, in spite of leaving from advanced assembly trenches, was checked in front of German machine-gun fire. To the right, the 16/Rifle Brigade lost its way in the morning mist, only a few men managing to enter the front line. Moving up to support the left flank of the 49th Division on the other side of the Ancre, the 5/Black Watch (118th Brig.) could do little in view of the failed attack of 146th Brigade. The Scots were, however, able to support the right flank of the 11/Royal Sussex of 116th Brigade. A lot of work had been put into the improvement of communications north of the Ancre but the newly-laid cables were soon destroyed and the only way of getting messages to the rear was by using runners. With no man's land being swept continuously with machine-gun and shell-fire, many runners were lost. The same barrage prevented supplying the men trying to hold on to their lodgements and when the last of the bombs had been used, there was little hope. Fire from St. Divion added to the increasing problems and the men began to withdraw, some Sussex and Black Watch hanging on until nightfall. On being informed of the outcome of the 49th Division's attack, Cuthbert considered it unwise to commit his 39th Division further. His Division had lost 1,850 all ranks including Capt. F. Skinner commanding the 14/Hampshire.

The policy of small raids and artillery bombardment, particularly to cut the enemy wire was continued while waiting to renew the attack in the Beaumont Hamel sector. General Haig was anxious to renew operations north of the Ancre as soon as Thiepval and its ridge were secured.

At the beginning of October, tunnellers, mainly of the 252nd Tunnelling Coy. R.E. in the Beaumont Hamel area were busy mining the crater on Hawthorn Ridge - it would be the second explosion on the ridge. At the same time the line west of the Beaumont Hamel/Serre road was being prepared for the coming attack. Russian saps were reconditioned and tunnellers were engaged in smaller mining operations along this front.

The preparations continued for some time and on the 28th October a very effective gas attack on Beaumont Hamel was made using 4-inch Stokes mortars to fire 1,126 lacrymatory bombs. Machine-guns and *Minenwerfer* (i) were silenced in two minutes. During the night Stokes mortars laid a barrage on the village of Beaumont Hamel, Y Ravine and into the valley before Redan Ridge. The frequency of raids and patrols was stepped up, the men finding these operations tiresome in the muddy and often foggy conditions. Some raids were quite successful, especially to the north of the sector where the Germans seemed less vigilant than at the garrison of Beaumont Hamel. Most patrols were executed at night and on the 31st October one such patrol entered the German front line south-east of Hébuterne finding parts which were not occupied. Similarly, patrols broke into the enemy line at Serre and entered the support line. A week later, about thirty Germans were killed in a further raid near Hébuterne.

With winter only a few weeks away and the front deep in mud, the question was whether to continue the offensive. Attacks had been planned only to be repeatedly postponed. This was bad for the moral of the troops. The senior commanders all wished to start a major attack, Beaumont Hamel, Redan Ridge and Serre had not seen any major offensive since the first of July. The artillery had kept up a harassing barrage throughout the summer keeping a number of German regiments tied up north of the Ancre, thus preventing the transfer of troops to other fronts. It had also recently been preparing the way for the next proposed attack and finally the decision was taken to make an attack on the 13th November, zero being at 5.45am. Changes in the pattern of artillery fire would, it was hoped, effect a tactical surprise. At 5.00am on the 11th November, Beaumont Hamel was again targeted with lacrymatory bombs and in the afternoon gas drums were launched into the village and Y Ravine to the south. Some tanks were available but the general feeling was that they would be of little use in the soft ground. The troops which had so often laboriously taken up positions in the assembly trenches only to hear the attack had been cancelled, were now taking up their assembly positions again, this time, in spite of the conditions, somewhat more optimistically. There had been no rain to speak of for the last four days and the 13th November was to remain dry with a maximum temperature of 54 degrees Fahrenheit. The seconds ticked away to zero hour, the pre-dawn mist limiting visibility to about 30 meters.

Maj.-Gen. Shute's 63rd Royal Navy Division(ii) formed the right of the V Corps and the 189th Brigade, commanded by Brig.-Gen. Philips, was assembled on the west side of the Ancre, more or less on a level with St. Pierre Divion near the opposite bank. Brig.-Gen. Prentice's 188th Brigade was next on the left. Both brigades had two leading battalions in line, two in support and two attached from Brig.-Gen. Heneker's 190th Brigade. This was to be the Division's first engagement on the Somme and officers and men had been well briefed. While waiting for zero, hostile fire caused some casualties, Lt.-Col. Saunders being killed while moving up to the assembly position. At zero the Hood Battalion, closely followed by the Drake, were quickly away on the right of 189th Brigade. Caught by enfilade fire from the left, both battalions suffered considerable losses but the advance did not lose its impetus and the German front line system was stormed and 300 prisoners sent back. There was some confusion in the morning fog as the battalions continued their advance having to negotiate waterlogged trenches and shell holes, but, being a naval brigade, the men were able to make good use of the compass. Rallied by Lt.-Col. Freyberg, the Hood and Drake continued their advance seizing first Station Trench, then on to Station Road and finally to Beaucourt Station at the eastern end of Station Road. Almost four hundred prisoners were captured in this operation. A company of the Honourable Artillery Company (HAC) of 190th Brigade was detailed to clear any isolated pockets of resistance, including "The Mound" to the immediate east of the railway line. The HAC cleared the numerous dugouts in the railway embankment. It involved a protracted struggle using bombs, machine-guns and trench mortars. Supplies found here were used for consolidation of the gains of the right flank. While the Hood and Drake had achieved their objectives, the Hawke and Nelson, the left battalions of 189th Brigade, were met by heavy machine-gun fire as they advanced towards the German front line. Officer casualties were heavy, the men splitting into small groups. The enemy trench was at last entered where a bitter struggle with bomb and rifle continued for some time. The Howe Battalion and 1/Royal Marines led the attack of 188th Brigade on the left, the Anson and 2/Royal Marines following in close support. The four battalions were held up by accurate machine-gun fire, the attack losing its cohesion. Mixed units from

Right: The ruins of Beaucourt-Hamel Station. [A Perret]

Below: Beaucourt Hamel Station on the D50 Albert/Miraumont road was secured by the 63rd Division (author)

the Howe, Anson and Nelson, led by the Anson commander Lt.-Com. Gilliland, fought their way over the crest to reach the next objective. On the extreme left the 1/Royal Marines lost its four company commanders before reaching the German front line, NCO's taking over the command of the many isolated groups. A similar fate awaited the 2/Royal Marines when heavy casualties brought the attack to a halt. Just a few Marines managed to get forward and make contact with the 51st Division on the left. It was established later that the Germans had placed low-sited machine-guns in medieval tunnels and catacombs. Almost out of site, the machine-gunners were able to pour continuous fire into the advancing troops, the judicious use of these guns preventing the capture of a good part of the original German front line. While the 188th Brigade organised bombing attacks on the tunnels, the 190th Brigade was ordered to continue the advance. The advance did not start well, the 7/Royal Fusiliers coming under fire from the front line. The 4/Bedford and 10/Royal Dublin Fusiliers, further to the north were having much difficulty in getting past Station Road at its western end. Mixed units eventually made some progress on the Beaumont Hamel spur. Some of the machine-guns in the tunnel entrances were silenced but there were so many exits and entries that the total clearance could not be achieved quickly, even with the most courageous of bombing attacks. Meanwhile, Lt.-Col. Freyberg led a mixed force of Hood and Drake towards the second objective, collecting some men from the HAC on the way. Advancing up Railway Road, close to the railway and up Railway Alley to the left, the mixed units reached the edge of Beaucourt. Friendly and hostile shells caused a withdrawal, but afterwards the men moved forward again and took over the shell holes. An attempt to use German wire was hampered by sniper-fire from the village. Lt.-Com. Gilliland's group, now at its second objective, had lost all communication with the rear and the party joined Freyberg on the edge of Beaucourt. Freyberg was refused permission to assault Beaucourt on the grounds that it was too risky a venture, there being no reinforcements for the 63rd Division, all units having been engaged. Bombing attacks continued throughout the day. Maj.-Gen. Shute ordered an attack on the German third line and although the preliminary barrage was started, the infantry on the front of 188th Brigade could not be organised in time.

    Lt.-Gen. Fanshawe, commanding the V Corps, allotted the 111th Brigade from 37th Division to reinforce the depleted 63rd and the Brigade made its way to the front ready to renew the attack on the following day. The 13/KRRC was one of the to arrive and was sent forward to support Freyberg who then extended his left flank to include Redoubt Alley. The 13/Royal Fusiliers and 13/Rifle Brigade were promptly sent forward to support the troops at the first objective. The fourth battalion of 111th Brigade, the 10/Royal Fusiliers arrived at the British front line from Hamel about 11.00pm.

    Maj.-Gen. Harper's 51st Division was next on the left, its first objectives being the Y Ravine salient and the village of Beaumont Hamel and the second being Frankfort Trench which ran north from about half way on the Beaumont Hamel/Beaucourt road. On the right, Brig.-Gen. Campbell's 153rd Brigade assembled with two battalions in line, one in support and the fourth to provide carrying parties. On the left, Brig.-Gen. Burn's 152nd Brigade was facing Beaumont Hamel from the west, its assembly being the same as the 153rd Brigade with the exception that the fourth battalion was to be kept in reserve. Brig.-Gen. Stewart's 154th Brigade was to be kept in divisional reserve east of Forceville. A few minutes before zero the men left the assembly trenches and cleared the British wire. Zero hour was announced by the explosion of 30,000 lbs. of ammonal under the previously-made crater on Hawthorn Ridge. On the right of 153rd Brigade the 7/Gordon stormed and seized the German front line. Keeping close to the creeping barrage and crossing the maze of trenches in the German front system, the Gordons passed the eastern end of Y Ravine and, joining ranks with a few Marines from 63rd Division, reached their first objective at 6.45am. The enemy was in flight before them, chased by the Gordons who took a number of prisoners before consolidating in Station Road. The attack by the right of the Gordon Battalion had been quick and successful but the left was engaged in a bitter struggle to get past the German front system causing some delay in getting forward. German machine-guns and snipers were silenced by the Gordon bombers and on they continued to be met with heavy fire from Y Ravine, a thirty-foot deep salient with almost vertical sides into which dug-outs had been excavated giving excellent protection from British fire. The 5/Gordon hurried forward in support at 7.00am and small parties were engaged in individual contests. The 6/Black Watch was involved in the struggle, some units managing to skirt the ravine to continue the advance northwards. Harper had ordered up the 4/Gordon from 154th Brigade in reserve in order to isolate Y Ravine by multi-

Above: the ruins of Beaumont Hamel Church 1916. (Michelin)

Left: Beaumont Hamel Church as it is today (author)

directional bombing attacks. In the event, the bombers were not needed, at least in Y Ravine as Germans from the *62nd Regiment, I Battalion* began to surrender in large numbers. There were still many Germans taking cover in the complex system of dug-outs but, realising they were surrounded, they gradually surrendered. The newly-arrived 4/Gordon joined up with its sister battalion the 6th and the 6/Black Watch, the mixed units advancing and entering the southern part of Beaumont Hamel. Meanwhile the battalions of the 152nd Brigade had been heavily engaged. The 5/Seaforth Highlanders and 8/A&SH led the attack and were held up at the start of their advance in the German front system. The wire was not well cut south of Hawthorn Crater, the Seaforths on the right groping about in the half-light trying to find a way through. However, the bad visability did protect the men to some degree as they cut passages through the wire. The Sutherland Highlanders on the left and north of the Beaumont/Auchonvillers road took considerable casualties before breaking through the front line. Continuing to the support line, the Highlanders encountered worsening ground conditions, the barrage gradually getting away from them. Fortunately, resistance in the support line was fairly quickly overcome and the men paused to take breath. The 6/Seaforths moved up to help its sister battalion while the 6/Gordon reserve battalion was also ordered forward, its bombers quickly engaged at the German reserve line. Mixed units now forced an entry into the western edge of Beaumont Hamel where, elated by their success, began a systematical clearance of the houses and cellars. Although some Germans surrendered quickly, others were more obstinate and more drastic methods were needed. Although there was some close-quarter work, bombs were mostly used. Even so, several hours were needed to clear the village. Meeting up with units from 153rd Brigade, a defensive link was established on the eastern edge of the village. The Germans had been perfecting the defences of Beaumont Hamel over a long period and its final capture was a fine feat of arms. The trophies included all types of arms and ammunition, canteen stores including aerated water, a piano and "incoming mail"(iii) It can be imagined that the food and water would be very welcome after the hard day's work. The capture had been completed before the arrival of two tanks, both of which ditched on their way forward.

The 8/Royal Scots, the divisional pioneers came up to help in the consolidation of the first objectives, the new line now being held by the 4/Gordon, 6/Black Watch, 6/Gordon and the 8/A&SH, the latter in touch with Maj.-Gen. Walker's 2nd Division. The pioneers were then engaged with the sappers in the essential work of improving roads, tracks and communcations. The 7/Gordon, both Seaforth battalions and half the 7/A&HS were withdrawn to the former German reserve line.

Maj.-Gen. Walker's Division was to the immediate north of the Highland Division, facing Redan Ridge with its 6th and 5th Brigades on the left and right respectively, the 99th Brigade being in reserve. After seizing the German front line trench, the plan was to secure Beaumont Trench which runs north from Beaumont Hamel to the Quadrilateral south-west of Serre, then to advance east over the Redan Ridge to the final objectives Munich and Frankfort Trenches.

The initial advance was made over open ground, the assembly trenches not allowing any sort of rapid movement through the mud. Right on the heels of the barrage, the 2/HLI and 24/Royal Fusiliers of Bullen-Smith's 5th Brigade caught the Germans by surprise, capturing around a hundred and fifty prisoners as they came out of the dug-outs. Thereafter, the enemy was prepared and his fire took a heavy toll of the attackers. Not slowed by the fire, the two leading battalions, now with the 2/OBLI in close support, advanced and forced an entry into Beaumont Trench . With initial delays by the brigades on each side, both flanks were in the air and barricades were quickly constructed to give the necessary protection. The Germans mounted periodical bombing attacks from the north but, helped by the blockades, these were repulsed.

Brigadier-General Daly's 6th Brigade was in difficulty from the start. The wire was for the most part still intact and as the 2/South Staffs and 13/Essex groped around in the mud trying to find gaps in the wire, the morning fog making it difficult to see the firmer ground, they were cut down by enfilade fire from the front and the Quadrilateral. Many of the German gunners could not see the attackers but the whole front, of which every meter was familiar, was raked with deadly fire. On the right the Essex, closely followed by the 1/King's, now well behind the creeping barrage, struggled on to Beaumont Trench. After forcing an entry a block was formed at the junction with Lager Alley. On the left, enemy fire forced the Stafford, now joined by the 17/Middlesex, off course and the battalions eventually met up with units of the 3rd Division to the north. In the resultant confusion, further forward movement was impossible against the concentrated close-range fire. Many men trying to get back on the designated line of attack were not able to reach even the front-line trench and were pinned down in water-logged shell-holes. There was no question of the 6th Brigade being able to help the 5th to any great extent with its assault on Munich and Frankfort trenches. And so, small units of the two support battalions of the 5th Brigade, the 2/OBLI and 17/Royal Fusiliers, made their attack on the right with just a few parties of Essex and King's of 6th Brigade on the left. Isolated small groups penetrated Frankfort Trench, but with insufficient numbers and gaps between the individual parties, their position was untenable. Under attack by bombers and fire from snipers, the groups made a fighting retreat to Munich Trench, 200 meters to the west. Continuous counter-attacks forced a further withdrawal, the right retiring via Crater Lane and Waggon Road to Beaumont Trench while the left withdrew directly down Crater Lane to Beaumont Trench where a defensive flank was formed at the junction with Lager Alley.

Maj.-Gen.Walker was anxious to renew the attack and the disorganised units of the 6th Brigade including elements of the 13/Essex and 1/King's from the 5th, were ordered to withdraw to the start line in order to regroup and reorganise. Brig.-Gen. Daly expressed the view that his troops were in no shape to attack again. Under pressure from V Corps headquarters, Walker issued orders at 9.30am for two battalions of the 99th Reserve Brigade to move up. Half the 23/Royal Fusiliers were sent to support the 5th Brigade in Beaumont Trench, there still being no contact with the 51st Division to its right. In the early afternoon the 22/Royal Fusiliers moved to take up position in Beaumont Trench to the north of the 5th Brigade, the left facing the untaken Quadrilateral to the north-west. As will be seen later, the 3rd Divison's attack failed against the German defences at Serre causing Lt.-Gen. Fanshawe to cancel the revised attack of the 2nd Division. By evening the 2nd Division was holding Beaumont Trench, Lager Alley and also part of the southern section of Serre Trench while the right had at last linked up with 51st Division. Parties of the divisional pioneers, the 10/DCLI, assisted by elements of the 226th Field Coy. RE went forward to establish strongpoints in Beaumont Trench. Further reinforcements on the way from Brig.-Gen. Robinson's 112th Brigade (37th Div. in reserve) had reached Mailly-Maillet by 9.00pm.

On the left of Fanshawe's V Corps, Maj.-Gen. Deverell's 3rd Division had the task of securing Serre Trench and ultimately, the village itself. The original plan to engage all three brigades had to be changed due to the awful ground conditions and finally, the 8th Brigade, commanded by Brig.-Gen. Williams, was assembled on the right with Brig.-Gen. Kentish's 76th just to the north. The 9th Brigade was in divisional reserve.

The 1/Royal Scots Fusiliers and 2/Royal Scots (8th Brig.) faced Serre. Initially advancing on time with the 8/East Yorks and 7/KSLI of the same Brigade close behind, in spite of partially-cut wire, secured the front trench. From thereon the ground had been churned to pulp by the constant bombardment and with virtually no firm open ground left, the men had to advance along the trenches through mud almost up to the waist in places. Any forward movement was painfully slow. To add to the problems, the fog at Serre on the high ground had lifted, giving the German gunners a perfect target. The combination of the horrendous conditions and deadly machine-gun fire created confusion amongst the troops who not only were at risk of death from enemy fire, but from drowning if falling wounded into the sticky loam. Men tried desperately to extricate themselves and take cover and it was at this time that a confused link was made with units of 6th Brigade of the 2nd Division which has been mentioned previously.

The 76th Brigade, leading with the 10/RWF and 2/Suffolk and with the 8/King's Own and 1/Gordon immediately behind, had the same problems getting forward. By sheer determination mixed units of Fusiliers and Gordons crossed the front and support lines, just a few small parties then forcing an entry into Walter Trench, a long trench running south from east of John and Luke Copses. Utterly wearied by their efforts, they were quickly overwhelmed. The Suffolk on the left struggled on past the first line but an attempt to hold a small gain in the support line was unsuccessful, the battalion suffering many casualties. The King's Own were prevented from keeping up with the Suffolk by a heavy hostile barrage which churned the already devastated ground as though the shells were acting like a misused large modern concrete-mixer preparing a mixture of mud, water, torsos, limbs and the detritus of war. The survivors began the dangerous and exhausting trail back although a few clung grimly to their personal shell-hole, firing when they could from rifles which would soon be unusable in the mud. Only bombs were of any use to these men but there was no possibility of further supplies.

Maj.-Gen. Deverell was anxious to get the attack going again and ordered up two battalions from the 9th Reserve Brigade. There were so many conditions and elements to be considered, of which very few could be met, that in the end, no further operations were effected by the V Corps on the 13th November.

On the right of XIII Corps, Maj.-Gen. Wanless-O'Gowan's 31st Division had assembled its 92nd Brigade (Brig.-Gen. O. de L.Williams) to the north-west of Serre facing Puisieux, 2,600 meters to the east. The Brigade was formed entirely by battalions raised in Hull which carried such names as 'Hull Commercials, Tradesmen, Commercials and T'Others'. The brigade had mercifully escaped the slaughter at Serre on the 1st July when it was decided not to engage the 92nd following

Waggon Road (Wagon Road), the sunken lane heading north from Beaumont-Hamel gave welcome shelter to the troops when, at first, they were repulsed during their attacks on Munich and Frankfort Trenches. (author)

the virtual destruction of the 93rd and 94th Brigades. Four and half months later the 92nd Brigade was facing Serre again, just a few hundred meters to the north of its assembly positions on the 1st July, its two objectives being the German front system on the right and the support system on the left.

Prior to zero hour, Lewis guns and snipers were established in no man's land and the leading battalions, the 12th and 13/East Yorks began their advance at 5.45am behind the barrage. Making good progress from the start over slightly better ground conditions than those to the south, the Yorkshiremen carried the front line and set about clearing the dug-outs with bombs. Some prisoners of the *66th Regiment* were sent to the rear. A crater adjacent to John Copse was occupied and fortified to protect the right flank but was subsequently destroyed by a hostile barrage. German resistance was much stronger at the support line and some hard close-quarter fighting was necessary before a section of the trench was secured. Some units of the 13/East Yorks on the right, passing over the support line to keep up with the barrage, reached their objective, the German reserve line. They held on all morning, vainly waiting for reinforcements and were gradually overwhelmed, only a few men managing to withdraw. The remainder of the 12th and 13th Battalions held on grimly to the support line(iv), the bombers extending right and left until the supply of bombs ran out. Casualties were heavy and supplies were desperately needed. Even with the help of 36 machine-guns firing from positions south of Hébuterne the advance could progress no further. However, the machine guns served a useful purpose in checking a counter-attack in the late evening. Carrying parties and two companies of the 11/East Yorks moving up to reinforce the weakened battalions were held up by an intense hostile barrage on no man's land. The hold on the German support line was rapidly weakening and with mounting casualties and without hope of receiving reinforcements, the gallant Yorkshiremen were ordered to withdraw. Retiring to the German front line, bringing the wounded with them, parties turned their guns on the enemy while the wounded withdrew to the British front line. Only then, did the others follow. The casualties of the 12th and 13/East Yorkshire amounted to nearly eight hundred officers and men and although the Brigade was back at its starting point, the Germans had been mauled and some hundred and thirty prisoners taken.

As the day wore on and the reports gathered in, the results of the day's fighting could be assessed. Although the attack on the northern sector of V Corps had failed, General Gough was reasonably pleased with the outcome of the right wing. Orders were issued for the renewal of the attack the next day, the 14th November, the orders being simply to secure the objectives set for the previous day.

Improved visibility on the 14th allowed the RFC to take photographs over the front of V Corps. Conditions over the open ground improved slightly but there was little change in the trenches and shell craters. Throughout the night of the 13/14th November the British heavy artillery had maintained its barrage on the enemy positions, increasing in intensity on Beaucourt and Munich Trenches some twenty minutes before zero hour.

The left of the 63rd Division's attack on Beaucourt Trench and Muck Trench beyond was undertaken by the 13/Royal Fusiliers and 13/Rifle Brigade (attached to 63rd Div. from 111th Brig. 37th Div.). Advancing from Station Road, the first objective lay about 900 meters ahead. With the right on Redoubt Alley, the two battalions advanced against enemy fire from Beaucourt village and Muck Trench, hostile fire forcing the troops a little to the left. The change of direction was corrected but the advance had lost its initial élan. Although Lt.-Col. Ardagh, G.O.C. of the Fusiliers was wounded, his men continued somewhat behind schedule and advanced about half way to Beaucourt Trench before being stopped by enemy fire. The Rifle Brigade and the 13/KRRC (111th Brig.) which had moved up in support, pushed their advance to within two hundred meters of Beaucourt Trench. On the right, the 190th Brigade of 63rd Division was to attack Beaucourt village. Advancing from near Beaucourt Station, mixed units of the HAC and 7/Royal Fusiliers managed to get forward and joined Freyberg's units on the edge of Beaucourt. A combined attack was now mounted on the village, Freyberg, although again wounded, led his troops to the centre of the village while the 13/KRRC and the battalions of 190th Brigade pushed forward from the south-east. The capture of Beaucourt was completed with little difficulty and almost five hundred prisoners were captured. The number of Germans killed in the encounter was substantial. The fall of Beaucourt allowed the Fusiliers and Rifle Brigade to renew their attack on Beaucourt Trench, most of which was now secured, the resistance much weaker than at the time of the first assault. A good number of prisoners from the *55th Reserve Regiment* were sent to the rear. Bombers of the Rifle Brigade advanced on Leave Avenue but there was no sign of the 51st Division's 153rd Brigade on the left.

Three tanks were detailed to move forward at daybreak from Auchonvillers to assist in the clearance of the German front system. One was damaged by a shell at the assembly point and was unable to leave while the other two found they could not cope in the muddy conditions - one bogged down before the German front line and the other, although stuck in the mud, opened fire with its Hotchkiss guns. The fire was sufficient to cause a breakdown of the enemy resistance, the 10/Royal Dublin Fusiliers (190th Brig.) rounding up over four hundred prisoners while the Howe Battalion of 188th Brig. collected a further two hundred from the Station Road sector.

Meanwhile, German troops were seen to be massing east of the Bois d'Hollande and a counter-attack seemed imminent. This eventuality was cancelled by an accurate British barrage to the east of Beaucourt. Lt.-Col. Freyberg was now able to return to have his wounds dressed and was subsequently awarded the Victoria Cross for his courageous and efficient leadership during the capture of Beaucourt. Assisted by the 14/Worcester, the divisional pioneers, mixed units established a defensive flank on the eastern edge of Beaucourt and in Beaucourt Trench to the west of the village. Six hundred meters to the east, the HAC held the railway bridge over the Ancre until relieved by the 37th Division which then began the full relief of the 63rd Division.

Not having received its orders, the 7/A&SH attached to 152nd Brigade from the 154th, was late in starting its attack on Munich Trench. Having lost the protection of the creeping barrage, the Highlanders sent forward patrols which were strongly resisted. By 7.30am, two companies were attacking the southern end of Munich Trench. The struggle through the mud and slush was more trying than the enemy resistance in this part of the trench and the Highlanders secured it with little difficulty. A further 200 meters to the east, Frankfort Trench was reported to be well fortified. At 11.00am, the Highlanders were caught in a friendly barrage and were forced to retire and take cover in shell holes in front of Munich Trench. Bombers of the 9/Royal Scots (attached from 154th Brig.) advanced up Leave Avenue with a view of making some progress towards Frankfort Trench from the south , but little could be done without support from the rest of the 152nd Brigade. Orders for another attack on Frankfort Trench had to be postponed. During the night of the 14/15th November, the 51st divisional pioneers, the 8/Royal Scots, along with the 2nd Highland Field Coy. RE started digging a new trench, aptly named 'New Munich Trench' running north/south some 200 meters west of Munich Trench.

At 6.20am on the 14th, the 1/KRRC and 1/Royal Berkshire (99th Brig. 2nd Div.) left Beaumont Trench on its assault on Munich. Being brought from the reserve positions, the battalions were unfamiliar with the terrain and some confusion was caused by friendly fire. The general direction of attack was lost, some parties of the KRRC on the right entering Leave Avenue under the impression that they had secured their objective, Munich Trench. On discovering the error, bombers were sent forward towards the objective but they could not reach Munich which was, unknown to them, still being shelled by the British artillery. Eventually, the Rifles withdrew to Waggon Road with over fifty prisoners. On the left, parties of the Berkshire reached Munich where enemy resistance varied considerably but the Berkshires, too few in number, could not hold and eventually withdrew to Waggon Road. Other parties of the same battalion, now reinforced with the 23/Royal Fusiliers, crossed Lager Alley, mistaking it for Munich Trench. However, on wheeling west they attacked the Germans in Serre Trench from the rear, taking a number of prisoners. Consolidation was undertaken to safeguard the extended hold on Serre Road. The 22/Royal Fusiliers, also of 99th Brigade had taken up a position on the defensive flank between the Quadrilateral and Lager Alley.

Divisional H.Q. was unaware that Munich Trench was still occupied by the enemy, and the freshly-arrived 11/Royal Warwick and 6/Bedford (112th Brig. attached to 2nd Div.), were ordered to attack Frankfort Trench. Expecting to use Munich as the starting point for their advance, they were surprised by accurate machine-gun fire from Munich. With no possibility of getting forward, the new arrivals withdrew to join the other mixed battalions in Waggon Road.

In the clearing sky, the British artillery was guided to good effect by aircraft reconnaissance. There were no sign of a massing of enemy troops or other activities which precede a counter-attack.

The 37th Division having arrived at Beaucourt, the troops of the 63rd in the area were withdrawn. The difficulty in establishing a link with the 51st Division on the left had presented problems throughout the 13th and one of the first priorities on the 14th was to make contact with the 153rd Brigade. On the left flank of 37th Division, bombers of the 13/Rifle Brigade (111th Brig.) worked

Site of Frankfort Trench where the very last action on the Somme in 1916 was enacted by the 16/Highland Light Infantry and the 11/Border. New Munich Trench Cemetery is on the left of the track. (author)

their way towards Munich via Beaucourt Trench at 9.00am but it was not until thirteen hours later that contact was made with the 51st Division. Meanwhile, Beaucourt village was subjected to a hostile barrage for most of the day. A night patrol to the north-west of Beaucourt found the enemy had abandoned the junction of Railway and Muck Trenches, the latter appropriately named, being full of mud. As previously mentioned, the HAC had occupied the railway bridge east of Beaucourt but the post was now subjected to heavy shell-fire, the HAC being obliged to withdraw to the west, the ground conditions in the marshland below the bridge being totally unsuitable for digging.

A German barrage caused casualties in the ranks of the 7/A&SH (154th Brig.) as it waited for zero at 9.00am on the 15th prior to its attack on Frankfort Trench. At last on the move, the men caught up with the British barrage and more casualties were sustained. Passing to the south of Munich and with numbers much reduced, a few parties succeeded in reaching Frankfort Trench where the bombers assaulted the enemy dug-outs. They could not hold against superior numbers and withdrew to their starting line in New Munich Trench.

The 10/Loyal North Lancs and 8/East Lancs of 112th Brigade and attached to Maj.-Gen. Walker's 2nd Division, were late in forming up in the assembly trenches and were also unfamiliar with the terrain. Advancing through the mist at 9.00am, they lost direction and came under heavy fire as they moved forward over ground well-known to the enemy. There were many officer casualties and the survivors withdrew to swell the ranks of the mixed units in Waggon road. Meanwhile the 6th Brigade on the left flank of 2nd Division enjoyed more success on Redan Ridge, its bombers advancing their position up the slope of the ridge and establishing a strongpoint in the Quadrilateral near the crest of Redan Ridge. Two tanks had been made available to assist in the assault but both ground to a halt in the thick mud.

Many of the objectives had not been taken on the 15th November. As well as Munich and Frankfort Trenches still being in the hands of the Germans, the plans to attack Serre, Grandcourt, and Puisieux Trench seemed to be jeopardized. Gough's orders for the next day in the V Corps sector were to extend the hold on Ancre Trench running east from Beaucourt, secure the Bois d'Hollande and establish a line west back to north Beaucourt and finally, the capture of Frankfort Trench.

There was no rain on the 16th and after a frosty start, the weather remained cold but fine. The only activity in this sector on the 16th was a night patrol by the 8/SLI (63rd Brig., attached to 37th Div.) which advanced up Ancre Trench and constructed a post in the Bois d'Hollande. At the same time the 10/Royal Fusiliers and 13/KRRC (111th Brig.),assisted by the 152nd Field Coy. RE, established posts in Railway and Muck Trenches.

During the night of the 15/16th Maj.-Gen. Rycroft's 32nd Division had begun the relief of the

2nd Division and Brig.-Gen. Compton's 14th Brigade immediately took up position to the south of 3rd Division. The 112th Brigade in Waggon Road was relieved twenty-four hours later by Brig.-Gen. Jardine's 97th Brigade whose right relieved the 51st Division in New Munich Trench and Leave Avenue. The frontage of the 32nd Division was now from Leave Avenue to the Quadrilateral. Another fine day dawned on the 17th allowing good aircraft reconnaissance. On reports that Puisieux Trench and the Grandcourt Line were unoccupied, General Gough issued orders for patrols to occupy the trenches at night if possible. Patrols later found the enemy working on the defences and laying wire in front of Grandcourt Line and that he was in fact well entrenched in both trenches.

By the 18th the 8/SLI had extended the line of posts from the Bois d'Hollande to the Puisieux road north of Beaucourt and also south to Ancre Trench. One company of the Somersets made an unsuccessful attempt to move up Puisieux Trench to the east of the Bois d'Hollande. Although well supported by machine-gun and Stokes mortar fire, no progress could be made against a series of well-fortified German posts. During the night of the 17/18th there was a good deal of activity north-west of Beaucourt when the 8/Lincoln, also of 63rd Brigade and the 10/Royal Fusiliers and 13/KRRC of the 111th Brigade established posts west along Muck Trench in order to provide covering fire for the coming attack on Frankfort Trench

At 6.10am the 13/KRRC on the right pushed patrols north from the new positions in Muck Trench encountering little opposition in Railway Trench which runs north to Leave Avenue. On the left, bombers of the 10/Royal Fusiliers continued pushing west along Muck Trench to secure the junction with Frankfort where it was intended to link up with the 32nd Division. It was not possible to secure the junction because, as we shall see, the advance of the 32nd Division had been halted by a combination of withering enemy fire and the terrible weather conditions.

Having relieved the 51st Division in New Munich Trench on the 16th, Brig.-Gen. Jardine's 97th Brigade now had the onerous task of assaulting Munich and Frankfort Trenches between Leave Avenue at the southern end and then north to the junction of Lager Alley. The importance placed upon the attack was such that an order was received from Corps headquarters to the effect that all four battalions were to be engaged. From left to right these were the 2/KOYLI, 11/Border, 16/HLI and 17/HLI. In rain and sleet, tapes pointing north had been laid over a frontage of one thousand meters and soon after dark on the 17th, the men made their way forward to the assembly positions. This, in itself, was a difficult and physically draining experience, the 16/HLI not arriving until 5.10am, an hour before zero. At 6.10am they moved forward, the 17/HLI on the right quickly halted by machine-gun and rifle fire from undestroyed German positions. Next on the left, the 16/HLI, suffered a similar fate, particularly on its right flank, the British barrage evidently having failed to silence the enemy guns. The gunners had done their best in the wintry conditions but the mist, rain and sleet had made it extremely difficult to obtain accurate reports of the barrage. The remaining companies, having sustained fewer casualties, stormed Munich and Frankfort Trenches capturing a number of German prisoners on the way. The Borders were quickly engaged in a bitter struggle getting through the machine-gun fire but, even so, parties entered Frankfort Trench where, after a lengthy and courageous contest with the enemy, most of the Borders, out of bombs, were forced to retire. On the left flank of Jardine's brigade, the right company of the 2/KOYLI came under fire from a strongpoint in Munich, the men diving for cover in the shell

The kilted Highlander of the 51st (Highland) Division stands guard over his fallen comrades at the northern extremity of Newfoundland Park. (author)

German Officers and other ranks in the southern tip of Leiling-Schlucht Y Ravine. (J Verdel)

craters where they remained, cold and wet until they could return after dark. The left companies pushed forward much easier, sweeping over the junction of Munich and Lager Alley, continuing north towards *Feste Soden*(v), a well-fortified group of trenches immediately north of Munich and Frankfort.

Brig.-Gen. Compton's 14th Brigade had been busy before zero hour, the 2/Manchester having worked its way down Lager Alley from its western extremity and at zero hour, the leading companies, with units of the 2/KOYLI of 97th Brigade, wheeled left to advance down the valley towards Serre. Four and half months earlier, in the height of summer with temperatures in the seventies, two brigades of the 31st Division were almost annihilated during their attack on Serre from the north. On Saturday the 18th November, advancing with their left flank on Serre Trench, heads down in the driving sleet, the Manchester and KOYLI attacked from the opposite end of the compass. It made no difference, most of the officers were early casualties leaving just a few small parties to continue in ever-reducing numbers to the outskirts of Serre. Apart from the weather, it was so reminiscent of the 1st July - isolated groups were seen entering Serre where they fought to the point of exhaustion and, out of ammunition, were overwhelmed. Many of these gallant north-countrymen were killed and German reports claim seventy prisoners were taken.

Bombers of the 15/HLI on the left of the Manchesters trying to bomb up the trenches were halted by machine-gun fire at close range. Apparently, the British barrage had overshot the target leaving the Highlanders the task of advancing against the full weight of the German defences. The 1/Dorset was brought up later to reinforce the Highlanders and take over from parties of the 2/Manchester which had remained in Lager Alley. With this support, the Highlanders extended the left flank from the north-western corner of the Quadrilateral.

The retaliatory barrage had destroyed much of the communication system and the only way of getting messages to the rear was by pigeon or runner. In the din and vortex of battle, the use of carrier pigeons could not be relied upon although some messages did get through and runners were much used. The conditions under which the runners performed their duty is today almost beyond imagination. Sending one runner was of little use and usually, several were sent with the same message in the hope that one or more would get through(vi). Even without a single round of enemy fire, the

The western entry to Y Ravine in Newfoundland Park. (author)

Y Ravine became known to the Germans as "Leiling-Schlucht" in order to honour Hauptmann Leiling, who displayed outstanding courage on the field of battle. He died in the Battle of Cambrai in October 1918. (J.Verdel).

runner stood every chance of falling through sheer exhaustion while struggling through the mud and slime. If he slipped in this state into a water-logged crater, he would likely succumb to his fatigue and drown. Add to this the real situation of carrying his message under hostile and sometimes friendly fire, being the much sought-after target of German snipers and to know that his chances of survival were only between five and ten per cent, then one just begins to have some idea of the task facing these unsung heroes. Five hours after zero, Jardine learned that his brigade's attack had failed and he instructed the wounded to be brought in and the survivors to rally in New Munich Trench and Waggon Road.

An attack on Puisieux Trench by the 8/SLI was delayed until 11.20am on the 18th, the preparations not being completed in time. As the men went forward they were caught by British artillery fire forcing them to take cover in the shell craters. A combined attack was later organised with the support of the 4/Middlesex of the 63rd Brigade. While the Middlesex was making its way forward from Beaucourt, Somerset bombers had gained entry into Puisieux Trench and advanced south. Joined now by the Middlesex, the trench was secured and consolidated as far as Ancre Trench. The northern reaches of Puisieux Trench were, however, still in German hands.

Following a long barrage by the heavy artillery in the morning of the 19th, the 10/Royal Fusiliers (111th Brig.) made an unsuccessful attack on the junctions of Munich and Frankfort Trenches with Leave Avenue. Poor visibility had prevented accurate reports of the artillery barrage which had in fact failed to silence the enemy defences in Leave Avenue, the Fusiliers being met by a curtain of machine-gun and rifle fire.

Further rain on the 19th rendered the terrain totally unsuitable for any further offensive and the immediate priority was to bring in the wounded, relieve the exhausted divisions and, with fresh units, consolidate the adjusted front line. Much work needed to be done in repair of artillery pieces, roads, trenches, communications as well as getting supplies and ammunition and food to the front. Although the offensive had ground to a halt, another attack was launched against Munich and Frankfort Trenches on the 23rd November by the 16/Lancs Fusiliers (96th Brig) but the German defences were still too strong and no gain was made.

The final chapter in this sector was written by the 16//HLI which, on the 18th November found itself isolated in Frankfort Trench. Having no idea of what was happening elsewhere on the front of its 97th Brigade, the principle aim now of the Highlanders was to hold this small section of trench until reinforced. Numbering about ninety including a few men from the Border Regiment and with a reasonable amount of ammunition available, the prospects looked fair. The Germans were already launching the first of many counter-attacks and runners sent back to ask for urgent reinforcements and supplies did not get through. Eventually two men from the Border reached the rear with the urgent message. A number of patrols were sent forward with supplies but each sustained heavy casualties and no contact could be made with the Highlanders. With both food and ammunition now desperately low and casualties mounting, the Scotsmen stuck grimly to their task. They had taken quite a heavy toll of the attackers and the pocket of resistance was seriously annoying the Germans, everywhere else in the sector having calmed down. Now under orders to secure the trench, the Germans came with bombs, machine-guns, rifle and bayonet and still the Highlanders held out. On Saturday the 25th November, the Highlanders were invited to surrender - no reply was given. The next day, the Germans came again in force and overwhelmed the pitifully few able-bodied men. The Germans counted thirty wounded, the remaining sixty had fought to the death.

(i) Literally, 'mine-thrower' the fore-runner of the trench mortar.
(ii) The Division was made up of two naval brigades, the third brigade being composed of one Territorial, one New Army and two Extra Reserve Battalions.
(iii) Official History, page 493
(iv) The Victoria Cross was awarded to Pte. J. Cunningham of the 12/East Yorkshire Regiment for conspicuous bravery in repelling single-handed two German counter-attacks. He survived the War and died in 1941, aged only forty-three.
(v) Not to be confused with the *Feste Soden,* renamed 'Goat Redoubt' which is 1,000 meters north of Mouquet Farm.
(vi) One of the Fourth Army divisions lost every one of its nineteen runners taking messages to the rear.

OBJECTIVE 36
DESTREMONT FARM, GIRD LINES WEST, LE SARS, FLERS TRENCH WEST, EAUCOURT ABBEY, BUTTE DE WARLENCOURT

The first action in this area was undertaken by the 47th Division of Lt.-Gen. W. Pulteney's III corps on the day after the general advance on the 15th September. Situated about half-way between Martinpuich and Flers, the German strongpoint with the somewhat unusual name of Cough Drop was the first objective, to be followed by the capture of Eaucourt Abbey(i) 1,800 meters to the north. Advancing from Crest Trench, the 23/London (142nd Brig.) on the right crossed the Switch Line, heading towards Cough Drop. Heavy artillery and machine-gun fire halted the advance, particularly on the right flank, some 400 meters short of the objective, the men occupying the Starfish, a group of trenches about 2,000 meters to the east of Martinpuich. On the left, units of the 6/London (140th Brig.) had made better progress and had forced an entry into Cough Drop and were later reinforced by parties from the left company of the 23/London.

In the early hours of the 18th September, bombers of 140th Brigade worked their way towards Flers Trench, protected to the north by the New Zealanders who had just secured Flers Support Trench almost as far as the junction with Goose Alley. Good progress was made towards the junction of Drop Alley and Flers Trench. In the early evening on the following day the 2nd Auckland (2nd N.Z. Brig.) completed the junction of Flers Support and Goose Alley while Hampden's 140th Brigade entered Flers Trench West but the tired men could not hold and were obliged to retire to the Cough Drop. Having been engaged at the final capture of High Wood and ground to the north, the 47th Division had suffered 4,554 casualties during its thirteen-day duty in the area. Early next day, the Londoners were relieved by Maj.-Gen. Strickland's 1st Division. Towards the end of the day, the 1/Black Watch of Brig.-Gen. Reddie's 1st Brigade assisted the New Zealanders in another attempt to gain entry into Flers Trench West. Advancing without artillery support at 8.30pm, the 2nd Canterbury surprised the enemy, driving him back towards Eaucourt Abbey. Meanwhile a company of the Black Watch moved up Drop Alley and linked up with the New Zealanders. Forward units of the Canterbury were almost cut off by a German counter-attack but the situation was saved by the prompt action of the remaining Canterburys who, after a hard struggle, pushed the Germans once more towards Eaucourt Abbey. The Black Watch occupied Drop Alley to its junction with Goose Alley and Flers Trench West.

After a rainy spell between the 18th and 21st bright and sunny weather started to improve conditions throughout the whole line. On the 25th, while the XV and XIV Corps were fully engaged on the right wing, the 23rd, 50th and 1st Divisions of III Corps took up positions ready for the assault north towards Le Sars. Brig.-Gen. Reddie's 1st Brigade was quickly engaged after zero in a short but bitter struggle in the Flers Line which resulted in a gain of some 300 meters. The sector was taken over by the New Zealand Division. Brig.-Gen. Price's 150th Brigade of 50th Division was formed up almost due south of Eaucourt Abbey and during the night of the 24/25th September had pushed out

222   *148 DAYS ON THE SOMME*

posts towards its objective. In the evening of the 25th, an advance post had been established northeast of Martinpuich where Crescent Alley heads for Eaucourt Abbey.

Maj.-Gen. Russell's New Zealand Division was to establish a defensive flank running from Goose Alley, east of Eaucourt Abbey, back to the Gird Lines. The German barrage caused some casualties whilst the 1st Brigade awaited zero hour in the secured part of Grove Alley but promptly at

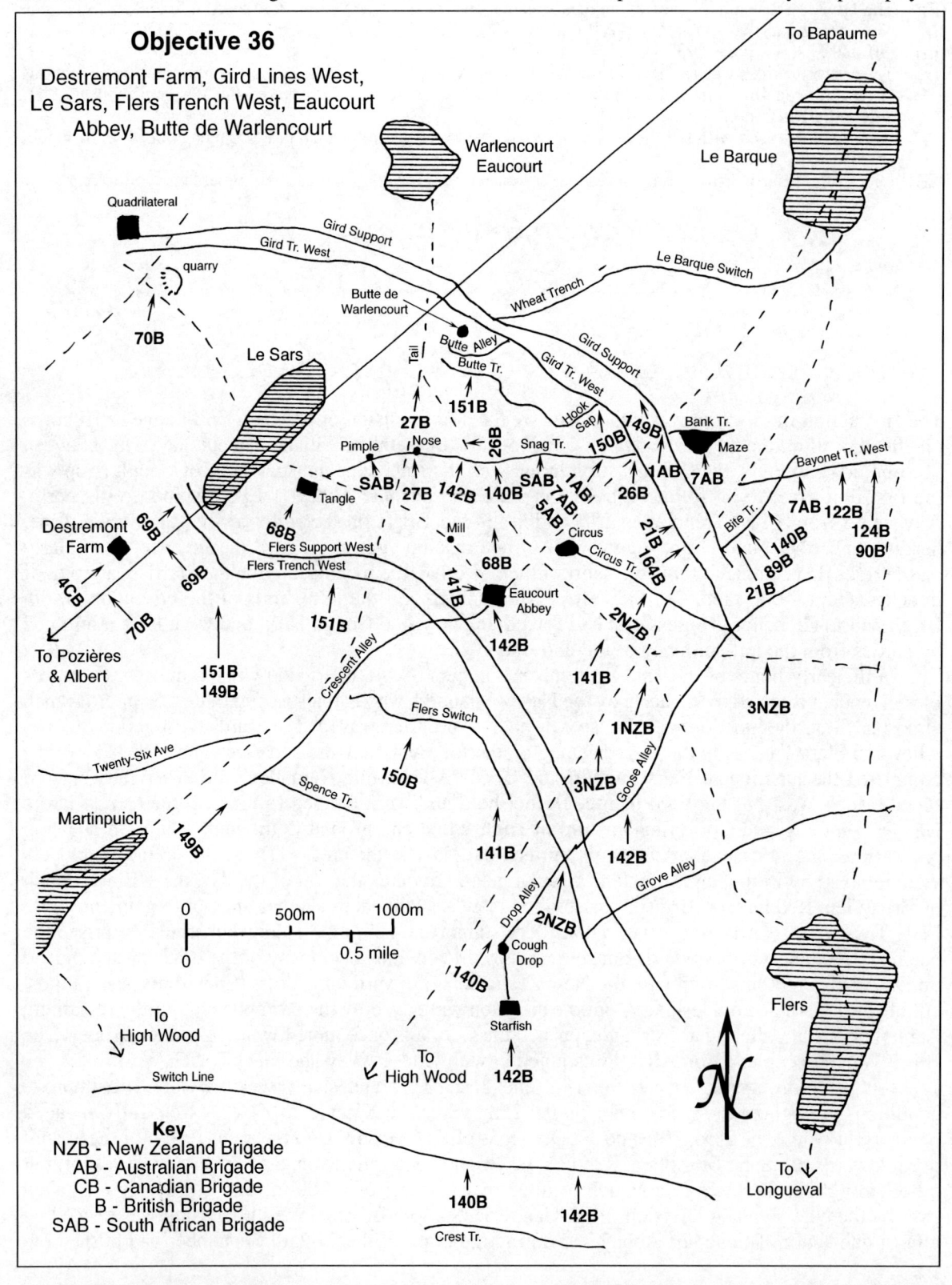

Objective 36
Destremont Farm, Gird Lines West, Le Sars, Flers Trench West, Eaucourt Abbey, Butte de Warlencourt

Key
NZB - New Zealand Brigade
AB - Australian Brigade
CB - Canadian Brigade
B - British Brigade
SAB - South African Brigade

zero the 1/Auckland, 1/Canterbury and 1/Otago pressed forward and, with little opposition, secured Factory Corner, the Canterburys taking some prisoners including battalion staff of the *13th Bavarians*. The southern part of Goose Alley was assaulted and secured by the Otagos where more prisoners and machine-guns were taken. The New Zealanders were now in Flers Support where links were made with 55th Division on the right and 1st Division of III Corps on the left. The timing of the New Zealand attack had worked to perfection. The results of XV Corp's attack had been somewhat less successful than on the XIV Corp's front. Nevertheless, the Gird Lines, if not fully secured, had been breached. Parts of Grove Alley and Goose Alley were also secured.

At 11.00pm on the 26th, the 1st and 50th Divisions attacked a section of the Flers Switch to the south of Eaucourt Abbey. The 1st Division on the right was caught by enfilade fire and a subsequent bombing attack in Flers Trench brought no positive results. Meanwhile, the Northumberland Fusiliers (149th Brig. 50th Div.) entered the western end of Flers Switch and bombed up Crescent Alley, continuing as far as Spence Trench.

The planned attack on the Gird Lines West on the 27th by the 164th Brigade of 55th Division was undertaken at 2.15pm by the 8/King's. The Liverpool Irish advanced with great élan, taking the allotted section of the Gird Trench with little loss. The British barrage having taken a heavy toll of the German defenders, the resistance of the survivors was much diminished. The 8/King's then linked up with the 1/Canterbury (1st N.Z. Brig.) on the Ligny road. After initial problems in cutting a way through uncut wire, the 1st Auckland, of the same Brigade, continued its advance to the north. Enemy shell and concentrated machine-gun fire from beyond Abbey Road caused horrendous casualties in the ranks of the 1st Otago Battalion. Only the left company managed to get forward to bomb its way up Goose Alley. North of the junction with Abbey Road, Goose Alley was hardly recognisable, being for the most part blown in by the British artillery. Strongpoints were made in this sector. The untaken part of Gird Trench West to the left of the King's delayed further movement to the north for the time being.

In the III Corps sector to the left of XV Corps, units of the 1st Division successfully assaulted Flers Switch but there was still strong opposition in Flers Trench West. During the night the 47th Division, now under the command of Brig.-Gen. Greenly, began to relieve the 1st Division. Meanwhile, patrols from the 50th Division were working their way up Crescent Alley towards Eaucourt Abbey. On the left Twenty-Sixth Avenue, the eastern approach to Courcelette, was entered through Spence Trench while a little further north, small gains were made towards the long Flers Line. The 41st Division relieved the 55th during the night of the 28/29th September.

A reconnaissance party from 23rd Division near Destremont Farm on the north of the Bapaume road between Courcelette and Le Sars was met by severe enemy fire. A further attack on the farm by the 70th Brigade (23rd Div.) on the 28th was also repulsed by German bombs and machine-gun fire. A combined tank and New Zealand infantry attack at dawn on the 28th on the still untaken junction of Goose Alley and Gird Trench had to be postponed. The tank had not arrived and hostile fire had reduced the ranks of the Wellingtons. Situated in a depression, the junction was out of site of the F.O.O's and was, therefore, almost immune from the British barrage. It was clearly necessary to assault the junction only after securing the part of Gird Trench on the higher ground. Meanwhile, reconnaissance had been made towards Destremont Farm from the eastern side of Courcelette and at 7.00am on the 28th, the 19th Battalion (Central Ontario, 4th Brig. 2nd Can. Div.) advanced towards the farm. German fire checked the advance but the men constructed a defensive flank facing northeast. The next day a further attack was made on the farm by a company of the 8/York & Lancaster of 70th Brigade (23rd Div.) which stormed and secured the farm, linking up with the Canadians in the Courcelette sector. Heavy rain fell during the day creating difficulties for the 18th London (141st Brig. 47th Div.) as it bombed its way up Flers Trench West. No rain fell on the 30th but ground conditions hardly changed. With the N.Z. Rifle Brigade now in Flers Support West, the Londoners engaged the enemy in Flers Switch, driving him to the north.

In accordance with plans made for the next attack, a continuous bombardment commenced at 7.00am on the 1st October, a fine day. Zero hour was fixed for 3.15pm. At 3.14pm, the Special Brigade RE discharged 36 prepared oil projectors providing a smoke and flame screen in readiness for the New Zealand attack with its 2nd Brigade. Leaving the assembly positions just east of Goose Alley the 2/Canterbury and 2/Otago (2nd Brig.) lost heavily from accurate machine-gun fire from the unsecured part of Gird Trench. However, the Canterbury secured the junction of Goose Alley and Gird

Trench which allowed entry into the eastern end of Circus Trench. The depleted Otago, on the left side of Goose Alley, moved north passing its objective destroyed by the barrage and passing the former strongpoint "The Circus", abandoned by the enemy. Arriving at Le Barque road and reinforced by the 2/Wellington of the same Brigade, the mixed units consolidated the new line and made contact on the left with the 19th London (141st Brig. 47th Div.). The New Zealand Brigade's losses had been heavy, especially at the start of the attack, but it had advanced rapidly, gaining all its objectives and collecting around 250 prisoners.

On the right flank of III Corps, Brig.-Gen. McDouall's 141st Brigade (47th Div.), with two tanks at its disposal was facing the German line east of Eaucourt Abbey. The leading battalion, the 19/London was halted by machine-gun fire and took cover in the shell craters while awaiting tank support. Both machines came up crushing the German defence and the Londoners moved forward again, the leading companies passing Eaucourt Abbey on the left and finally making contact with the New Zealand 2nd Brigade. The rear companies began consolidation of Flers Support. Eaucourt Abbey which, in reality, was two large farms built on the site of an Augustine Abbey, was the objective of the 20/London. Preceded by the tanks which had created a passage across Flers Trench West, the Londoners rushed the farm buildings and continued north to link up and extend the line with the 19/London. On the Brigade's left flank, the 17/London was halted by uncut wire and machine-gun fire, the Germans quickly launching a counter-attack down the trenches. Both tanks having ditched west of Eaucourt Abbey, they were in danger of being captured and the crews abandoned the machines after having set fire to them.

To the left of the 47th Division, the 50th, in the centre of III Corps sector, engaged Brig.-Gen. Cameron's 151st Brigade. The leading battalion, the 6/DLI, was a composite battalion of the remnants of the 8/DLI and 5/Border - the losses of both battalions not yet having been replaced - and the 5/NF, attached from 149th Brigade. The flank of the 6/DLI was in the air due to the hold up on the left of the 17/London, and the Durhams lost heavily before eventually forcing an entry into Flers Trench West. Lt.-Col. Bradford brought up his 9/DLI under heavy fire to support the weakened 6th whose commanding officer had become a casualty before zero hour. Bradford's(ii) leadership and organisational skills ensured renewal of the attack and the securing of the part of Flers Trench West facing him with units from the composite battalion and the Northumberland Fusiliers.

Maj.-Gen. Babington's 23rd Division formed the left flank of III Corps. The 11/Sherwood Foresters and 8/KOYLI of Brig.-Gen. Gordon's 70th Brigade had already moved forward from their assembly trenches prior to zero.The leading companies advanced north-east, storming and securing Flers Trench and most of Flers Support and also made contact with 151st Brigade on the right. Stiffer resistance was met on the opposite side of the Bapaume road and it was some while before the Germans could be pushed back. When this was achieved, the 8/KOYLI linked up with the Canadians. The 9/York & Lancaster of the same Brigade reinforced the two battalions. Probes towards Le Sars were quickly repelled by German fire.

The following morning, the 23/London (142nd Brig.) advanced along the line taken by the 17/London on the previous day. The 23rd, already weak in numbers, fared no better, sustaining about 170 casualties from machine-gun fire. The survivors were withdrawn and Maj.-Gen. G. Gorringe,

Goose Alley which cut across the Gueudecourt to Flers road and here heading southwest towards High Wood. It was successfully assaulted by the 55th Division and the 1st New Zealand Brigade. (author)

freshly arrived from Mesopotamia, took over command of the 47th Division from Brig.-Gen. Greenly. The same morning, under the command of Lt.-Col. Bradford, the remaining Germans were driven from Flers Support by the 6th and 9/DLI. A defensive flank was established on the right to cover the area not secured by the 141st Brigade and numerous counter-attacks were repulsed. The 151st Brigade was relieved by the 68th Brigade of 23rd Division. At noon on the 2nd, patrols of the 18/London were sent forward in the rain to ascertain the position in the Flers Trench covering Eaucourt Abbey. Following a favourable report, the Battalion occupied without resistance a position just to the northwest of Eaucourt Abbey and in touch with 20/London on the right and the freshy arrived 68th Brigade on the left. By the 4th October the remainder of Flers Support in the 47th Division's sector was secured and by the following evening the new line included the ruined mill 400 meters north-west of Eaucourt Abbey.The 69th Brigade had taken over near Le Sars from the 70th and two days later, an assault by bombers up Flers Support to dislodge the Germans at the Bapaume road immediately south of Le Sars, achieved little gain. The situation of a company of the 11/NF of 68th Brigade which had occupied the group of trenches called "The Tangle" a few hundred meters south-east of Le Sars became untenable under very heavy fire and the Fusiliers were withdrawn.

    The New Zealanders had been in action for sixteen consecutive days and with losses of a little under 7,000 were relieved on the 3rd October by Maj.-Gen. Lawford's 41st Division. Rain fell at some time during the next few days causing a postponement in operations, the next attack being rescheduled for the 7th, the preliminary bombardment beginning at 3.15pm on the 6th. Brig.-Clemson's 124th Brigade on the right faced Bayonet Trench West(iii). Accurate fire from the German trench took a heavy toll of the leading companies of the 26/Royal Fusiliers but the rest pushed forward, forcing an entry. The support battalions, the 21/KRRC and 10/Queen's moved quickly forward, also losing many men, but succeeded in joining the Fusiliers. The Brigade had started its attack well under strength but now it was down to between 25 and 30 per cent of its normal establishment. The left brigade, 122nd, fared only slightly better. All four battalions were committed. Leading the attack, the 15/Hampshire and 11/Royal West Kent sustained heavy casualties from Gird and Bite Trenches. Reinforced by the 18/KRRC and 12/East Surrey, the objective could still not be taken. Drums of burning oil had failed to discharge but bombers of the Kents made some slight progress in Gird Trench. On the left flank, a link was made with 140th Brigade in III Corps sector.

    Next on the left, the 47th Division took up its assembly positions covering the area west of the track from High Wood to Le Barque as far as Le Sars on the Bapaume road. The first objective facing Brig.-Gen. Hampden's 140th Brigade was Diagonal Trench, later renamed Snag Trench, situated some 800 meters north of Eaucourt Abbey. The 8/London led the attack and, like the divisions in XV Corps to the east, suffered horrendous casualties. The support battalions, the 15th and 7/London were also cut down by the same fire. The survivors dug in near Le Barque road in touch with 41st Division. It was grimly evident the German defences from Le Sars to Gueudecourt including the long Gird Lines, Bayonet Trench and other forward trenches had escaped severe damage from the preliminary barrage. Machine-guns awaited every infantry attack.

    On the left flank of III Corps, Maj.-Gen. Babington's 23rd Division was assembled just east of the Bapaume road. Leading the attack, the 12/DLI (68th Brig.) enjoyed the services of a tank which made a large contribution in the clearance of The Tangle. Immediately north of The Tangle, the tank turned left on to the Eaucourt Abbey/Le Sars road where it was destroyed by a shell. Machine-gun fire from the same direction held up the Durhams for a while. Meanwhile, the 9/Green Howards of 69th Brigade set off from the west side of the main road to secure the southern half of Le Sars as far as the Pys/Eaucourt Abbey road which bisects the village. Excellent progress was made, the Green Howards arriving at the central cross-roads. They were now joined by the 13/DLI (68th Brig.) which had moved up between the other two battalions to join the Green Howards in the centre of Le Sars. The enemy resisted strongly but he was overwhelmed and the village secured. Outposts were at once established around the village. A request to continue and attack the Butte de Warlencourt was refused. A little after 2.00pm the 11/West Yorks of 69th Brigade, on the west of the Bapaume Road attacked the upper reaches of Flers Support but was halted by hostile shelling, rifle and machine-gun fire. Assisted later by the 10/DWR, bombers eventually gained some 300 meters west up Flers Trench and Flers Support in the 4th Canadian Division sector.

    At 4.50am on the 8th, two companies of the 8/York & Lancaster (70th Brig. 23rd Div.) continued clearance of the Flers trenches and established an outpost at the quarry some 800 meters north-west of Le Sars. Elsewhere along the whole of the Fourth Army front, the wounded were brought in, a tiring job for the stretcher-bearers in the rain which was making the bad conditions even worse. Many of the walking-wounded who would normally make their own way back, were having great difficulty struggling through the mud and needed help to get to the rear. In the evening, the 21st and 22/London of 142nd Brigade crept forward as far as they could towards Snag Trench and dashed the last few meters when the barrage lifted. Entry was made into the western end of the trench where the Londoners came under enfilade fire from the opposite flank which was still strongly held. The Londoners were withdrawn to the Eaucourt/Warlencourt road where forward posts were made and a link made with 23rd Division.

It was essential to relieve the tired troops and get supplies forward. This was effected with much difficulty between the 8th and 11th. On the XV Corps front, Shea's 30th Division replaced the 41st while to the left in the III Corps sector, Furse's 9th Division replaced the 47th and McCracken's 15th Division took over from the 23rd. General Rawlinson was anxious to renew the attack - a number of the 7th October objectives were still not secured. Zero was set for 2.05pm on the 12th.

The newly-arrived 30th Division now formed the left wing of XV Corps. At zero hour, the 2/Royal Scots Fusiliers and 17/Manchester of 90th Brigade assaulted Bayonet Trench West. The Fusiliers were met by a curtain of machine-gun fire and were forced back. Accurate fire harassed the Manchesters as they went forward and losses were heavy. Nevertheless, small parties gained entry into Bayonet Trench but their hold quickly became untenable. Bombers of the 89th Brigade approached from the Gird Line meeting stiff opposition, the 2/Bedford just managing to secure a small section of Bite Trench which runs north-east from the Gird Lines parallel with the High Wood/Le Barque road. Meanwhile, the advance of the 17/King's (89th Brig.) was also checked by heavy enemy fire and although the assault was launched with great determination, no appreciable gain was made.

The 15th Division on the left of III Corps had been busy establishing forward posts to assist the attack of the 9th Division on its right whose final objectives were the Butte de Warlencourt (hereafter, called the Butte) and the Warlencourt line. No approach to the Butte could be made from the south without first securing Snag Trench and other obstacles including The Tail which runs from Snag Trench towards the Butte, the Nose, adjacent to Snag Trench and the Pimple, a fortified position at the western extremity of Snag Trench. The objectives were targeted by the artillery while Royal Engineers set up a smoke screen from the Butte to Little Wood to the north of Le Sars. The British barrage had left Snag Trench virtually untouched and the 7/Seaforth Highlanders, supported by the 10/A&SH (both of 26th Brig.) were met by a curtain of enemy fire. Suffering also from the British barrage, the two battalions could only get forward between a hundred and fifty to two hundred meters, some parties linking up with the South African Brigade on the left. The 2nd and 4th South African Regiments were also raked by machine-gun fire and dug in about half way to Snag Trench where they remained until being ordered to withdraw the following morning

The III Corps attack had failed and some responsibilty must lie with the failed artillery bombardment. The men attacked with great determination and staying power but were well under strength and, for the most part, inexperienced. The South Africans in particular had sustained horrendous casualties at Delville Wood and although replacements had been brought in, the men were only half-trained, the battalions were still only at half strength, which was indeed the case with so many units of Rawlinson's Fourth Army. Another cause was the long-distance fire from hidden German machine-guns which so effectively raked the leading battalions. Again, the wounded were brought in while headquarters planned the next attack, provisionally set for the 18th October. Before the general attack, XIV Corps was to secure certain 'Meteorological' trenches while the III was ordered to secure Snag Trench and the Gird Lines east of the Eaucourt Abbey/Le Barque road. Parts, but not all of these preliminary objectives were taken, the 3rd South African Regiment advanced under cover of night on the 14th seizing The Pimple and securing some 80 meters east along Snag Trench.

On the 18th the 2/Green Howards (21st Brig. 30th Div.) advanced towards Bayonet Trench West but the Battalion was checked a short distance from the trench by heavy and accurate grenade attacks causing many officer casualties. The Green Howard bombers then approached the objective up Bite Trench, a part of which had been previously secured by the 2/Bedford. As the Green Howards approached the northern end of Bite Trench they sustained heavy casualties from accurate machine-

On the right of the Albert-Bapaume road at Le Sars stood this elaborate German monument which, although severely damaged, gives an idea of permanency which the Germans felt in this area and which led to the construction of a number of monuments. Note that the lower half of the statue appears to be propped up by a rifle! (Michelin)

gun fire. The muddy ground, deteriorating each hour, delayed any hope of immediate reinforcement. To the left, the 18/King's and 2/Wiltshire attacked the Gird Lines, the King's encountering uncut wire, one of the most dreaded obstacles, and could not get forward. Meanwhile, the Wiltshire forced an entry in its sector of the Gird Lines but were nearly wiped out by enfilade fire.

Following the infantry attack, a tank was sent up from Flers and, after having crossed the British front line, stopped at Gird Trench where its Hotchkiss guns took a toll of the enemy, many taking fright and fleeing to the north-east. The commander opened the hatch to call the infantry forward, the way had been cleared, but the foot soldiers were few in number and physically exhausted. The tank continued on its own as far as Le Barque Road from where it made its way back. The performance of this single tank was a most refreshing and moral-boosting event - that it had managed to keep going in the awful ground conditions was almost a miracle in itself. A second tank had been available but was soon stuck in the mud.

Pulteney's III Corps was to attack the heavily-defended Snag Trench and the 15th Division discharged smoke and lacrymatory bombs to keep the enemy down during the attack by Furse's 9th Division. The 5/Cameron Highlanders of 26th Brigade somehow got forward against the enemy fire, entering Snag Trench at the junction with Le Barque road, then, joined by a few 2/Wiltshire (21st Brig.) bombed along the trench to within 200 meters of The Nose, this being the junction of Snag Trench and The Tail. The enemy regained a footing in the trench during a counter-attack delivered in the afternoon but was later dislodged. Meanwhile, the South African Brigade on the left of the Highlanders attacked with the same resolution displayed on each of its engagements on the Somme and although most of the men were inexperienced, two companies crossed the western end of Snag Trench, continuing ever towards their objective. Many of these gallant men perished from machine-gun fire from the Butte, the survivors clinging desperately to their hold in Snag Trench. Their work was not yet finished. At dawn on the 19th, the South Africans bombed east up Snag Trench from the Pimple but the enemy, still holding a good hundred meters of the trench each side of the Nose, prevented them from making much progress. They tried again in the early evening and eventually made some gains. A few hours previously, the 8/Black Watch (26th Brig.) had relieved the Cameron Highlanders. Shortly afterwards, German bombers equipped with *Flammenwerfer*, entered Snag Trench via the Tail and bombed east, pushing the South Africans towards the Black Watch. A continuous barrage was placed on the Nose and Tail positions but the South Africans, having used all available troops, were in no shape to continue the attack. During the night of the 19/20th October the 27th Brigade relieved both the 26th and South African Brigades. The weary troops were at the limit of physical endurance, many unwounded men having to be helped back. As for the wounded, they had to be carried. It is difficult for us today to imagine the stress and hardship of relieving troops in these conditions. Stretcher bearers were quickly at the point of complete exhaustion and when it seemed as though they could do no more, they rested a few minutes and, lighting up a cigarette, struggled to their feet to continue the arduous but essential and humane work. It has been recorded(iv) that the kilts of the Highlanders were so heavy with mud that the men could not bear the weight and so discarded them. In spite of these atrocious conditions, the 6/KOSB of the newly arrived 27th

Brigade was ordered to attack at 4.00pm. The Borderers, without flinching, attacked and seized the position called the Nose but lost the gain during a counter-attack. A further attack re-secured the Nose and, joined a little later by a company of the 11/Royal Scots, the Borderers strengthened their hold in Snag Trench while the Royal Scots bombed and secured 250 meters up the Tail towards the Butte.

Due to the deteriorating ground conditions, no further attack was made in this sector until the 5th November. After having served with the British 61st Division at Fromelles well to the north of the Somme in an action which was far from satisfactory, Maj.-Gen. McCay's 5th Australian Division came into the Somme on the 22nd October taking over the line of the 30th Division. Although not engaged in any major attack, McCay's Division was busy working on trenches, roads and communications as well as helping to bring up supplies and was eventually relieved by the 2nd Australian Division early in November.

On the 5th, the 27th Battalion (7th Brig. 2nd Aus. Div.) which, in fact, was complimented by mixed units from the 25th, 26th and 28th Battalions, attacked on the right. For some reason, the composite battalion was three minutes late in leaving, ground conditions making it impossible to catch up with the creeping barrage. Bayonet Trench was entered at a number of points and the struggle to retain the hold continued until dark when the men were forced to withdraw. Some of the units entered the Maze, 1,500 meters north-east of Eaucourt Abbey, which they defended against a number of counter-attacks. On the left the 28th Battalion came under a hail of fire from Gird Trench and took cover in the water-logged shell holes. A second attack produced no result and the men were forced to the ground again. Cold and wet, they waited until dark to make their way back.

On the left of the Australians, the 151st Brigade of Maj.-Gen. Wilkinson's 50th Division was assembled in III Corps sector to the west of the Eaucourt Abbey/Le Barque road. The attack was made by the 6th, 8th and 9th Battalions of the Durham Light Infantry. On the right the 8/DLI, instead of keeping up with the creeping barrage, the men found themselves helping one another forward in the slimy, sticky mess. It took a superhuman effort simply to advance anywhere near the German front line. Utterly exhausted, they struggled on and almost reaching the enemy front line, the waiting machine-gunners executed their deadly work from both flanks. The survivors made their way back as best they could. The right of the 6/DLI was caught by the same fire and no entry could be made in the enemy front line. However, parties on the left managed to force an entry with units of the 9/DLI. The latter had made an extraordinary advance over two lines of trenches and, on reaching the Butte, established a strongpoint on the Bapaume road, entered the Warlencourt line and, fighting all day, occupied the quarry immediately west of the Butte. It is not the length of the advance which makes the Durham attack 'extraordinary' but the unimaginably horrendous conditions in which it was made. The inevitable counter-attacks were at first repulsed but as these became stronger, the defenders weakened and in the end, the weary Durhams had to abandon their hard-earned gains. These sturdy men from the north of England had performed feats of valour and stamina worthy of the highest recommendation.

The artillery of Rawlinson's Fourth Army placed a barrage on the German positions designed to retain the attention of the enemy while General Gough was completing his preparatives for his attack astride the River Ancre. It was becoming more and more difficult to effect a continuous bombardment - supplying the forward pieces with ammunition through the mud was a major problem. In addition,

Eaucourt Abbey Farm, captured by the 47th Division. (author)

Kings Park Memorial, Perth, Western Australia, noted for its black swans. (D. R. Wood)

the guns had been used continuously over a long period, many were out of commission and others awaited spare parts.

A combined assault by the New South Wales Brigade of the 2nd Australian Division and 149th Brigade of 50th Division was planned for the 14th November against objectives astride the Eaucourt Abbey/Le Barque road. Dealing first with the Australian attack, the 26th Queensland & Tasmania Battalion on the right (on loan with the 25th Bn. from 7th Brig.) quickly lost the support of the creeping barrage while struggling through the mud towards the Maze, some 1,250 meters north-east of Eaucourt Abbey. Enemy machine-gun fire from the objective prevented the Australians from advancing over the open ground which, although not dry, was infinitely better than conditions in the trenches and shell craters and the advance was checked. The same conditions halted the 25th Queensland in the centre but fortunately, the 19th N.S.W. Bn. on the left kept pace with the 5/NF on the right of 149th Brigade, both battalions pushing forward as far as Gird Support. Consolidation was out of the question in the damaged and muddy trench, the New South Wales and Northumberland Fusiliers withdrawing to Gird Trench, 200 meters to the rear. Lewis gun and bombing posts were established in the trench astride the Le Barque road and about 50 prisoners sent to the rear. On the left of 149th Brigade, the 7/NF advanced well, entering Hook Sap, a short trench running north from Butte Trench to Gird Trench, but here, the Fusiliers were subjected to a terible curtain of concentrated fire. With communcation and reinforcement impossible, the Fusiliers held on as best they could. They were not seen again and it is assumed they were overwhelmed.

A number of counter-attacks were successfully repulsed during the day but a subsequent attack by two companies of the 20th N.S.W. Battalion on the Maze was checked by the same fire which had halted the 26th Battalion during the dawn attack. Parties of the 4th and 5/NF tried to get forward towards Hook Sap before midnight but no progress was made. The only gain at the end of the day was the section of Gird Trench astride Le Barque road. A strong counter-attack from both flanks was checked by the Australians and Northumberland Fusiliers. On the 15th November, bombers of the mixed units in Gird Trench endeavoured to work their way west towards the Butte but every movement through the mud exacted its toll and the men were soon exhausted. The venture was abandoned. In the late afternoon the 28th Bn. (West Australia) of 7th Brigade relieved the 19th Bn. while on the British side, two companies of the 4/East Yorks (150th Brig.) replaced the Northumberland Fusiliers. The problem of communication and supplies worsened by the hour.

The 16th November brought the last action in this sector. Towards the end of the afternoon and following a heavy bombardment, the Germans attacked in force from the front and both flanks. The enemy barrage had caused many casualties in the ranks of the British and Australian units. With bombers coming in from both sides and against the frontal fire, the men had to relinquish their hold in Gird Trench. The survivors made their way back as best they could - a sad end in the filth, mud and desolation of the Somme battleground.

NOTES
(i) Shown on Map 2407 Est as l'Abbaye d'Eaucourt Fme
(ii) Taking over command of both battalions, Lt.-Col. R.B. Bradford was awarded the Victoria Cross for his leadership and subsequent acts of bravery. He was killed on the 30th November 1917 by a stray shell which hit brigade H.Q. and is buried at Hermies British Cemetery.
(iii) Bayonet Trench East is included in Objective 32.
(iv) Official History, page 447, footnote 3

CONCLUSION

And so the battle of the Somme ground to a halt in the mud to the north-east of Beaumont Hamel and on the Transloy Ridge. Both British and German troops were exhausted and the British infantry settled down to its cold winter quarters trying to make the best of the awful ground conditions. There was little chance of drainage in the scarred terrain which was in fact a vast expanse of water-filled shell craters. The sole chance of any slight improvement could only occur in the event of a severe frost which would at least firm up the trenches. Many of the wounded were waiting for help or trying to get back to their lines and these could now be brought in, for the most part, unmolested. Both sides needed time to recover.

The map below shows the approximate Allied gains between the 1st July and the end of November. High Wood had held out against innumerable attacks for two months and one day. Delville Wood also had been a constant thorn in the side of any serious advance to the north. The capture of Pozières with a population today of under three hundred, the nearby Mouquet Farm and Pozières Ridge, all objectives on the 1st July, incurred casualties in excess of 30,000 Australian, British and Canadian troops. The high water level in the area of the Meteorological trenches(i) prevented any further movement towards Le Transloy, Ligny Thilloy or Le Barque. No advantage had been gained by the bitterly contested capture of the Meteorological trenches, the Germans withdrew to more favourable ground leaving the British troops in the mud.

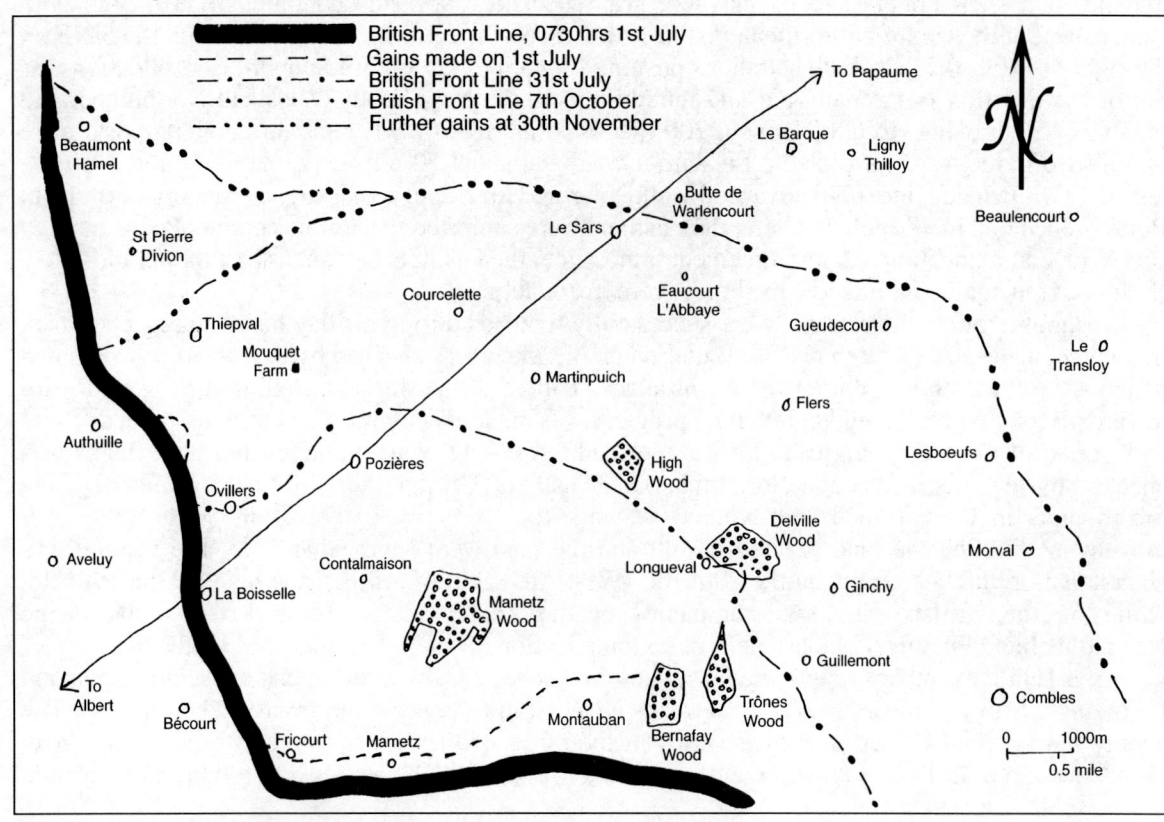

The best estimations of casualties during the Battle of the Somme in 1916 are that the British casualties amounted to 419,000; the French 204,000 while the figure for German losses range from 419,989(ii) to between 660,000 and 680,000(iii). The large disparity in the casualty figures has been the subject of discussion for many decades and a fuller study of these figures is outside the scope of this guide. However, even taking the smaller estimates, the total casualties of the belligerents amounted to over a million. The area gained by the Allies during the 1916 offensive was approximately 100 square kilometers or 38.61 square miles. With casualties of 623,000, this means that for each square kilometer gained, the cost in casualties was 6,230 or, for each square mile, the corresponding figure is 16,135. These are indeed costly gains and it is from statistics such as these that Generals Haig and Rawlinson became the butt of much criticism by eminent historians such as B. H. Liddell Hart and others. There was much talk of "Lions led by Donkeys", although I do not personally subscribe to this theory. With the advantage of hindsight it is not difficult to understand such claims when we consider the future of the survivors who returned to England to a land "fit for heroes". Many jobs were retained by those who did not join one of the armed forces - most quite legitimately, being exonerated from military service due to essential war work on munitions, arms, coal and other essential areas. The demobbed soldiers had much difficulty in finding work which caused much bitterness by those who had fought and were wounded, many of them seriously. The dead could not talk and this haunted the survivors for many years - almost a sense of guilt about surviving the war. But a commander's single duty is to win his battle. There is no room for a soft-hearted commander in war. The days of the gentleman's war at Agincourt and Crécy had long finished. How the brigade and battalion commanders must have felt when transmitting the dreaded order from Corps Headquarters "....to be taken at all costs", cannot be imagined. It will be recalled that Major-General Sir C. St. L. Barter, commander of the 47th Division, received such an order at High Wood which he transmitted to his Brigade and Battalion commanders. High Wood was taken but Barter was sacked for "wanton waste of life". Where is the justification? The answers to such questions have been considered by many eminent scholars and I recommend particularly John Terraine's Essays on the Leadership of War. Mr. Terraine asks the question 'Who won the War ...?"

The Germans had held the high ground on the Thiepval ridge until finally thrown back in October, the British commanders being well pleased to have command of this advantageous position. But the natural industrious nature of the German mind was very much in evidence. As soon as they were aware they no longer occupied advantageous positions, preparations were put in hand to construct a new line of defences a few miles to the east. These were designed to be even more impregnable than those facing the British troops on the 1st July. The new line, aptly named, the Hindenburg Line, started from Arras to the north of Bapaume, then headed south-east towards Cambrai before finally heading more or less south towards St. Quentin, the whole forming a rough triangle between Arras, Cambrai and St. Quentin. While this work was being executed, German troops remained in the Somme area in order to hide the intentions of an eventual withdrawal to the new Hindenburg Line. When this took place between the 15th and 18th March 1917 following a period of total destruction of the area to be evacuated, the Germans, in doing so, conceded the untaken objectives of the Allied forces on the Somme Front. The village of Serre, with the skeletons of the pals battalions who had attacked on the 1st July still entangled in the barbed-wire, was ceded to the British as a master throws a bone to his dog. Similarly, the Butte de Warlencourt was abandonned to the British. It begs the question - should the Butte have been attacked. The answer has to be in the affirmative - had not the Meteorological trenches been captured in a morass of mud and sludge? Without such pressures, the Germans may not have evacuated the Somme. It is true also that the construction of the Hindenburg Line would present particular difficulties to the Allied attacks in 1917.

And now, to bring this guide to a close, it would seem appropriate to conclude with a poem written by Harry Fellows who has been mentioned in the events enacted at Mamezt Wood. However, before doing so, a few biographical details would perhaps serve as an introduction. Harry was an interesting character. Son of a Nottingham coal miner, he started work at the Raleigh cycle factory at the age of thirteen and earned five shillings a week. Both his parents died in their forties and Harry's brothers were looked after by relatives in Melton Mowbray while his sister went into domestic service. At the outbreak of war Harry volunteered with his pal 'Pip' Henson and they had the choice of joining the Sherwood Foresters, the Duke of Cornwall's Light Infantry or the Northumberland Fusiliers.

Without hesitation they opted for the Northumberland Fusiliers because there was no football team in Cornwall and although Nottingham and Derby both had teams, the name of Newcastle United was a household name and then, there was the long free ride on the train.... And so Harry and Pip, inseparable pals, became privates in the 12th Battalion of the Northumberland Fusiliers in 'C' Coy. of 62nd Brigade, 21st Division. The Battalion left for Loos in September 1915, Harry now a fully-qualified Lewis gunner. The two pals arrived at the infamous Hill 70. The gunners were ordered to return to the transport depot to collect their guns and ammunition. On arrival they faced the dreadful sight of dead and wounded soldiers and mules, a chance enemy shell having wreaked havoc when striking an ammunition limber. While the others returned to the trench, Harry was ordered to wait for a message. Harry records, "I waited, Pip left me to walk back to the trench ...and out of my life...I never saw him again."

Many men in the battalion had the opportunity in early 1916 to return to England to take up their former jobs as miners or dockers and so help with essential home work but few took up the offer. They preferred to stay together. Robert 'Geordie' Appleton, although in his mid-forties, refused to leave. He was killed on May 21st just before the great Somme offensive. Harry immortalises him is his poem 'Geordie'. It is sufficient to quote just one verse:-

> In the Spring of '16 some men left for home
> To work in the shipyards and mines,
> A concession freely granted to them
> Through the Conscription Act of that time.
> Geordie, a miner, was tops in his trade
> Overman of local renown,
> When asked why he decided to stay
> "I don't want to let the lads down!"

The losses at Hill 70 were replaced and the men were sent to Méaulte for training in preparation for the attack on Fricourt on the 1st July. The 10/Green Howards were to lead the 62nd Brigade's attack while the 12th and 13/NF were scheduled to move up in support at noon. However, the Howards losses were so great, including Major S.W.

Above: Harry Fellows, 12/Northumberland Fusiliers, Machine Gun section. (M Fellows).

Left: Harry Fellows with his son Mick (M. Fellows)

Harry laying a wreath at the grave of Major S. W. Loudoun-Shand at Norfolk Cemetery (M Fellows)

Loudoun-Shand who was to be awarded the Victoria Cross, that the Fusiliers were not sent forward at the appointed time and did not move forward until the evening. Harry considers it is very probable that this saved his life. Less than a fortnight later Harry was engaged in Mametz Wood. His battalion was also engaged in action in the Flers and Gueudecourt sectors in September.

During the Battle of Arras in March 1917 Harry lost two more close friends, John Dyson and George Gunnell while his pal Dick Turnbull suffered horrendous leg injuries. Harry seemed to be leading a charmed life but, three months later, received a serious head wound. He realised he had got his 'Blighty'. He was eventually repatriated and underwent surgery at Lincoln. After a period of convalescence, he was posted to Etaples on burial duty - this brought back memories of 'the poor boys' in Mametz Wood. However, he was now to turn to more pleasant work, his previous training at the Raleigh factory as a wheel truer providing him with the job of repairing damaged motor cycle wheels near St. Omer. Later transferred close to the Channel ports during the last great German offensive, he was posted to Rouen until the end of the war. After demobilisation in March 1919, Harry returned to Raleigh, eventually being promoted to General Foreman. After 53 years service with the firm, he retired in 1962. Thirteen years later, while reading a book about the Battle of Loos, Harry was surprised to read so many inconsistent and contradictory comments made by some of the contributors to the book. And so Harry, now aged eighty-one, but still able to recall vividly his actions all those years ago, decided to write his memoires.

With his second-hand typewriter he set to work on the mammoth task. His style is simple, clear and lucid without exaggeration. I particularly like his small collection of poems which bring to light a style of literary significance which is both sensitive and moving and yet at times satirical in its criticism of the Establishment. They are essentially concerned with people, particularly his pals Geordie and Pip but also pay homage to NCO's Sgt. Walter Smith, MM, Cpl. McIntosh and to Major Loudoun-Shand of the 10/Green Howards. Sixty five years after being demobbed, Harry suddenly felt the need to return to Flanders. With his son Mick as chauffeur and accompanied by a guide, the three pilgrims retraced the steps of the 12th Northumberland Fusiliers through to Loos, Hill 70 and to Dud Corner Cemetery where Pip is commemorated. The journey continued south to the Somme and to Norfolk Cemetery where Harry paused and reflected at the grave of Major Loudoun-Shand, VC and at that of his pal Geordie Appleton, two dead soldiers from opposite ends of the social sphere. At 7.30am on the 1st July 1984 at the annual ceremony at Lochnagar Crater, La Boisselle, Harry made the acquaintance of Monsieur le Marquis de Thézy, owner of Mametz Wood and the two men became firm friends. Each related their earlier experiences in Mametz Wood, Harry with the burying of the dead and Monsieur de Thézy relating how, when helping his father to replant the wood, they found the remains of many soldiers. He told Harry, "We let them lie, after all, it is their wood, they paid for it with their lives" The old veteran returned as often as he could, always placing a wreath on the grave of Loudoun-Shand, always in contact with Monsieur de Thézy until, on the 1st September 1987, the Great Reaper called Harry to the fold. But that is not quite the end of the story. Through the kindness and generosity of M. de Thézy, Harry's ashes were interred in Mametz Wood on the 13th March 1988 and a memorial stone now stands in the wood which had haunted Harry throughout his life.

Harry Fellows' memorial in Mametz Wood. (author)

From Harry's collection of poems I have chosen "They Grow Not Old" as its moving words typify the thoughts of so many survivors who recall so vividly the young faces of pals who paid the supreme sacrifice. The passing of years did not dim their vision of the youthful, innocent faces of those who were never to return. Harry's philosophical style seems a fitting end to this book.

THEY GROW NOT OLD

As I sit in my chair, content with my pipe
In smoke clouds some faces come clear
Of those who are gone, cut off in their youth
But whose memory I hold so dear
Like the flow of a stream, time passes by,
It's seventy years on I am told
But the faces I see still seem to be young
But I, who was left, have grown old.

My pal, "Pip" Henson a North Country lad,
Our dialects so far apart
He called me "marrer", I called him "mate",
But our friendship came from the heart
"I'll be gannin' marrer!"
He walked from my side
The Great Reaper took him to his fold
But the face that I see is still just nineteen
But I, who was left, have grown old.

Grey hairs, baldness, arthritis,
Glasses and similar aids
These things never entered their minds
Their thoughts centred more on the maids,
The mud and the rats of the trenches
And other discomforts untold
Seem never to alter the visage of youth
But I, who was left, have grown old.

The walls of the Thiepval Memorial
Carry thousands of names which tell
Of those who were lost, with no known grave
Just dumped in the ground where they fell.
In my smoke clouds I see them all marching,
Singing, with manner so bold
Full of vim and vigour of youth
But I, who was left, have grown old

They fought and died for their country
To make a land in which heroes could live
Just like pipe dreams of those at Westminster,
Had they lived would they ever forgive?
We returned to find nothing had altered
Some men had a base greed for gold
And we still had to fight - for a living
It's a wonder we lived to grow old.

Harry Fellows

ILS NE VIEILLISSENT PAS

Comme je m'assieds dans mon fauteuil content avec ma pipe
Dans un nuage de fumée, quelques visages clairement apparaissent
De ceux qui sont partis, en pleine jeunesse
Mais dont le souvenir est en moi si cher
Comme coule le ruisseau, le temps passe.
Il y a soixante-dix ans on m'a dit
Mais les visages que je vois, semblent encore être jeunes
Mais moi, qui suis encore là, j'ai vieilli.

Mon ami, "Pip" Henson, un gars du nord,
Nos dialectes si opposés
Il m'appelait "pote", je l'appelais "copain",
Mais notre amitié venait du coeur.
"Je dois y aller, mon pote!"
Il s'est éloigné de moi et
La mort l'a pris sous son aile
Mais le visage que je vois, semble avoir encore juste dix-neuf ans
Mais moi, qui suis encore là, j'ai vieilli.

Cheveux gris, calvitie, arthrite,
Lunettes et autres aides
Ces choses, ils n'y ont jamais pensé.
Leurs pensées étaient plutôt vers les filles,
La boue et les rats des tranchées
Et autres inconforts passés sous silence
Ne semblent jamais changer le visage de la jeunesse
Mais moi, qui suis encore là, j'ai vieilli.

Les murs du Mémorial de Thiepval
Portant les milliers de noms
De ceux qui étaient perdus, de tombes inconnues
Juste jetés dans la terre où ils étaient tombés.
Dans mon nuage de fumée, je les vois tous marchant,
Chantant, d'allure téméraire
Pleins de vigueur et d'énergie de la jeunesse
Mais moi, qui suis encore là, j'ai vieilli.

Ils se sont battus et sont morts pour leur pays
Pour créer une terre où pourraient vivre les héros.
Juste un rêve pour ceux de Westminster,
S'ils avaient vécu, leur auraient-ils jamais pardonné?
Nous sommes retournés pour trouver que rien n'avait changé
Quelques hommes étaient avides d'or
Et nous avions toujours à lutter pour vivre
C'est étonnant que nous ayons pu vieillir.

Harry Fellows

At the dawn of a new century, the Battle of the Somme in 1916 seems very distant and yet there are still a few survivors of this epic campaign. These veterans have for many years been much sought after by authors and the media in general. What is remarkable is the lucidity of interviews given by these grand old men who, for possibly over forty years after demobilisation, spoke rarely of their experiences during the Great War and, as twilight approached, found themselves stars of numerous television programmes. As Europe honours them, they, for their part, with images undimmed with the passing of time, still lament the loss of their friends over eighty years ago. Time has not effaced the memories, good and bad, of the conflict which was forever to mark the history of Europe. Soon, they will be no longer of this earth and as we have remembered those who fell on the field of battle, we shall remember those recently departed whose fate was that they would survive to stand witness to the emotive and tumultuous events of the Battle of the Somme.

(i)   Here, where the water table was, and is still high, a water-pumping station was built at the junction of the Flers/Thilloy and Gueudecourt/Eaucourt Abbey cross-roads to supply water to the large sugar factory to the south-east of Longueval, a distance of almost 5,000 meters. The ruins can be seen today.
(ii)   When the Barrage Lifts, page 411, G. Gliddon, Alan Sutton Publishing Ltd 1987
(iii) Official History, Vol. II, page 553, IWM, Battery Press 1992

ACKNOWLEDGEMENTS

The compilation of the statistics of infantry units and trenches could not have been undertaken without the help of a good number of people. Although knowing exactly the required format of the book, I was faced with the immediate problem of how to go about it. How does one go about the ambitious task of allocating the infantry and pioneer battalions to objectives of my own grouping throughout the Somme campaign? The obvious answer is that I needed a source containing the necessary information. Somewhat apprehensively, my eyes wandered along the shelves of my library and I took out Ray Westlake's book, 'British Battalions on the Somme'. This fine book went way above what I needed, giving details of the battalions before, during and after their tour of duty on the Somme and without this source of information, my humble book could not have come to fruition. To Ray, I offer my sincere thanks not only for giving me the means to undertake my work but also for his extra fast and friendly service from his archives on questions such as the amalgamation of battalions and other related matters. The trench list has been compiled from numerous sources including trench maps purchased from the WFA and elsewhere. A good number also originated from books in my library or from maps lent by friends, too numerous to name but I wish to mention especially Leslie Syree by name for his help in finding the exact location of a number of small sites. To Lesie and the others, I thank them all. I am indebted to Richard Golland, Keeper of the Department of Printed Books at the IWM for permission to extract casualty figures, the names of commanders and other relevant detail from Volume II of the Official History. I wish to acknowledge Mick Fellows for permission to reproduce freely from Harry Fellow's memoires and recordings. For photographs and cards from British or Commonwealth sources, other than my own, I thank Mick Fellows, Ann Warren, Denis Wood from Spearwood, Western Australia and Derek Heaney for permission to reproduce the unusual Australian WW1 characters sculptured from the stumps of Red Gum trees in East Victoria, Australia. To Glynne Payze, a computer buff and friend of many years who has been as ever willing to help in disk management. Almost all visitors to the Somme visit the museum and visitor's centre at Delville Wood where they will meet Janet and Tom Fairgrieve who provide a cheery welcome and a seat to weary walkers along with a drink while they browse through the many books for sale on the shelves. Janet has put me in contact with many people. Those who are in the northern sector of the Somme can call in at Avril Williams, who offers bed and breakfast at Auchonvillers - passers by are always welcome to call in for a drink and, preferably by prior arrangement, can visit the interesting cellars. A similar welcome awaits the tourist who requires bed and breakfast, guided tours or simply a drink at Colin & Lisa Gillard at Martinpuich, Diane & Vic Piuk at Hardecourt-aux-Bois and Christine and Jean-Pierre Matte at Bernafay Wood. Special thanks go to Jim Fallon who regales everyone with his interesting odd facts gained over nearly forty years of visiting the Somme. To my brother Raymond who, living in England, has easier access to addresses, telephone and fax numbers. And lastly, to my willing and very patient wife Sylvia, who yet again has had to give up her best table for the study of maps and has given freely of her time in all matters concerning the writing of the book.

I also feel it would be appropriate to acknowledge in their own language a number of French people who have contributed in the preparation of this guide.

A l'aube d'un nouveau siècle, le nombre de personnes qui ont vécu la Grande Guerre diminue chaque année et aujourd'hui il en reste peu. Mais en France les fils et les petits-fils des participants ont gardé le souvenir de la vie dans cette région où, trois fois en soixante-dix ans, ils ont dû faire face à l'envahisseur. Ils ont vu leur terre déchirée par les obus, et, l'ennemi refoulé, ces braves gens de Picardie ont repris leur bien tant aimé qui, aujourd'hui regorge de centaines de milliers de disparus. Il n'est donc pas étonnant que les fermiers et autres gens se rappellent bien de ces évènements et qu'ils ont raconté aux enfants l'histoire de leur pays. Grâce à l'amabilité de ces descendants, aux lettres, aux mémoires et un grand choix de livres, nous avons une source précieuse de renseignements. On peut apprendre beaucoup des livres, mais pour mieux connaître le champ d'honneur, il faut le voir, le sentir sous les pieds, parler à ceux qui y vivent. La plupart des fermiers de la Somme et du Pas de Calais supportent gentiment de nombreux Anglais et autres personnes qui veulent, tout en respectant leurs propriétés privées, faire des recherches ou simplement, découvrir le champ de bataille où la perte des belligérants dépassa un million deux cents mille hommes. Je veux juste exprimer l'énorme plaisir, que j'éprouve toujours, lors de rencontres et de discussions avec les cultivateurs et autres gens parmi lesquels naissent de sincères rapports d'amitié.

Dans mon village d'Attignat, je voudrais remercier Monsieur Roger PIGUET pour m'avoir prêté un vieux Guide Michelin. Ce beau livre, édité en 1920, a été retrouvé couvert de poussière dans un grenier où personne ne l'avait ouvert depuis plus de soixante-quinze ans. Mes remerciements vont aussi à Madame Nicole GUERRY et Monsieur Jean-Paul GONZALES, d'Attignat, qui m'ont aidé de diverses façons. A Philippe CARON, Rédacteur en Chef au quotidien, La Voix du Nord, à Lille, pour l'autorisation de reproduire la photographie de la démolition de l'église de Morval dans l'édition du 27 septembre 1987. A Monsieur le maire de Morval, Jean-Pierre POUTRAIN et l'ancien maire, Monsieur Alfred DELMOTTE pour leur coopération concernant les évènements qui ont aboutis, par sécurité, à la démolition de l'église de leur village. A Monsieur Jacques BOMMELEYN pour l'autorisation de reproduire de nombreuses cartes d'époque de sa grande collection et qui vont enrichir ce guide. A Jean-Pierre MATTE, dont la vie est étroitement liée à la Grande Guerre, demeurant au Bois de Bernafay et travaillant au Musée des Abris à Albert. Je le remercie sincèrement de m'avoir fourni quelques photographies, et aussi, pour l'accueil chaleureux qui m'est toujours réservé chez lui. Je recommande fortement la visite du Musée des Abris sous la Basilique au centre ville d'Albert. A Monsieur Jean VERDEL, de Miraumont, qui m'a accordé le droit de reproduire deux photographies allemandes de sa belle collection. A Monsieur Bernard MAES pour m'avoir prêté une des rares cartes postales de l'ancienne Ferme du Bois à Fricourt. A Alain PERRET qui m'a prêté quelques vieilles photographies.A Monsieur Arthur LEECH et son fils, Jean-Philippe. Apres avoir passé presque cinquante ans avec la *Commonwealth War Graves Commission* et terminé sa carrière comme *Senior Head Gardener*, Arthur jouit de sa retraite à Hamel. La pelouse de sa propriété est le témoin de son ancien métier. Fils de 27354 Ben Leech qui a participé avec son Régiment de Manchesters à la bataille de la Somme en 1916, Arthur m'a toujours fourni des renseignements quand j'en avais besoin. Son fils, Jean-Phillipe, est directeur de la Tour d'Ulster. Au *Visitor's Centre,* se trouve un musée où il est possible de visionner un film en anglais ou en français, de l'attaque de la 36ème Division le premier juillet 1916. Sur place, des repas légers sont disponibles. Vous y trouverez aussi une bonne selection de livres, principalement sur l'histoire de la 36ème Division.

Enfin, mes reconnaissances à la famille MATHON, propriétaire du Bois des Fourcaux, ainsi qu'à Messieurs Henri LEMAIRE, Jacques FOURDINIER, Dominique ZANARDI et Gérard DESAILLY. J'ai fortement apprécié, au cimetière britannique de Grandcourt Road, la rencontre de trois des graveurs de la *Commonwealth War Graves Commission* qui ont pris le temps de répondre à mes questions en ce qui concerne la restauration des tombes. Je les félicite ainsi que les jardiniers de la *CWGC* de leur beau travail d'entretien dans tous les cimetières, y compris ceux où les visiteurs sont rares.

Les photographies MICHELIN sont © MICHELIN, d'après les Guides Illustrés Michelin des Champs de Bataille - Edition 1920 - Autorisation n° 9911476.

BIBLIOGRAPHY AND FURTHER READING

The statistical information and other information contained in this guide have been extracted from a number of sources, the principle ones are listed below, (R = reprint)

British Battalions on the Somme	Ray Westlake	Leo Cooper 1994
Official History Vol. I*	Br.-Gen. Edmonds	Battery Press 1993 R
Official History Vol. II*	Capt. Miles	Battery Press 1992 R
Somme, A Day by Day account	Chris. McCarthy	Arms & Armour Press 1992
Somme, Then & Now	John Giles	Battle of Britain Prints 1991 R
The British Campaign in France and Flanders - 1916	Arthur Conan Doyle	Hodder & Stoughton 1918
Topography of Armageddon	Peter Chasseaud	Mapbooks 1991
Les Batailles de la Somme Somme - Vol. 1**	Michelin et Cie	Michelin 1920
La Bataille de la Somme 1916	André Laurent	Martelle 1996
The Hell they called High Wood	Terry Norman	Patrick Stephens Ltd 1989 R
When the Barrage lifts	Gerald Gliddon	Allan Sutton 1990 R
Somme	Lyn Macdonald	Papermac 1990 R
VC's of the Somme	Gerald Gliddon	Budding Books 1995 R
Essays on Leadership & War	John Terraine	W.F.A. 1998
Mametz	Colin Hughes	Gliddon Books 1990 R
WW1 1914-1918	Peter Simkins	Tiger Books 1994
Somme Battlefields	Martin & Mary Middlebrook	Penguin Books 1991
Before Endeavours Fade	Rose Coombs	Battle of Britain Prints 1994 R
Battlefield Guide to the Somme	Major & Mrs. Holt	Leo Cooper 1996
Walking the Somme	Paul Reed	Leo Cooper 1997

And last, but by no means least, a number of fine books, each well illustrated and dealing with a specific area, all published by Leo Cooper, Imprints of Pen & Word Books Limited:-

Title	Author	Year
Beaumont Hamel	Nigel Cave	1994
La Boisselle	Michael Stedman	1997
Courcelette	Paul Reed	1998
Delville Wood	Nigel Cave	1999
Fricourt, Mametz	Michael Stedman	1997
Guillemont	Michael Stedman	1998
Mametz Wood	Michael Renshaw	1999
Montauban	Graham Maddocks	1999
Pozières	Graham Keech	1998
Serre	Jack Horsfall & Nigel Cave	1996
Thiepval	Michael Stedman	1995

* Reprints of the twenty Official Histories, Military Operations in France & Flanders are available from the Imperial War Museum, Ray Westlake and other booksellers.

** Michelin & Co. published eighteen guides between 1917 and 1920 in English dealing with the campaigns in France and Flanders. They make interesting reading and are well illustrated These have long been out of print by Michelin but can occasionally be found at bookfairs or markets. The last guide in English, Verdun - Argonne, was published in 1931. However, a number of titles in English are now available as reprints at specialist booksellers.

APPENDIX ONE
SELECTION OF BRITISH & ALLIED CEMETERIES ON THE SOMME

CEMETERY	LOCATION	MAP	MAP REF
AIF Burial Ground	900m W of Gueudecourt	7 E	63540/256278
Acheux Brit. Cem.	660m WSW of village	7 W	61344/256392
Adanac Mil. Cem.	1700m N of Courcelette	7 E	62924/256432
Albert Communal Cem. Extn.	500m SSE of Town Hall	8 W	62284/255574
Albert Nat. (Fr.) Cem.	1100m ESE of Town Hall	8 W	62344/255570
Ancre Brit. Cem.	660m NNE of Hamel	7 W	62368/256362
Assevillers New Brit. Cem.	Eastern exit of village	8 E	63640/254458
Auchonvillers Communal Cem.	South east Auchonvillers	7 W	62128/256476
Auchonvillers Mil. Cem.	West Auchonvillers	7 W	62084/256502
Authuille Mil. Cem.	West Authuille	7 W	62362/256056
Aveluy Communal Cem. Extension	360m WNW Aveluy church	8 W	62298/255882
Aveluy Wood (Lancashire Dump) Cem.	On D50 Aveluy Wood East	7 W	62324/256106
Bapaume Post Mil. Cem.	2000m NE of Albert	8 W	62422/255732
Bazentin-le-Petit Comm. Cem. Extn.	420m E of Baz-le-Petit	8 E	63090/255964
Bazentin-le-Petit Mil. Cem.	Just W of Baz-le-Petit	8 E	63038/255970
Beacon Cem.	2800m NNE of Sailly-Lau.	8 W	62012/254898
Beaumont Hamel Brit. Cem.	540m NW of Beau. Hamel	7 W	62242/256562
Bécourt Mil. Cem.	Western exit of village	8 W	62490/255648
Bernafay Wood Brit.	NW side of Bernafay Wd	8 E	63274/255740
Bertrancourt Mil. Cem.	520m WSW of village	7 W	61536/256610
Blighty Valley Cem.	760m N of Cruc. Corner	8 W	62386/255974
Bouzincourt Comm. Cem. & Extn.	NW exit of village	8 W	61948/255926
Bouzincourt Ridge Cem.	2000m ESE of Bouzincourt	8 W	62156/255846
Bray Hill Brit. Cem.	2200m N of Bray	8 E	62732/255148
Bray Mil. Cem.	900m N of Bray	8 E	62752/255020
Bray Vale Brit. Cem.	1900m NNW of Bray	8 E	62666/255102
Bray-sur-Somme German Cem.	West Bray	8 E	62690/254944
Bray-sur-Somme Nat. (Fr.) Cem.	400m N of Bray	8 E	62730/254970
Bronfay Farm Mil. Cem.	Adjacent Bronfay Farm	8 E	62920/255242
Buire-sur-Ancre Comm. Cem.	400m N of village	8 W	61834/255256
Bulls Road Cem.	460m E of Flers	7 E	63524/256154
Canadian (2nd) Cem.	840m SSE of Pozières	8 E	62822/255964
Carnoy Mil. Cem.	Just SE of Carnoy	8 E	63000/255400
Caterpillar Valley Cem.	860m W of Longueval	8 E	63266/255892
Cerisy Nat. (Fr.) Cem.	400m W of village XR	8 W	62130/254550
Cerisy-Gailly Mil. Cem.	600m W of village XR	8 W	62120/254536
Chipilly Comm. Cem. & Extn.	Northern exit of village	8 W	62256/254606
Citadel New Mil. Cem.	1800m SW of Mansell Cop.	8 E	62744/255344
Colincamps Communal Cem.	500m S of village	7 W	61892/256670
Combles Communal Cem. Extn.	East Combles	8 E	63842/255732
Connaught Cem.	740m WNW of Thiepval	7 W	62468/256260
Contalmaison Château Cem.	Centre of Contalmaison	8 E	62826/255878
Côte 80 Nat. (Fr.) Cem.	1300m N of Etinehem	8 W	62566/254926
Courcelette Brit. Cem.	900m WSW of Courcelette	7 E	62850/256224
Courcelles-au-Bois Com.Cem.Extn.	Northern exit of village	7 W	61780/256780
Dantzig Alley Brit. Cem.	On D64 ENE of Mametz	8 E	62924/255604
Dartmoor Cem.	360m NE of Bécordel B.	8 W	62528/255518
Delville Wood Cem.	Facing Delville Wd	8 E	63416/255874
Dernancourt Com. Cem. & Extn.	360m NW of village	8 W	62068/255340
Devonshire Cem.	540m SE of Mametz Halt	8 E	62864/255472
Dive Copse Brit. Cem.	1900m N of Sailly-Laur.	8 W	61930/254814
Dompierre-Becquincourt (Fr.) Cem.	Western exit of village	8 E	63336/254544
Eclusier Communal Cem.	500m WNW Eclusier-Vaux	8 E	63186/254926
Englebelmer Comm. Cem. & Extn.	Just SW of village	7 W	61912/256236
Euston Road Cem.	1300m E of Colincamps	7 W	62028/256732
Faffemont Farm (near)	400m W of Farm(3 burials)	8 E	63644/255654
Flatiron Copse Cem.	Just E of Mametz Wd	8 E	63024/255828

A nest of seventeen partridge eggs at Dernancourt Cemetery. (Ann Warren)

The author talking to one of the CWGC engravers at Grandcourt Road Cemetery. These hardy men do this work throughout the year. (Ann Warren)

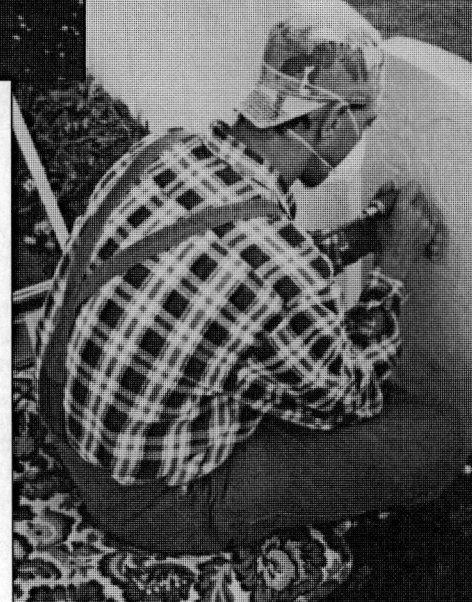

A CWGC engraver at work. Note he is working on a headstone with two regimental crests (author)

Foncquevillers Mil. Cem.	500m WNW of F.villers centre	7 W	62072/257262
Forceville Comm. Cem. Extn.	300m WNW of village	7 W	61546/256290
Frankfurt Trench Brit. Cem.	860m NE of Beaum. Hamel	7 W	62366/256588
French National Cem.	1280m WSW of Serre	7 W	62286/256712
Fricourt Brit. Cem. (Bray Road)	Bray Road, Fricourt	8 E	62690/255550
Fricourt New Cem.	NW of Tambour Craters	8 E	62682/255616
German Cem. Fricourt	700m N of Fricourt	8 E	62716/255652
Gommecourt Brit. Cem No.2	1200m ENE of Hébuterne	7 W	62246/257054
Gommecourt Wood New Cem.	440m NW of Gommecourt	7 W	62166/257196
Gordon Cem.	740m SE of Mametz Halt	8 E	62884/255462
Gordon Dump Cem.	900m NE of Lochnagar cr.	8 E	62658/255836
Grandcourt Road Cem.	1800m SSW of Grandcourt	7 E	62632/256318
Grove Town Cem.	1100m SW of Bray Aeorodr	8 W	62504/255204
Guards Cem.	SW exit of Combles	8 E	63754/255680
Guards Cem., nr. Lesboeufs	2200m NE of Ginchy	8 E	63704/256034
Guillemont Road Cem.	700m WSW of Guillemont	8 E	63440/255728
Hamel Mil. Cem.	300m S of Hamel	7 W	62332/256274
Hawthorn Ridge Cem. No.1	600m WSW of Beaum. Hamel	7 W	62240/256512
Hawthorn Ridge Cem. No.2	Newfoundland Mem. Park	7 W	62234/256472
Hébuterne Communal Cem.	South Hébuterne	7 W	62162/256936
Hébuterne Mil. Cem.	460m W of Hébuterne	7 W	62112/257006
Hédauville Comm. Cem. Extn.	Northern exit of village	7 W	61642/256128
Heilly Station Cem.	360m SSE of Heilly Stn.	8 W	61476/254938
Hem Farm Mil. Cem.	600m W of Hem Monacu	8 E	63556/255090
Herbecourt Brit. Cem.	Just W of Herbecourt	8 E	63588/254756
Hunter's Cem.	Newfoundland Mem. Park	7 W	62244/256472
Knightsbridge Cem.	1840m SE Auchonvillers	7 W	62206/256356
Lancashire Dump Cem.	See Aveluy Wood Cem.	7 W	62324/256106
Laviéville Communal Cem.	400m NNW of village	8 W	61702/255536
London Cem. & Extension	Immed. W of High Wood	8 E	63190/256034
Longueval Road Cem.	820m SSW of Longueval	8 E	63320/255822
Lonsdale Cem.	1700m SSW of Thiepval XR	7 W	62476/256050
Luke Copse Brit. Cem.	660m S of New Touvent Fm	7 W	62312/256806
Mailly Wood Cem.	600m SW of Mailly-M.	7 W	61880/256438
Mailly-Maillet Comm. Cem. Extn.	800m NW of village	7 W	61856/256558
Martinpuich Brit. Cem.	320m S of Martinpuich	7 E	63052/256118
Martinsart Brit. Cem.	Just S of Martinsart	8 W	62142/256026
Maurepas Nat. (French) Cem.	460m SE of Maurepas	8 E	63710/255484
Méaulte Mil. Cem.	540m SE of Méaulte	8 W	62330/255346
Méricourt l'Abbé Comm. Cem. Extn.	400m E of village	8 W	61686/255076
Mesnil Communal Cem. Extn.	700m SW of Mesnil-Martinsart	7 W	62194/256148
Mesnil Ridge Cem.	1800m SE Auchonvillers	7 W	62194/256346
Mill Road Cem.	740m NW of Thiepval	7 W	62490/256284
Millencourt Comm. Cem. Extn.	West Millencourt	8 W	61760/255636
Morlancourt Brit. Cem. No.1	540m W of Morlancourt	8 W	62044/255060
Morlancourt Brit. Cem. No.2	600m WNW of Morlancourt	8 W	62050/255080
Morval Brit. Cem.	180m W of D74 Morval	8 E	63826/255964
Munich Trench Brit. Cem.	1240m NNE of Beaum.Hamel	7 W	62340/256656
New Munich Trench Brit. Cem.	780m ENE of Beaum. Hamel	7 W	62368/256568
Norfolk Cem.	700m S of Bécourt	8 W	62562/255590
Ovillers Cem.	Western exit of Ovillers	8 W	62550/255924
Owl Trench Cem.	Just S of Rossignol Wood	7 W	62382/257036
Peake Wood Cem.	800m SW of Contalmaison	8 E	62770/255804
Péronne Road Cem.	W on D938 Maricourt	8 E	63200/255364
Point 110 New Mil. Cem.	1660m S of Fricourt	8 E	62748/255422
Point 110 Old Mil. Cem.	1440m S of Fricourt	8 E	62746/255444
Pozières Memorial & Cem.	940m SW of Pozières	8 E	62718/255984
Quarry Cem.	860m N of Montauban	8 E	63196/255766
Queen's Cemetery	920m S of New Touvent Fm	7 W	62306/256780
Railway Hollow Cem.	900m SSW of New Touvent Fm	7 W	62284/256786
Redan Ridge Cem. No.1	1000m NNW of Beaum.Hamel	7 W	62260/256638
Redan Ridge Cem. No.2	700m NNW of Beaum. Hamel	7 W	62278/256592
Redan Ridge Cem. No.3	840m NNW of Beaum. Hamel	7 W	62274/256620
Regina Trench Cem.	1500m NW of Courcelette	7 E	62814/256332
Ribemont-sur-Ancre Comm. Cem.Extn.	600m NW of village centre	8 W	61610/255210

Rossignol Wood Cem.	400m W of Rossignol Wood	7 W	62356/257046
Senlis Comm. Cem. Extn.	500m NW of village	8 W	61696/255924
Serre Road Cem. No.1	1100m SW of Serre	7 W	62294/256720
Serre Road Cem. No.2	1700m WSW of Serre	7 W	62264/256670
Serre Road Cem. No.3	1180m SSW of New Touvent Fm	7 W	62292/256758
Stump Road Cem.	1340m S of Grandcourt	7 E	62682/256362
Sucrerie Mil. Cem.	1800m ESE of Colincamps	7 W	62050/256666
Sunken Road Cem.	1000m N of Contalmaison	8 E	62834/255962
Suzanne Communal Cem. Extn.	400m NW of Suzanne	8 E	63054/255072
Suzanne Mil. Cem. No. 3	1400m NE of Suzanne	8 E	63166/255174
Ten Tree Alley Cemetery	1060m SSE of Serre	7 W	62432/256670
Thiepval Mem. and Anglo/French Cem.	Facing west of Memorial	7 W	62516/256166
Thistle Dump Cem.	760m S of High Wd	8 E	63220/255938
Varennes Mil. Cem.	800m NW of village	7 W	61362/256208
Ville-sur-Ancre Comm. Cem.& Extn.	SW of village	8 W	61954/255138
Waggon Road Cem.	1040m NNE of Beaum. Hamel	7 W	62348/256628
Warlencourt Brit. Cem.	1700m NE of Le Sars	7 E	63322/256504
Warloy-Baillon Comm. Cem. & Extn.	On NE exit of village	8 W	61380/255750
Y Ravine Cem.	Newfoundland Mem. Park	7 W	62264/256452

THIEPVAL MEMORIAL

The imposing memorial at Thiepval carries the names of over 73,000 soldiers who have no known grave and who fell between October 1915 and March 1918. Nearly all died in the battle of the Somme between July and November 1916 and included are the names of 858 South Africans. The Australians, Canadians, Newfoundlanders and New Zealanders have their own memorials elsewhere. Designed by Sir Edwin Lutyens, the memorial was inaugurated by the Prince of Wales in 1932.

Thousands of visitors come each year to Thiepval, some just to look, wonder and reflect on the endless list of names. Others come to place a wreath, a bunch of flowers or a simple poppy at the base of one of the large faces. Finding a Regiment or an individual name can take quite some time and the attached listings are designed to make this task easier. The names on the memorial are listed alphabetically by regiment and rank order. Only the London regiments are listed by numbered battalions.

On the attached sheets the regiments and London battalions are listed alphabetically showing the number of the pier and the corresponding face. For example, if you are looking for the name of a soldier from the Devon Regiment, this can be found on Pier 1, Face C. Some of the battalions suffered

photo: author

148 DAYS ON THE SOMME

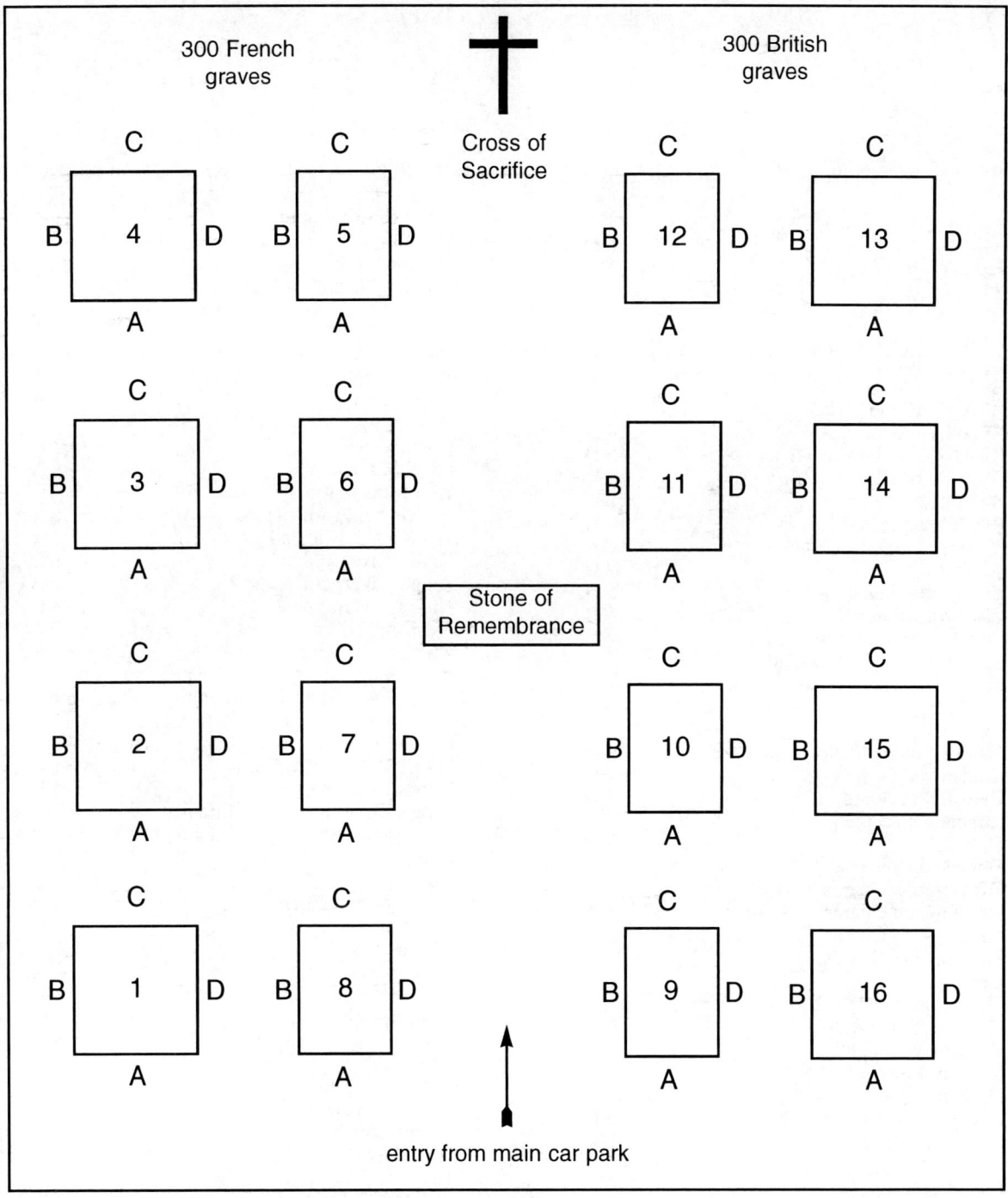

300 French graves — Piers 1–8
Cross of Sacrifice
300 British graves — Piers 9–16
Stone of Remembrance
entry from main car park

Each pier has faces labelled A (south), B (west), C (north), D (east).

very heavy losses and the Northumberland Fusiliers, for example, can be found on Face B of Piers 10, 11 and 12.

Many visitors come to find the name of a soldier known to them only by the familiar name of the battalion - Accrington Pals, Sherwood Foresters, Green Howards, Buffs, etc. On page 46 to 48 there is a separate list of these battalions showing the familiar name as well as the official name of the regiment and, from there, the following list will guide you to the appropriate pier and face on the memorial. For example, the Green Howards are known officially as the Yorkshire Regiment and are recorded under this name on Pier 3, Faces A and D.

The names of all the missing can be found in the numerous registers at the memorial but, due to a spate of register thefts, these are not always available to the visitor. However, they can normally be viewed at the gardner's workshop to the left of the main entrance of the memorial.

APPENDIX TWO
LOCATION OF REGIMENTS ON THE THIEPVAL MEMORIAL

No. REGIMENT/BATTALION	PIER/FACE	No. REGIMENT/BATTALION	PIER/FACE
Argyll & Sutherland	15 A	Hampshire Carabineers	1 A
Argyll & Sutherland	16 C	Herefordshire	12 C
Army Cyclist Corps	12 C	Hertfordshire	12 C
Artists Rifles	12 C	Highland Light Infantry	15 C
BLANK FACE	1 B	Honorable Artillery Company	8 A
BLANK FACE	2 B	Household Battalion	1 A
BLANK FACE	3 B	Huntingdon Cyclist Battalion	12 C
BLANK FACE	4 B	10th Hussars	1 A
BLANK FACE	13 D	20th Hussars	1 A
BLANK FACE	14 D	Inniskilling Dragoons	1 A
BLANK FACE	15 D	Irish Guards	7 D
BLANK FACE	16 D	8th K.R.I. Hussars	1 A
Bedfordshire	2 C	13th Kensington	9 C
Black Watch	10 A	1st King Edward's Horse	1 A
Border	6 A	2nd King Edward's Horse	1 A
Border	7 C	King's Liverpool	1 D
Buffs (East Kent)	5 D	King's Liverpool	8 B
Cambridgeshire	16 B	King's Liverpool	8 C
Cameron Highlanders	15 B	3rd King's Own Hussars	1 A
Cameronians (Scottish Rifles)	4 D	4th King's Own Hussars	1 A
Cavalry	1 A	King's Own Scottish Borderers	4 A
Cheshire	3 C	King's Own Scottish Borderers	4 D
Cheshire	4 A	King's Own Yorkshire Light Infantry	11 C
City of London Rifles	9 D	King's Own Yorkshire Light Infantry	12 A
Coldstream Guards	7 D	King's Royal Rifle Corps	13 A
Coldstream Guards	8 D	King's Royal Rifle Corps	13 B
Connaught Rangers	15 A	King's Shropshire Light Infantry	12 A
Devon	1 C	King's Shropshire Light Infantry	12 D
Dorsetshire	7 B	Labour Corps	5 C
3rd Dragoon Guards	1 A	Lancashire Fusiliers	3 C
5th Dragoon Guards	1 A	Lancashire Fusiliers	3 D
6th Dragoon Guards	1 A	17th Lancers	1 A
7th Dragoon Guards	1 A	5th Lancers	1 A
Duke of Cornwall	6 B	9th Lancers	1 A
Duke of Lancashire Yeomanry	1 A	Leicester Yeomanry	1 A
Duke of Wellington	6 A	Leicestershire	2 C
Duke of Wellington	6 B	Leicestershire	3 A
Durham Light Infantry	14 A	Leinster	16 C
Durham Light Infantry	15 C	1st Life Guards	1 A
East Kent (see Buffs)	5 D	2nd Life Guards	1 A
East Lancashire	6 C	Lincolnshire	1 C
East Surrey	6 B	7th London	9 D
East Surrey	6 C	10th London	9 C
East Yorkshire	2 C	11th London	9 C
21st East of India Lancers	1 A	12th London Rangers	9 C
Essex	10 D	17th London	13 C
Essex Yeomanry	1 A	19th London	13 C
First Surrey Rifles	13 C	20th London	13 C
General List	4 C	22nd London	12 C
Gloucestershire	5 A	22nd London	13 C
Gloucestershire	5 B	23rd London	12 C
Gordon Highlanders	15 B	24th London	12 C
Gordon Highlanders	15 C	London Irish Rifles	13 C
Grenadier Guards	8 D	London Rifle Brigade	9 D
Hampshire	7 B	1st London Royal Fusiliers	16 B
Hampshire	7 C	2nd London Royal Fusiliers	16 B

No. REGIMENT/BATTALION	PIER/FACE	No. REGIMENT/BATTALION	PIER/FACE
3rd London Royal Fusiliers	9 D	Royal Inniskilling Fusiliers	4 D
4th London Royal Fusiliers	9 D	Royal Inniskilling Fusiliers	5 B
London Cyclist Bn.	12 C	Royal Irish	3 A
London Scottish	9 C	Royal Irish Fusiliers	15 A
London Scottish	13 C	Royal Irish Rifles	15 A
Loyal North Lancashire	11 A	Royal Irish Rifles	15 B
Machine Gun Corps	12 C	Royal Lancaster (King's Own)	5 D
Machine Gun Corps (Cavalry)	5 C	Royal Lancaster (King's Own)	12 B
Machine Gun Corps (Infantry)	5 C	Royal Marines	1 A
Machine Gun Corps (Motor)	5 C	Royal Munster Fusiliers	16 C
Manchester	13 A	Royal Navy	1 A
Manchester	14 C	Royal Naval Volunteer Reserve	1 A
Middlesex	12 D	Royal North Devon Hussars	1 A
Middlesex	13 B	Royal Scots	6 D
Monmouthshire	16 B	Royal Scots	7 D
Norfolk	1 C	Royal Scots Fusiliers	3 C
Norfolk	1 D	Royal Scots Greys	1 A
Norfolk Yeomanry	1 A	Royal Sussex	7 C
North Staffordshire	14 B	Royal Warwickshire	9 A
North Staffordshire	14 C	Royal Warwickshire	9 B
Northamptonshire	11 A	Royal Warwickshire	10 B
Northamptonshire	11 D	Royal Welsh Fusiliers	4 A
Northern Cyclist Battalion	12 C	Royal West Kent (Queen's Own)	11 C
Northumberland Fusiliers	10 B	Royal Wilts Yeomanry	1 A
Northumberland Fusiliers	11 B	Scots Guards	7 D
Northumberland Fusiliers	12 B	Scottish Rifles - see Cameronians	4 D
Northumberland Hussars	1 A	Seaforth Highlanders	15 C
Nottinghamshire & Derbyshire	10 C	Shropshire Yeomanry	1 A
Nottinghamshire & Derbyshire	10 D	Somerset Light Infantry	2 A
Nottinghamshire & Derbyshire	11 A	1st South African Regiment Infantry	4 C
Ox & Bucks Light Infantry	10 A	2nd South African Regiment Infantry	4 C
Ox & Bucks Light Infantry	10 D	3rd South African Regiment Infantry	4 C
P.W.O. Civil Service Rifles	13 C	4th South African Regiment Infantry	4 C
Post Office Rifles	9 C	South Irish Horse	1 A
Post Office Rifles	9 D	South Lancashire	7 A
Queen Victoria Rifles	9 C	South Lancashire	7 B
Queen's Bays	1 A	South Staffordshire	7 B
Queen's Own Oxford Hussars	1 A	South Wales Borderers	4 A
Queen's Westminster	13 C	Suffolk	1 C
Queens (West Surrey)	5 D	Suffolk	2 A
Queens (West Surrey)	6 D	Sussex Yeomanry	1 A
Rifle Brigade	16 B	Welch	7 A
Rifle Brigade	16 C	Welch	10 A
Royal Army Chaplain's Department	4 C	Welsh Guards	7 D
Royal Army Medical Corps	4 C	West Kent Yeomanry	1 A
Royal Army Ordnance Corps	4 C	West Yorkshire	2 A
Royal Army Service Corps	4 C	West Yorkshire	2 C
Royal Army Service Corps	5 C	West Yorkshire	2 D
Royal Army Vetinary Corps	4 C	Wiltshire	13 A
Royal Berkshire	11 D	Worcestershire	5 A
Royal Dragoons	1 A	Worcestershire	6 C
Royal Dublin Fusiliers	16 C	York & Lancaster	14 A
Royal Engineers	8 A	York & Lancaster	14 B
Royal Engineers	8 D	Yorkshire (Green Howards)	3 A
Royal Field Artillery	1 A	Yorkshire (Green Howards)	3 D
Royal Field Artillery	8 A		
Royal Fusiliers	8 C		
Royal Fusiliers	9 A		
Royal Fusiliers	16 A		
Royal Garrison Artillery	8 A		
Royal Horse Artillery	1 A		
Royal Horse Artillery	8 A		

A number of additonal names were engraved after completion of the memorial. These addenda are shown in random order on the following piers and faces: 4 C, 5 D, 8 D, 9 B and 12 B.

The Author

Since settling in France in 1991 following his retirement as a teacher of French and Business Studies at a comprehensive school in Nottinghamshire, the author has been able to devote much of his leisure time in the pursuit of data concerning the Battle of the Somme in 1916. He spends four months each year researching and interviewing people from all walks of life who live or visit this corner of Picardy which was invaded three times between the years 1870 and 1940. Being physically part of the great continent of Europe, France has, like so many countries within its boundaries, had to defend its frontiers against the invasion of enemy troops. It is the resultant change in politics, history, literature and culture which interests the author and incites him to further study.

This is the author's second book on the Battle of the Somme, the first dealing with the events of the 'Big Push' on the 1st July 1916. This work is concerned with the events from the 2nd July to the end of the campaign in 1916 and reveals the fortitude, courage and patience of the British "Tommy" and fellow soldiers drawn from throughout the British Empire - which enabled them to stand up to the might of the Kaiser's disciplined armies. It is noticeable that the belligerents showed their fortitude in different ways, the German in the Teutonic disciplined tradition, the Frenchman in his ardour to crush anyone who dares to invade the sacred soil of France, the dogged spirit of the Australian and New Zealanders, the flare and grit of the Canadian and South African. And finally, the British "Tommy" who fought in the old traditions of our long history and could endure the most horrendous conditions with a grumble but often with humour. These are some of the facets of the great offensive in 1916 which have fascinated the author and it is not really surprising he chose to settle in the country which has attracted and held his attention for over forty years.